# UBIQUITOUS
# MULTIMEDIA
# COMPUTING

# Chapman & Hall/CRC
# Studies in Informatics Series

## SERIES EDITOR

### G. Q. Zhang
Case Western Reserve University
Department of EECS
Cleveland, Ohio, U.S.A.

## PUBLISHED TITLES

Stochastic Relations: Foundations for Markov Transition Systems
**Ernst-Erich Doberkat**

Conceptual Structures in Practice
**Pascal Hitzler and Henrik Schärfe**

Context-Aware Computing and Self-Managing Systems
**Waltenegus Dargie**

Introduction to Mathematics of Satisfiability
**Victor W. Marek**

Ubiquitous Multimedia Computing
**Qing Li and Timothy K. Shih**

Chapman & Hall/CRC
Studies in Informatics Series

# UBIQUITOUS MULTIMEDIA COMPUTING

EDITED BY

## QING LI

CITY UNIVERSITY OF HONG KONG

HONG KONG SAR, CHINA

## TIMOTHY K. SHIH

ASIA UNIVERSITY

WUFENG, TAIWAN

CRC Press
Taylor & Francis Group
Boca Raton   London   New York

CRC Press is an imprint of the
Taylor & Francis Group, an **informa** business

A CHAPMAN & HALL BOOK

Chapman & Hall/CRC
Taylor & Francis Group
6000 Broken Sound Parkway NW, Suite 300
Boca Raton, FL 33487-2742

First issued in paperback 2017

© 2010 by Taylor and Francis Group, LLC
Chapman & Hall/CRC is an imprint of Taylor & Francis Group, an Informa business

No claim to original U.S. Government works

ISBN 13: 978-1-138-11274-2 (pbk)
ISBN 13: 978-1-4200-9338-4 (hbk)

### Library of Congress Cataloging-in-Publication Data

Ubiquitous multimedia computing / Qing Li and Timothy K. Shih.
     p. cm. -- (Chapman & Hall/CRC studies in informatics series)
    Includes bibliographical references and index.
    ISBN 978-1-4200-9338-4 (alk. paper)
    1. Multimedia systems. 2. Ubiquitous computing. I. Li, Qing, 1971- II. Shih, Timothy K., 1961- III. Title. IV. Series.

QA76.575.U25 2009
006.7--dc22
                                                 2009022437

**Visit the Taylor & Francis Web site at**
**http://www.taylorandfrancis.com**

**and the CRC Press Web site at**
**http://www.crcpress.com**

# Contents

## Part I   Ubi-Media Infrastructure

## Part II   Ubi-Media Middleware

## Part III   Ubi-Media Applications

# Foreword

"Computing is ubiquitous" has become a cliché that is now used without much thought. The phrase, while undoubtedly true, hides a considerable amount of complexity. Ubiquity is not only due to the fact that we use computing in our daily lives, but it is also a function of the characteristic of the computing environment. Today's computing environment is rich, consisting of multiple media types accessed over distributed environments.

Consider what "modern" computing meant only a decade ago. We were excited that we could do textual searches over the World Wide Web, even if the precision of the results received were not high and even if it was slow (thus the phrase "World Wide Wait"). Image processing techniques that would scale up to large data sets were still uncommon, and delivery of video over wide area networks was thought to be only possible with special broadband networks such as Asynchronous Transfer Mode (ATM), in many cases separate from the more commonplace Internet connections. Although we talked about multimedia, what we had was, at best, multiple systems dealing with different media types. Certainly integrated systems that could handle multiple media types and provide users with a clean interface were not common.

Consider the environment today. Many of the computing systems that we use every day ubiquitously handle multiple media types, and the ability to search across (at least) text and images in a uniform way is being built into browsers whose accuracies have improved considerably. Sites such as YouTube have become repositories of user-supplied video that is now delivered ubiquitously over the Internet. The networks that now make up the Internet have widely implemented the above mentioned broadband network protocols in their core, making it possible to run many applications with heavy media demands (e.g., remote sensing, remote surgery, and the like). Social networking sites such as Facebook, Picassa, and Flickr are repositories of large amounts of images that can be shared, tagged, and searched. Most certainly the computing environment of today is richer involving multimedia more ubiquitously.

This should not be taken to mean that all is done and all of the major problems are solved. Far from it; there is still much work to do on a wide variety of fronts. We still need to find better ways of storing and accessing multimedia data, better precision and recall of retrieval systems, better user interfaces to access this media including ways to consider user preferences and context, better handling of mobile devices in this environment, and the list goes on.

This book contains chapters that address many of these issues. The wide coverage of the book is an indication of the complexity of finding an integrated solution to these problems. The book will undoubtedly be very welcome by both practitioners and researchers in this area that continues to occupy many of us.

**M. Tamer Özsu**

# Introduction

Events in cyberspace can be modeled in a spatiotemporal continuity by computing devices, communication channels, and multimodal interactions. Contemporary ubiquitous devices unleash the boundary of one-to-one human–computer interaction. Someone using a device somewhere through multimodal interactions has now become a *de facto* style of facilitating social events. How people access multimodal media in different contexts is an interesting yet challenging problem. Ubi-Media Computing, as it is bravely defined, brings together technologies for location–context adaptation, inter-device interaction–reaction, and media–data communication.

This book aims at revealing the fundamental technologies and potential research focuses of Ubi-Media Computing. The notion of context is not new. Context may include semantics information as defined in information retrieval. However, with the rise of mobile personal device use, the meaning of context could be extended to include location information, types of personal devices and their capacities, personal preferences, as well as the scenario where events occur between machines and the real world (i.e., searching for nearby restaurants). The new scenario of *interaction* under different types of contexts is more complicated. The multimodality need should be defined between different contexts and types of interactions. On the other hand, interaction among a group of people relies on typical *multimedia* communication technologies. Although communication via audio became popular on cellular phones, the need for short messages, file transfer, and video streaming was recognized by mobile device vendors. How to efficiently use media for communications and how to design a generic-friendly, multimodal user interface for easy use remains a challenge. The development of a new style of multimodal multimedia interaction will need sophisticated adaptation techniques for different mobile devices under different contexts.

The book discusses Ubi-Media Computing on three levels: infrastructures, where fundamental technologies need to be developed; middleware, where the integration of technologies and software systems need to be defined, and applications, where usage cases in the real world need to be realized.

Part I (Ubi-Media Infrastructure) contains five chapters. Chapter 1 is a survey to discuss various architectures for delivering multimedia content to users. The discussion starts from a uni-cast architecture. Challenges in multicast architecture follow, with a comparison to traditional uni-casting. The authors finally point out the advantages of using peer-to-peer (P2P) streaming technologies to reduce server load and network bottlenecks. With a focus on audio communication, Chapter 2 studies parameter values of control schemes on two-party VoIP (Voice over Internet Protocol) systems. The study and experiments suggest that the subjective preferences of users could be tested against different parameters and evaluated to review the impact-to-user satisfaction. The accuracy and efficiency of conducting subjective tests are also addressed. Chapter 3 is about architecture for video communication over wireless networks. Especially, the standard of 3G circuit-switched mobile video (3G-324M) is discussed. In addition, Chapter 3 discusses usages of video sensor devices to be connected to networks like UMTS and WiMAX. As a result, the hybrid network is able to deliver sensor video streaming to terminals anywhere. In contrast to the first three chapters focusing on communication, Chapter 4 discusses the security issues of peer-to-peer networks. A reputation-based mechanism, without centralized control, points out the importance of automatically recording, analyzing, and even adjusting the metrics of reputation, trust, and credibility among peers. Chapter 5 points out the three main tasks of a sensor (i.e., sensing, processing, and transmitting data), by providing a task-flow graph as a theoretical model to evaluate the upper and the lower bounds of using a different number of data collection sensors under different scheduling algorithms.

Part II of this book (Ubi-Media Middleware) has six chapters on various software technologies for interaction, message embedding, and indexing. Chapter 6 presents a project that involves the use of a wearable computer and static sensor networks as remote sensing devices in a large-scale factory, especially in security control. Access points of the wireless network are used to connect sensors and wearable computers. In addition, a middleware developed under Linux is presented to enable seamless connections. The next two chapters demonstrate the usage of finger gestures and advanced devices to provide friendly human–computer interactions. Chapter 7 presents a set of finger gesture commands on multitouch screens to control document accessing among multiple users, as well as providing a set of interactive commands for resizing documents and creating user working areas. A practical implementation called JuTable is also presented. In addition to using a gesture interface of a stick shape object (enhanced based on a Wii controller), the authors of Chapter 8 create a two-dimensional sound display called Sound Table (with 16 speakers) for supporting collaborative work. Feedback to users can also be presented using video. Applications of this system include creating music by manipulating a number of sound objects. A digitally enhanced printout concept, called *newsputers*, is introduced in Chapter 9. Newsputers are applicable to conventional printed materials by providing embedded image information, which does not interfere with ordinary reading. The interaction through a scanner allows users to access multimedia information. It is also possible to extend the technique to include semantic information such that an object with newsputers has self-awareness functionality. Chapter 10 discusses mechanisms to embed secret messages into innocuous looking objects, known as covers. An efficient message embedding method for minimizing the embedding impact is presented. Chapter 11 is a survey of using color, shape, texture, motion, and additional information in searching for multimedia objects. Applications of such a retrieval technique include a Web image search on mobile devices.

Part III (Ubi-Media Applications) contains six chapters of applications using ubiquitous and multimedia technologies. Chapter 12 presents a system that allows users to save, retrieve, and share so-called serendipitous moments by means of location-aware multimedia objects. The application can be developed based on technologies such as P2P communication on mobile devices and the usage of Google Map. Chapter 13 presents a similar intelligent travel book management system using a PDA or cellular phone. Multimedia information supported in the system includes photos, audios, videos, and notes taken on the journey. A traffic congestion alleviation system discussed in Chapter 14 uses Vehicular Ad hoc NETworks (VANET) and GPS-enabled devices. Two types of communication modes: vehicle-to-roadside and vehicle-to-vehicle, are addressed. Chapter 15 presents a system using RFID in emergency room operations. The chapter reports on the findings of a case-based research investigation in a hospital. The RFID is applied in emergency room workflow as a new service to ease patient waiting time, and accuracy for patient treatment, and prevent human errors. The last two chapters discuss practical usages of ubiquitous and multimedia technologies in education. Chapter 16 presents a multiagent architecture for supporting users learning the ubiquitous environment. The agents sense users' physical locations automatically, find the suitable learning map, plan the learning route, and guide the users in the real world. Chapter 17 presents alternatives of using intelligent tutoring techniques with adaptive media. The authors conclude that the learning experiences will come to users in response to their strengths and prior learning, interests, and aspirations.

With the continued and increasingly attracted attention on ubiquitous multimedia computing, we foresee that this fast growing field will flourish as successfully as what the Web has achieved over the past decade and a half. We hope the readers will like this book and enjoy the journey of studying the fundamental technologies and possible research focuses of Ubi-Media Computing. While the book is intended to be a timely handbook for researchers and senior-postgraduate students, it can also be a useful reference book for other professionals and industrial practitioners who are interested in learning the latest technology and application development in this increasingly important field.

# Part I

# Ubi-Media Infrastructure

# 1

# Peer-to-Peer Streaming Systems

Victor Gau, Peng-Jung Wu, Yi-Hsien Wang, and Jenq-Neng Hwang

## CONTENTS

## 1.1 Introduction

For more than a decade, content providers have been trying to deliver multimedia content to end users via the Internet. Before the introduction of streaming technology, users needed to first download complete multimedia content from the Internet down to their own storage before being able to play back the content. Owing to the development of streaming technology, users can playback content while downloading.

In an earlier age, multimedia content was only delivered via client/server architecture. One of the reasons is that at that time the computing power of a regular PC was insufficient to be a server. A server needs to serve requests from many users and requires much computing power. Most end users could not afford to buy computers with such server capabilities. As users' demand for multimedia content grew and the power of their machines increased, Internet Service Providers (ISP) and Internet Content Providers (ICP) figured that there will always be scalability and cost problems from only deploying client/server architectures. Subsequently, Peer-to-Peer (P2P) technologies started to receive the spotlight.

The idea of P2P networks is in contrast to client/server architectures. In a client/server architecture, dedicated servers are deployed to serve clients. However, as common PCs have increasingly more computing power, a lot more processing load can be distributed to the client side. Therefore, the roles of clients and servers have changed. In P2P networks, there are not clear roles for servers and clients; instead, peers share duties with both. At times, peers act as servers, and at times, as clients. By using P2P technologies, the load and cost of the servers are distributed to and shared by peers. Especially after the impact of Skype and BitTorrent on telephony and file sharing, ISPs and ICPs are more and more interested in P2P technologies.

### 1.1.1 Peer-to-Peer Streaming Systems

P2P streaming systems have been recently garnering more and more attention, because most people believe that IPTV is the next killer application and that P2P technologies will be the key to solve the scalability and cost problem. Basically, there are two ways of applying P2P technologies to reduce the cost and load on the server side. First, the media content can be distributed into peer networks, rather than being stored in centralized repositories. In such cases, peers are used as shared storage and requests for the content are directed to the peers that keep the content. Second, the media content can be relayed by peers instead of servers.

Typical modules within P2P streaming systems are shown in Figure 1.1. The *content lookup substrate* is in charge of content lookup or content discovery. The *streaming control module* is in charge of peer management and multicast

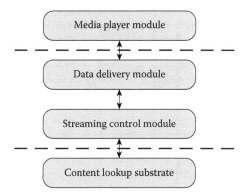

**FIGURE 1.1**
Modules within a peer of a P2P streaming system.

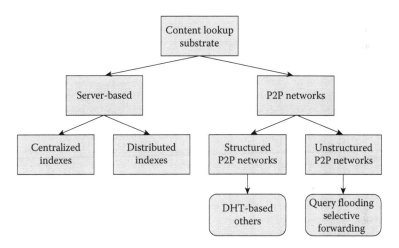

**FIGURE 1.2**
Categories of content lookup substrates.

topology maintenance. The *data delivery module* is in charge of data exchange among peers. The *media player module* is in charge of media playback.

Based on the way the protocols distribute and discover the content in the content lookup substrate, P2P networks can be categorized into structured P2P networks and unstructured P2P networks (see Figure 1.2). In structured P2P networks almost all protocols use a Distributed Hash Table (DHT) to manage content distribution and discovery. In unstructured P2P networks, most protocols simply flood the network with requests to discover content.

Based on the data delivery mechanisms of the networks controlled by the streaming control layer, the P2P networks can be categorized by whether they use tree-push, mesh-pull, or push-pull methods (see Figure 1.3).

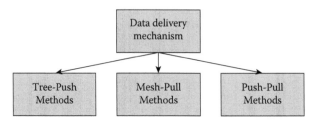

**FIGURE 1.3**
Categories of data delivery mechanisms.

The recent progress in P2P live streaming is prone to using mesh-pull streaming approaches where each peer keeps a buffer of video chunks. Peers exchange video chunks based on gossip-like protocols. The beauty of the mesh-pull approach is that it utilizes the upload bandwidth of all peers to achieve better video quality. The data transmission rate can be up to 300–400 Kbps or higher.

Notice that most P2P live streaming systems do not use P2P content lookup substrates; instead, they use server-based centralized indexes. Since the live channels are not as varied as on-demand media content, a dedicated server or several servers would be enough to handle all the search requests of the content. For some on-demand systems, multicast topology would not be formed. Since the media content would have been distributed in peers already and there would only be few peers interested in the same media content, peers with the content would have the ability to stream the content to all requested users.

The rest of the chapter is organized as follows. In Section 1.2, we introduce the architecture of video over IP networks. In Section 1.3, content lookup mechanisms in P2P networks are discussed. After that, methods of data delivery in P2P live streaming systems are presented in Section 1.4. Then, the peer dynamics in a streaming system are modeled, and packet level Forward Error Correction (FEC) is proposed to improve the robustness of the systems in Section 1.5. Section 1.6 concludes this chapter.

## 1.2 Delivery Technologies for Video over IP Networks

To deliver multimedia content to multiple users, a server could choose to use multiple unicasts (i.e., multiple one-to-one connections), or use multicast (i.e., one-to-many connections). In this section, we first introduce client/server-based architectures that use unicasts for video streaming. After that, we describe multicast and how it is used to lessen bandwidth bottlenecks and server load [1–4].

## 1.2.1 Unicasting Video Streams

Unicast is the traditional method for delivering content to users. There are two different setups for unicast video delivery, commonly called media server farms and Content Delivery Networks (CDN).

### 1.2.1.1 Media Server Farms

In a media server farm [2], several expensive network servers are housed in one location. There is a load balancer distributing user requests to different servers in the farm. In addition to load balancing, another advantage of such a setup is robustness. If any server breaks down, others will take over. All servers in the farm share the same upload capacity. Applications like YouTube and Webs-TV employ media server farms for content distribution.

A conceptual diagram of a media server farm is shown in Figure 1.4. In such a setup, a company would need to purchase or lease high-end servers along with sufficient upload bandwidth. The cost of bandwidth in total would be a lot higher than the fixed equipment. As population grows, a company would need to purchase more servers and pay for more bandwidth to serve the users.

### 1.2.1.2 Content Delivery Networks (CDN)

Content providers who do not want to spend money on maintaining server infrastructure use a CDN [16]. CDN companies, such as Akamai, own the infrastructure, and the content providers, such as CNN, pay for the service of delivering the video content to the users with the least possible delays.

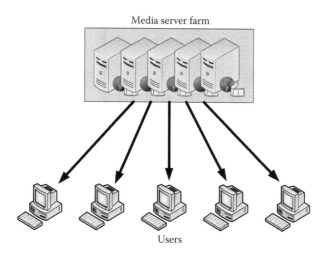

**FIGURE 1.4**
The media server farm.

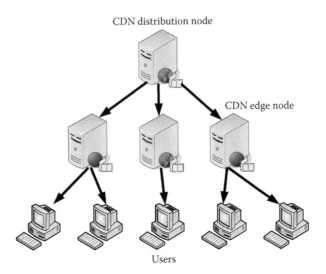

**FIGURE 1.5**
Content delivery network (CDN).

The CDN typically installs hundreds of servers in data centers throughout the Internet. The content providers can upload their contents to a specific server and the contents are then replicated and distributed to CDN edge nodes. When a user requests a video, the CDN server uses DNS-based redirection to redirect the request to the CDN edge node closest to the user. Thus, the delivery delay is reduced (see Figure 1.5).

A CDN architecture is more scalable than a Media Server Farm. However, when a big event happens and everyone wants to get to the same edge server for the same event, that server will still be overloaded. Even though the scalability problem would be solved to some extent, the cost and maintenance of the servers are still very high. And the cost is passed on to the content providers who need to pay more for better service.

Figure 1.6 shows the traditional IPTV architectures. The aforementioned Media Server Farms and CDNs use multiple unicasts to deal with multiple user requests. The arrows show the identical copies of the streams. From the figure, we can easily tell the disadvantages of delivering video streaming with this approach. As the number of connections increase, the server load and the link stress also increase.

It is obvious that multiple unicasts is not sufficiently scalable. Therefore, multicast technologies have been proposed to address this scalability problem.

### 1.2.2 Multicasting Video Streams

Multicast technologies can be loosely classified into IP multicast, Application Layer Multicast (ALM), and hybrid IP/Application-Layer Multicast. Figure 1.7 shows the taxonomy of multicast technologies.

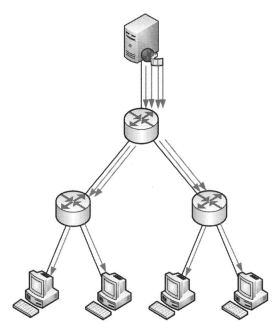

**FIGURE 1.6**
Multiple unicasts to the users.

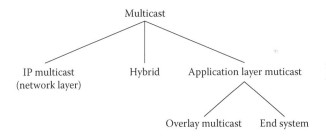

**FIGURE 1.7**
Taxonomy of multicast technologies. (From Buford, J., Presentation Slides, Scalable Adaptive Multicast Research Group, July 2006. With permission.)

### 1.2.2.1 IP Multicast

The IP multicast is infrastructure dependent. This means that the routing devices need to have the functionality to support IP multicast. In IP multicast, an end host uses the Internet Group Management Protocol (IGMP) to communicate with the edge router when attempting to join or leave a particular IP multicast group. The edge router then negotiates with other routers using a multicast routing protocol to join a group-shared tree or source-based tree and then relay the video content to the end user (Figure 1.8) [17].

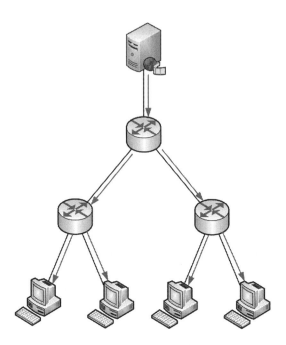

**FIGURE 1.8**
IP multicast.

Some well-known multicast routing protocols include the Distance Vector Multicast Routing Protocol (DVMRP), the Protocol-Independent Multicast in Sparse Mode (PIM-SM), Protocol-Independent Multicast in Dense Mode (PIM-DM), the Core-Base Tree, and the Multicast Open Shortest Path First Protocol (MOSPF). The DVMRP constructs a source-based tree using Reverse Path Forwarding (RPF). The PIM-DM employs a simpler mechanism similar to DVMRP to build a source-based tree. The MOSPF utilizes a protocol extending from the Open Shortest Path First (OSPF) Protocol. The CBT and PIM-SM organizes their multicast trees using a core or Rendezvous Point (RP).

Even though the IP multicast provides bandwidth efficiency, it has several disadvantages [10]. First, as mentioned earlier, IP multicast is infrastructure dependent. Without the support of the routing devices, an IP multicast cannot work. Second, because the edge router needs to maintain the membership information of the end hosts, it lacks scalability as the number of multicast groups increase. Third, it is not easy to implement access control or pricing policy over IP multicasts. Because of some of these disadvantages, even if most routers have the ability to support multicast, ISPs will generally not activate this function.

### 1.2.2.2 Application Layer Multicast

Figure 1.9 shows the conceptual diagram of ALM. In ALM, the server delivers the video content to certain users and the users that receive the content

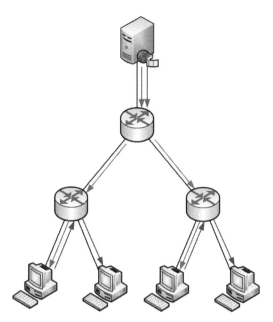

**FIGURE 1.9**
Application layer multicast.

act as relays, forwarding the content to other users in the group. Application layer multicasting is increasingly promising because of its scalability, flexibility, and ease of deployment. There are basically two kinds of ALM: end system multicast and overlay multicast.

With end system multicast, the membership management and multicast trees are maintained by participating end hosts. It does not require extra infrastructure to support ALM. All peers share the load and cost of the whole service.

With overlay multicast, proxies and intermediate nodes are deployed into the network as the "backbone overlay" for content delivery [19,20,24]. Overlay multicast tries to take advantage of both IP and ALM. Since the topological information of the proxies is known, a high performance overlay can be achieved by careful design. In addition, the proxies are well controlled and are less likely to fail or improperly leave the overlay. Therefore, it is more reliable than ALM. Another advantage of overlay multicast is that unlike ALM, overlay multicast can host several multicast groups. However, all these advantages are built upon the cost of deploying and maintaining the proxies over the Internet. Typical overlay multicast protocols include Scattercast [20] and OverCast [19]. Figure 1.10 shows a conceptual diagram for Scattercast, where SCX stands for Scattercast proXies.

The difference between overlay multicast and a CDN is that in a CDN, the client node requests data from CDN edge nodes; in contrast, a client node could request data either from the overlay proxy or from other nodes.

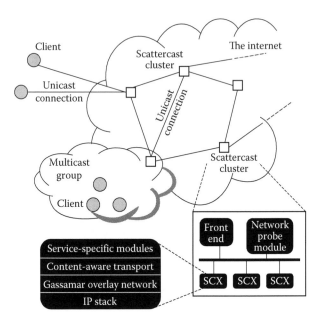

**FIGURE 1.10**
The architecture of Scattercast. (From Chawathe, Y., *Multimedia Systems*, 9, July 2003. With permission.)

Even though ALM is very flexible and very easily deployed, the efficiency of ALM depends highly on the method used to control and manage membership. There are several challenges in designing ALM protocols. First, the ALM protocol should be topology-aware to efficiently deliver data. Lack of topological awareness will cause low efficiency and longer delays in content delivery. Second, while control packets for membership management are inevitable, we need to minimize the control overhead to prevent self-induced congestion of the network. In addition, the dynamic nature of the end hosts affects the performance of downstream members significantly. These challenges become very difficult to deal with when multicast groups grow very large, and they hinder the quality-of-service (QoS) provided to the end users.

### 1.2.2.3 Hybrid IP/Application Layer Multicast

Hybrid IP/Application Layer Multicast tries to integrate IP multicast and ALM to create an efficient multicast architecture. If the current network environment supports IP multicast, then IP multicast will be used. Otherwise, ALM will be used to multicast contents to the users. Typical hybrid multicast approaches include the Host Multicast Tree Protocol (HMTP) and Universal Multicast (UM; Figure 1.11) [21,22].

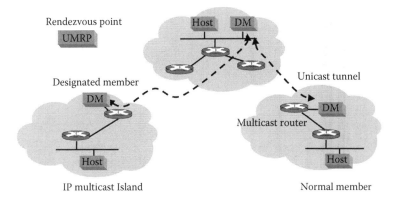

Rendezvous point

Host DM

UMRP

Designated member

Unicast tunnel

DM

DM

Multicast router

Host

Host

IP multicast Island

Normal member

**FIGURE 1.11**
The architecture of universal multicast. (From Zhang, B., Wang, W., Jamin, S., Massey, D., and Zhang, L., *Computer Networks*, 50, April 2006. With permission.)

In HMTP, an IP-multicast island is defined as a network of any size that supports IP multicast. One member host in an island is elected as the Designated Member (DM) for that island. Different islands are connected by user datagram protocol (UDP) tunnels. In HMTP, only DMs participate in tree construction and maintenance. Data communication among DMs use ALM, and, within each IP-multicast island, it takes advantage of IP-multicast. The HMTP is a protocol to build a group-shared tree where multisenders share this group-shared tree for bulk data transfer. UM employs HTMP for inter-island routing and the Host Group Management Protocol (HGMP) for intra-island multicasting. Disadvantages of hybrid multicasting include the complexity of the system and performance variation due to the dynamic nature of the end host.

## 1.3 Content Lookup in Peer-to-Peer Networks

As previously mentioned, P2P networks can be used to share the server load, and one of the methods is to distribute the content to peers. To do so, we need to have a protocol to distribute that content, and a method to locate the content afterward. In this section, we first introduce structured P2P networks, in which pieces of content are distributed and searched by using a DHT. We also introduce some systems that employ structured P2P networks as their content lookup substrate. In addition to the content discovery, we will see that in a structured P2P network, the route to the content can also be employed to form a multicast tree for content delivery. After that, we briefly introduce unstructured P2P networks and a system that uses an unstructured P2P network as its content lookup substrate.

### 1.3.1 Structured Peer-to-Peer Networks

Structured P2P networks are mostly based on different arrangements of a DHT. Hash tables are data structures designed for efficient search [18]. The DHTs are a distributed version of hash tables, where all participating hosts keep part of all the entries of the hash table [25–30].

#### 1.3.1.1 Distributed Hash Table (DHT)

In the DHT implementation, an end host would pass a message $X$ (e.g., IP address) to a hash function to get a key $H(X)$. The content object $Y$ would also be passed to the hash function to get a key $H(Y)$. The content object $Y$ would then be placed into the end host $X$ with the keys $H(X)$ and $H(Y)$ closest to each other.

Figure 1.12 shows a simple example of DHT-based protocol. There are four computers (i.e., A, B, C, and D) interconnected to each other based on DHT-based protocol. Each computer can send a message to a hash function to get a key (e.g., host A has the key numbered 507). Different content objects can

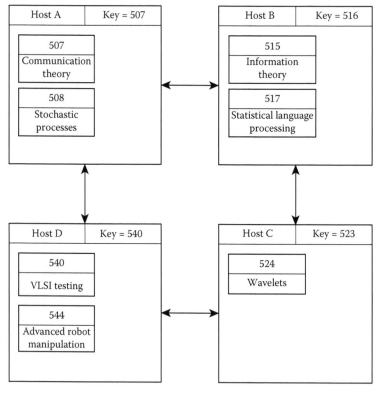

**FIGURE 1.12**
A simple example of a DHT-based concept.

use the same way to get their keys. The content objects are then placed into the host with the key closest to themselves.

DHT-based protocols differ in the ways of how they organize the membership of the participating hosts and how they route to locate the content objects.

### 1.3.1.2 Pastry

Pastry [25] is a popular large-scale DHT-based P2P network protocol developed by Microsoft Research and Rice University. Each participating node gets its unique 128-bit node identifier (nodeId) by a hash function (e.g., SHA-1). Pastry uses a circular nodeId space for organizing the membership of the nodes. Figure 1.13 shows a sample nodeId space used by Pastry. By reading from the prefix of a nodeId, we can quickly locate a node on the circular nodeId space. The nodeId can be viewed as a sequence of digits with base $2^b$. When $b = 4$, then the first digit in the nodeId divides the circular nodeId space into 16 equal pieces. The second digit divides 1/16 of the circular nodeId space into another 16 equal pieces. Therefore, we can locate a node by checking the prefix of its nodeId digit by digit (see Figure 1.14).

Figure 1.15 shows the routing table of a node in Pastry. When a node is trying to join Pastry, it first contacts a bootstrap node that is already in the nodeId space. The bootstrap node then routes the newly joined node to the corresponding location in the nodeId space. During the routing, the newly joined node collects the routing tables from the nodes it visits to create its own routing table.

When a node tries to publish a message to the Pastry network, it first calculates the hash key of the message. The message is then routed to the node

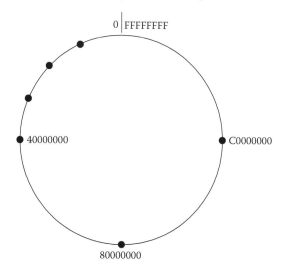

**FIGURE 1.13**
A sample nodeId space in Pastry.

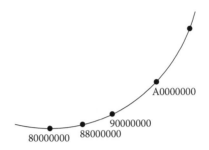

A0000000

90000000
80000000    88000000

**FIGURE 1.14**
A sample nodeId space enlarged from 80000000 to A0000000.

| 0 | 1 | 2 | 3 | 4 | 5 | 6 | 7 | 8 | 9 | a | b | c | d | e | f |
|---|---|---|---|---|---|---|---|---|---|---|---|---|---|---|---|
| x | x | x | x | x | x |   | x | x | x | x | x | x | x | x | x |
| 60x | 61x | 62x | 63x | 64x |  |  | 67x | 68x | 69x | 6ax | 6bx | 6cx | 6dx | 6ex | 6fx |
| 650x | 651x | 652x | 653x | 654x | 655x | 656x | 657x | 658x | 659x |  | 65bx | 65cx | 65dx | 65ex | 65fx |
| 65a0x |  | 65a2x | 65a3x | 65a4x | 65a5x | 65a6x | 65a7x | 65a8x | 65a9x | 65aax | 65abx | 65acx | 65adx | 65aex | 65afx |

**FIGURE 1.15**
Routing table of a Pastry node with nodeId *65a1x*, $b = 4$. Digits are in base 16, *x* represents an arbitrary suffix. (From Castro, M., Druschel, P., Kermarrec, A.-M., and Rowstron, A., Proceedings of SIGOPS European Workshop, France, September 2002. With permission.)

with the nodeId closest to the message key. Figure 1.16 shows the process of routing a message.

The DHT-based P2P networks have some disadvantages. First, they do not adapt well to node churns. The overhead of joining and leaving the network is usually high; therefore, as nodes join and leave the network frequently, the performance degrades. Second, DHT-based P2P networks do not support substring search. Due to the nature of the hash function, if there would be any modification of the message, the key generated by the hash function will be totally different. Therefore, the user would not be able to just use the substring of the key to search for a message.

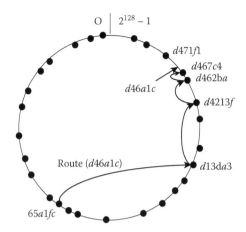

**FIGURE 1.16**
Routing a message from node 65a1fc with key d46a1c. The dots depict live nodes in Pastry's circular namespace. (From Castro, M., Druschel, P., Kermarrec, A.-M., and Rowstron, A., Proceedings of SIGOPS European Workshop, France, September 2002. With permission.)

### 1.3.1.3 Applications of Pastry

Scribe [34] is a large-scale event notification network based on Pastry. In Scribe, the publisher publishes a *topicId* to the network. The subscribers route to the topicId to subscribe for the topic. The Scribe then combines each route back to the subscribers to get an ALM tree. When the publisher updates topic related content, the content will be delivered through the multicast tree to the subscribers. Some of the nodes in the multicast tree do not subscribe to the topic for themselves—they serve as relays forwarding the content to the subscribers.

SplitStream employs interior-node-disjoint multicast trees for streaming. In most tree-based streaming protocols, few interior nodes within the multicast tree are in charge of forwarding the streaming content. This not only imposes a disproportionate load on those nodes, but also makes the streaming vulnerable to node failures. SplitStream divides the streaming into $k$ different stripes, and organizes nodes into $k$ or so-called interior-node-disjoint multicast trees. Participating nodes are interior nodes in one of the $k$ trees, and leaf nodes in the rest of $k$–1 trees. SplitStream uses Scribe and Pastry to build each multicast tree. For each tree, a stripeId with a different starting digit is assigned so that the participating nodes are automatically placed into leaf or interior nodes within each tree. Since there are $k$ trees, different error correction coding can be applied to SplitStream to increase the robustness of the streaming (see Figure 1.17).

The disadvantages of SplitStream are that it assumes that nodes uninterested in the streaming are still willing to be relays to help forward streaming content to other nodes. Additionally, there will be synchronization problems when applying error correction coding to different $k$ trees.

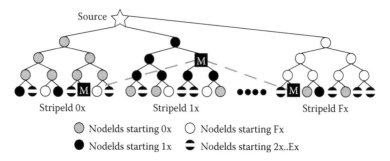

**FIGURE 1.17**
SplitStream's forest construction. Each stripeId starts with a different digit. The nodeIds of interior nodes share a prefix with the stripeId, thus they must be leaves in the other trees. (From Castro, M., Druschel, P., Kermarrec, A.-M., Nandi, A., Rowstron, A., and Singh, A., Proceedings of the 19th ACM Symposium on Operating Systems Principles, October 19–22, 2003. With permission.)

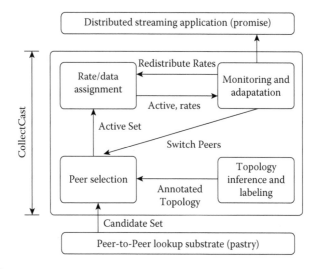

**FIGURE 1.18**
Components of CollectCast. (From Hefeeda, M., Habib, A., Xu, D., Bhargava, B., and Botev, B., *ACM Multimedia Systems J.*, 11(1), 68–81, November 2005. With permission.)

PROMISE is a P2P media streaming system intended for on-demand video playback that enables one receiver to collect data from multiple senders. PROMISE is based on a P2P service called CollectCast. CollectCast bridges PROMISE and DHT-based structures and performs three major functions: selecting peers based on the inferred underlying network topology, monitoring the status of the peers, and dynamic switching of the active and standby peers.

Figure 1.18 shows the components of CollectCast and the interactions among them. The P2P lookup substrate could be Pastry, CAN, or Tapestry. PROMISE outlines some good policies for peer selections and dynamic adaptation. However, it is mainly used for on-demand video playback. Even though their small-scale experiment appears promising, the large-scale performance is unknown. The time complexity of its topology-aware operation would make it very difficult to be scalable for large-scale applications.

### 1.3.2 Unstructured Peer-to-Peer Networks

In contrast with structured P2P networks, peers in unstructured P2P networks flood search requests to discover the media content.

Gnutella is a typical unstructured P2P network protocol for file sharing. The peers in Gnutella network use ping-pong messages to discover hosts in the network, and use query flooding to discover the content. The Time-To-Live (TTL) field of packets is used in the query messages to limit the number of times each query message is forwarded. It is generally believed that all nodes could be reached in six hops due to the famous study of the six degrees of separation. The value of TTL is typically set to seven.

GnuStream [54] is a P2P streaming system build atop Gnutella. The architecture of GnuStream consists of three layers: the Network Abstraction Layer (NAL), the Control Streaming Layer (CSL), and the Media Player Layer (MPL). The NAL is for locating, routing, and data retrieving from the network. The CSL is for load balancing and network condition adaptation. The MPL is in charge of playing back the video. GnuStream actually delegates the content lookup service to Gnutella.

## 1.4 Application Layer Multicasts in Peer-to-Peer Streaming Systems

As mentioned in an earlier section, multiple unicasts of the same content make a server a bottleneck in the unicast systems. Therefore, if there is a flash crowd of users who would like to watch the same video channel, we tend to arrange these users into different topologies to take advantage of ALM for data delivery. For P2P live streaming systems, there are basically three different data delivery methods: tree-push, mesh-pull, and push-pull [9–15,38–39,56]. In this section, we present these three data delivery methods and classify a few P2P live streaming systems based on these data delivery methods.

### 1.4.1 Tree-Push Methods

Most of the earliest ALM protocols employ tree-push methods to deliver media data [32,33]. The media content is first delivered from the source node to several intermediate nodes. The intermediate nodes then forward the data to the remaining nodes. The content is actively pushed from the root of the tree to all other nodes in the tree (see Figure 1.19).

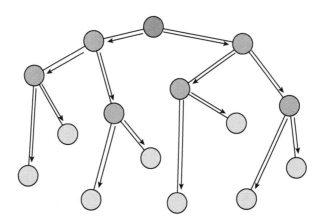

**FIGURE 1.19**
Tree-push streaming.

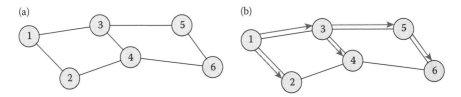

**FIGURE 1.20**
(a) Narada maintains an efficient mesh overlay among peers. (b) A source specific tree is built atop this efficient mesh overlay afterward.

### 1.4.1.1 Narada

Narada [31] is one of the earliest ALM protocols. It is intended for multi-source, multicasting. Narada is a mesh-first approach, that is, the main goal of the protocol is to maintain an efficient mesh overlay. The multicast tree is build atop this efficient mesh overlay. More specifically, when nodes have media data to broadcast, a source specific multicast tree will be built based on RPF. Even though Narada is very robust because of its mesh nature, it is not designed for large groups due to the high overhead of maintaining the underlying mesh overlay structure (see Figure 1.20).

### 1.4.1.2 NICE

NICE [40] arranges sets of end hosts into a hierarchy; the basic operations of the protocol are to create and maintain the hierarchy. The hierarchy implicitly defines the multicast overlay data paths.

NICE minimizes the end-to-end delay between the leaders to their cluster members by electing the leader as the center node of the cluster. The cluster size is bound by $[k, 3k-1]$, where $k$ is a universal constant of this protocol. Clusters

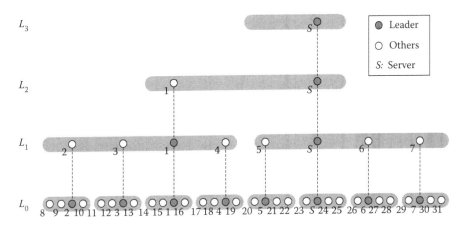

**FIGURE 1.21**
Control topology of NICE.

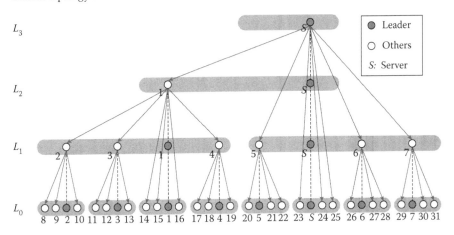

**FIGURE 1.22**
Multicast tree of NICE.

are maintained periodically. If cluster size is smaller than $k$ or larger than $3k-1$, follow-up cluster merges or splits will be performed (see Figure 1.21).

A cluster leader at layer $Lj$, is also a cluster leader of a certain cluster at a lower layer $Li$, where $Li < Lj$. In NICE, cluster leadership may change as members join or leave. If there is a change in leadership of a cluster $C$, in layer $Lj$, the current leader of $C$ removes itself from all layers $> Lj$. Each affected layers then must choose a new leader. New leaders join their super-cluster. If the state of the super-cluster is not locally available, the node will need to contact Rendezvous Point (RP; see Figure 1.22).

There is a possible bottleneck because in this hierarchical architecture. The node at the top layer would need to deliver data to about $k \times (\log n)$ members. As the scale increases, this will become the bottleneck of a system.

One of the goals of tree-push protocols is to construct an efficient multicast tree. However, it is observed that even though the data dissemination paths are efficient, the intermediate nodes tend to overload with increasing scale. Additionally, most tree-push architectures do not adapt well to node churns. When nodes join and leave frequently, the service would be interrupted [41,42].

### 1.4.2 Mesh-Pull Methods

Many popular P2P live streaming systems make use of protocols modified from P2P file sharing protocols like BitTorrent to deliver media content. The peers are randomly organized into a mesh structure. The selection of active peers for data exchange is commonly decided based on the round trip time (RTT) between peers. Peers exchange information about what video chunks they have before exchanging media data. This is known as the gossiping process. After that, in the pulling process, peers request video chunks that they do not have from other peers who do (see Figure 1.23) [5–7].

To explain the mesh-pull P2P streaming applications, we first briefly describe the BitTorrent file sharing protocol. There are basically three components within the BitTorrent protocol: a web server keeping the torrent files (the metadata of the sharing files); a tracker keeping track of the information of all participating peers; and a swarm of seeds and leechers, where seeds are peers that have finished downloading the file but are still active for uploading to others and leechers are peers that are still downloading (see Figure 1.24).

When a new leecher tries to join the torrent, it first connects to a web server to download the torrent file. The torrent file includes the detailed file and tracker location. After that, the new leecher registers its information with the tracker and gets an initial peer list. After the leecher has this initial peer list, it starts connecting to other peers to request exchanges of file chunks.

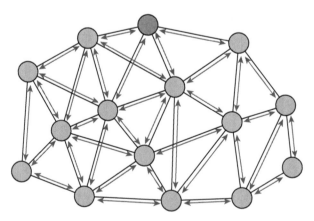

**FIGURE 1.23**
Mesh-pull streaming.

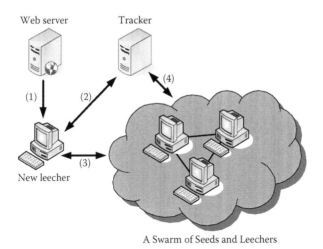

**FIGURE 1.24**
BitTorrent for file sharing.

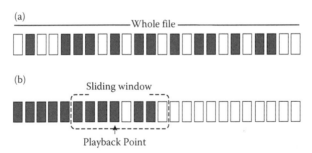

**FIGURE 1.25**
Buffer snapshots of (a) BitTorrent and (b) CoolStreaming. (From Liu, J., Rao, S., Li, B., and Zhang, H., *Proceedings of the IEEE*, 96, 11–24, 2008. With permission.)

BitTorrent and mesh-pull P2P streaming systems like CoolStreaming differ mainly in their buffer management. Figure 1.25 shows the buffer snapshots of BitTorrent and CoolStreaming.

### 1.4.2.1 CoolStreaming (DONet)

CoolStreaming, developed by Hong Kong University of Science and Technology, is one of the earliest P2P streaming systems. The first version of CoolStreaming uses a mesh-pull mechanism to collect video chunks from peers. Each video chunk contains about one second of content. A peer keeps about 120 video chunks in the buffer. A buffer map with a 2-byte offset and 120 bits for identifying the video chunks is used to exchange existing video chunks among peers. Peers try to identify the video chunks they want based on the buffer maps and send out requests for the video chunks once identified.

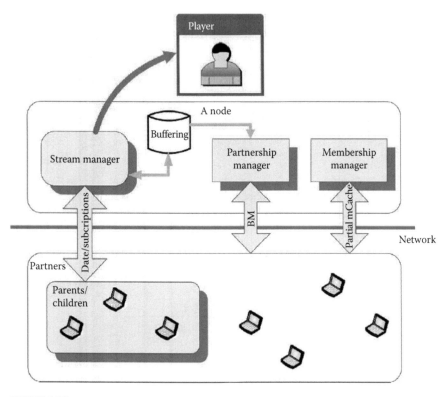

**FIGURE 1.26**
CoolStreaming system diagram. (From Hei, X., Liang, C., Liang, J., Liu, Y., and Ross, K., *Multimedia, IEEE Transactions*, 9, 1672–87, 2007. With permission.)

Figure 1.26 shows the conceptual diagram of CoolStreaming. There are three major components in CoolStreaming: Membership Manager, Partnership Manager, and Streaming Manager. The membership manager gets an initial list from a bootstrap node, and then tries to collect more peers using the scalable membership protocol (SCAMP). The membership manager keeps a partial list of the participating members in a membership cache (mCache). The partnership manager selects peers from the mCache based on RTT and puts the active peers into the partner list. The streaming manager is in charge of exchanging video chunks with other partners. In the most recent version of CoolStreaming, the push-pull mechanism has been adopted to reduce the playback delay.

### 1.4.2.2 PPLive

PPLive is a popular P2P streaming system that uses proprietary technologies for media delivery. Hei et al., 2007 [47,48] conducted a measurement study on PPLive. They showed that PPLive uses a protocol similar to the BitTorrent file sharing protocol. When joining the network, a peer first connects to a tracker to get the channel list. After deciding what channel to watch, the peer then registers its information with the tracker server and downloads an initial peer list. While trying to connect to peers in the initial peer list, it also exchanges peer list information with other peers, so as to collect more peers for best parent selection.

PPLive divides the streamed video into smaller video chunks. A buffer map records the existing video chunks in a peer. Peers then exchange buffer map information before pulling data from other peers. Figure 1.28 shows a conceptual diagram of the buffer map.

Figure 1.29 shows a more detailed working mechanism of PPLive. It shows that the protocol used in PPLive is very similar to BitTorrent with the difference being the time constraint of the buffer data. That is, the P2P streaming

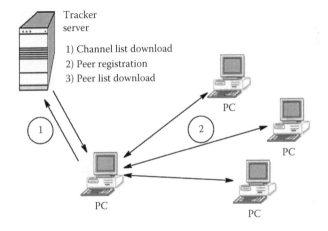

**FIGURE 1.27**
Channel and peer discovery in PPLive. (From Hei, X., Liang, C., Liang, J., Liu, Y., and Ross, K., *Multimedia, IEEE Transactions*, 9, 1672–87, 2007. With permission.)

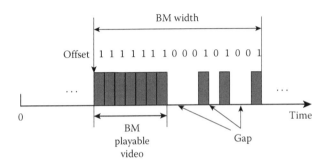

**FIGURE 1.28**
Buffer Map. (From Hei, X., Liang, C., Liang, J., Liu, Y., and Ross, K., *Multimedia, IEEE Transactions*, 9, 1672–87, 2007. With permission.)

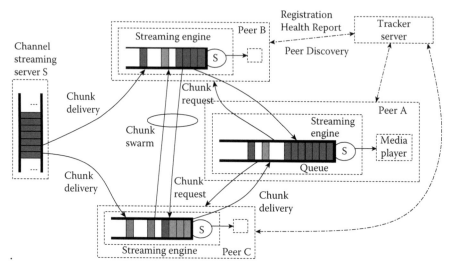

**FIGURE 1.29**
Conceptual diagram of PPLive. (From Hei, X., Liang, C., Liang, J., Liu, Y., and Ross, K., *Multimedia, IEEE Transactions,* 9, 1672–87, 2007. With permission.)

protocol needs to efficiently exchange the video chunks within designated time constraints.

## 1.4.3 Push-Pull Methods

As previously mentioned, the advantage of tree-push methods is their efficiency of data delivery, since once the multicast tree is built, the media data is actively delivered from the parents to the children and additional information exchange (gossiping) is not necessary. The advantage of mesh-pull methods is their robustness to node failure along with the efficient bandwidth usage of all nodes. Push-pull methods are proposed to leverage the respective advantages of tree-push and mesh-pull methods (see Figure 1.30).

### 1.4.3.1 Gridmedia

In mesh-pull P2P streaming systems, the sender needs to update the buffer map to all its neighbors, and the receiver needs to send out requests after identifying the video chunks it needs. Figure 1.31 shows that in mesh-pull methods, a receiver at least needs to wait for three one-way transmission delays before it gets the desired video chunks. By employing a push mechanism, the receiver requests the desired video chunks from the sender based on the sequence numbers of the video chunks, and the sender then pushes

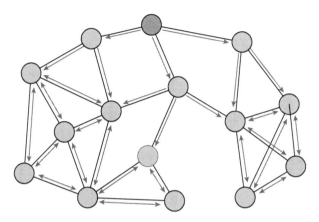

**FIGURE 1.30**
Push-pull streaming.

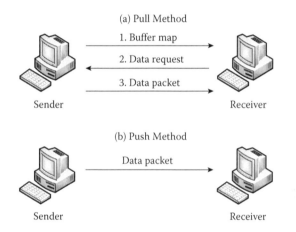

(a) Pull Method

1. Buffer map

2. Data request

3. Data packet

Sender                          Receiver

(b) Push Method

Data packet

Sender                          Receiver

**FIGURE 1.31**
Differences between push and pull methods.

the video chunks to the receiver once it has received them. A pull mechanism is then used as a packet loss resilience mechanism.

Gridmedia [49,50] tries to reduce the delay incurred during the information exchange and proposes a push-pull mechanism. Once a peer finds that it regularly pulls video chunks from the same peer, it will then try to switch to the push mode to reduce the delay.

Gridmedia uses a centralized tracker server. The server records the history data of the connected peers. When a new peer tries to request the initial peer list, the tracker server tries to respond with a peer list where half of the peers are closer to the new peer and the remaining peers are randomly selected by

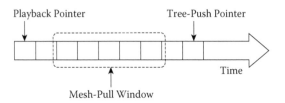

**FIGURE 1.32**
Push-Pull switching buffer in mTreebone. (From Wang, F., Xiong, Y., and Liu, J., Proceedings of International Conference on Distributed Computing Systems, Toronto, Ontario, Canada, June 2007. With permission.)

the tracker. With this mechanism, the new peer should quickly identify good peers and shorten the time for exchanging video chunks.

### 1.4.3.2 mTreebone

The idea of mTreebone is to design a protocol that combines the advantages of both tree-push and mesh-pull methods. A tree-push based backbone (called a treebone) is built from identified stable nodes in the network. A mesh overlay is built upon all the other nodes. The stable node identification is based on observing user behavior so that the nodes with a higher age tend to stay in the network longer. The media data is delivered through the treebone, and the lost packets are recovered by the mesh structure (see Figure 1.32).

## 1.5 Dynamic Behavior of Peers in Peer-to-Peer Systems

The dynamic behavior of peers in P2P systems is one of the major challenges. Since peers are transient in nature, once a parent peer departs from the system, the receivers receiving streaming content from that parent peer might suffer from temporarily unstable transmission. In P2P streaming systems that organize peers into a specific topology (e.g., tree-based topology), this temporarily unstable transmission could lead to burst packet loss [57,58,67]. To reduce the impact of parent departure, many researchers [35,37,45,65,66] focus on detecting the failure and locating a new parent as quickly as possible. However, even if the peer failure can be detected and quickly recovered from, the associated child peers still suffer burst packet losses before locating a new parent peer. Moreover, in a tree-based P2P system, once an ancestor peer loses packets, all of its descendant peers have no way to receive the missing packets. Subsequently, the packet losses accumulate along the forwarding path. Addressing the above problems, we first discuss the dynamic behavior of parents in P2P systems and then describe a multisource structure with *Forward Error Correction* (FEC), which can be used to overcome the burst packet loss problem.

## 1.5.1 Modeling Parent Behavior

A *Continuous-Time Markov Chain* (CTMC) is used to model the behavior of parents that are currently forwarding packets to a child peer [58,67]. More specifically, consider an *n*-source P2P system where each peer always tries to maintain connections with *n* parents and concurrently receives packets from these parents. Even though a child peer tries to maintain connections with *n* parents, due to parent departures the number of active parents of the child peer may not always be *n*. Every time a parent departs, the child peer invokes a parent recovery process to replace the departed. Note that the remaining parents might depart from the system during the parent recovery process; in that case, the child receiver will invoke another parent recovery process to find a new parent. Assume that multiple parent recovery processes can be executed simultaneously (e.g., with a concurrent technique). Let $S_i$ be the state of the CTMC representing that *i* parents of a *child peer* that has left the system, namely, *i* parent recovery processes are running. Suppose the potential time that a child peer needs to locate a new parent and the potential time that a parent (peer) stays in the system are exponentially distributed with means $1/\mu$, $1/\lambda$, respectively. The whole system can thus be seen as an M/M/c *queuing system* with arrival rate $\lambda$ and departure rate $\mu$. By the memoryless property of the exponential distribution, we have the following transition diagram of the system (see Figure 1.33).

Based on the principle that the rate at which the process enters state $S_i$ equals the rate at which it leaves state $S_i$ [59], we have the following equation

$$
\begin{array}{cc}
\text{state} & \text{Rate leave = rate enter} \\
0 & n\lambda P_0 = \mu P_1 \\
1 & (\mu + (n-1)\lambda)P_1 = n\lambda P_0 + 2\mu P_2 \\
2 & (2\mu + (n-2)\lambda)P_2 = (n-1)\lambda P_1 + 3\mu P_3 \\
i, \ 0 < i < n & (i\mu + (n-i)\lambda)P_i = (n-i+1)\lambda P_{i-1} + (i+1)\mu P_{i+1} \\
n & n\mu P_n = \lambda P_{n-1}
\end{array}
\qquad (1.1)
$$

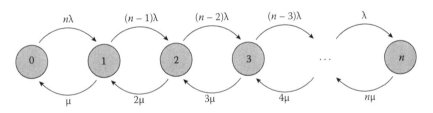

**FIGURE 1.33**
The transition diagram of the system.

By using

$$\sum_{x=0}^{n} P_x = 1,$$

the steady-state probability $P_i$ can then be obtained. Equation 1.2 shows the calculation result of $P_0$ where $\rho = \lambda/\mu$. Since each element of $P_i$ is directly related to $P_0$ this can be used to derive $P_i$. For more detail of the calculation, please refer to Wu, Lee, Gau, and Hwang [58].

$$P_0 = \begin{cases} \dfrac{1-\rho}{1-\rho^n}, & \text{if } \rho \neq 1 \\ \dfrac{1}{1+n}, & \text{if } \rho = 1 \end{cases}. \tag{1.2}$$

### 1.5.2 Overcoming Burst Packet Loss by a Multisource Peer-to-Peer System with Forward Error Correction Scheme

To overcome burst packet loss caused by parent departure, a multisource P2P live streaming system with a packet-level FEC scheme is proposed in Wu, Lee, Gau, and Hwang [58], as shown in Figure 1.34, where each parent is responsible for forwarding only partial streaming content, to reduce the impact of parent departure. Packet level FEC is a receiver-driven error correction scheme that overcomes the packet loss problem by adding a certain number of redundant packets that can be used to reconstruct the originally transmitted data packets after packet losses. For a packet level FEC scheme [60–64], FEC($n$, $k$), when $n$ packets are transmitted, the receiver can fully recover all $k$ information packets if $n - k$ or fewer packets are lost during transmission. In a multisource system with packet level FEC, as long as the remaining parents can provide sufficient packets, the lost packets can thus be recovered by FEC. In the example shown in Figure 1.34, an eight-source multicast structure with packet level FEC(8,5) is used. For every five data packets, three redundant FEC packets are added. For

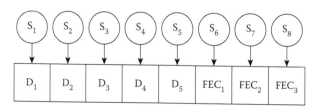

**FIGURE 1.34**
An example of eight sources structure with an FEC(8,5) scheme.

convenience, these eight packets altogether are called an *ensemble*. In this case, the FEC scheme is capable of recovering at most three missing packets out of an ensemble as shown in Figure 1.35. As shown in the example, even though three out of eight parents have failed, the receiver can still recover all of the original packets as well.

Figure 1.36 shows a possible topology where each peer has three different parents. Each parent peer can then forward partial streaming content to a child peer (i.e., one packet per ensemble).

Figure 1.37 shows the analytical packet loss probability of the multisource, multicast structure with a packet drop rate of 0.1 under different FEC protection levels. The average recovery time is 30 seconds and the average service time is 10 minutes. As one can see, the packet loss accumulates rapidly with increasing depth when there is no FEC protection. With a low level FEC protection, such as FEC(5,4), the packet loss performance improves, but still, the packet loss accumulates as the depth increases. However, an interesting result is that when a sufficiently strong FEC protection is used, the packet loss accumulation stops. That means when an appropriate FEC protection is incorporated into a multisource system, the impact of a packet loss due to the dynamic behavior of peers can be overcome. Further detailed mathematical analysis and discussion of multisource, multicast structures can be found in Wu, Hwang, Lee, Gau, and Kao [67].

The benefits of using multiple sources and a packet level FEC is that once a few parents fail or the connections to some parents are suffering from bad conditions, the remaining parents can still provide most of packets that can be used

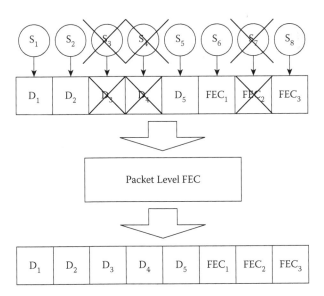

**FIGURE 1.35**
An example of packets recovery with three sources failed in an FEC(8,5) system.

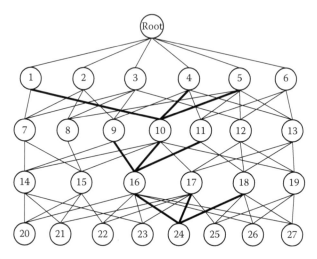

**FIGURE 1.36**
An example of a three source P2P video streaming system.

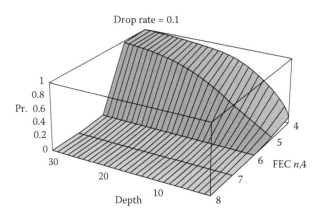

**FIGURE 1.37**
The packet loss probability of the multisource, multicast structure with different depths and FEC parameters.

to construct the original data. In addition, the concept of path diversity can also be employed by multisource P2P systems. More specifically, in Apostolopoulos [68], a multisource video communication system is proposed where the Multiple Description Coding (MDC) is used. Multiple video servers transmit video substreams to a receiver through different network paths. The concept of path diversity is that by transmitting packets through different network paths, once a path fails or encounters traffic congestion, the receiver can still obtain most remaining video substreams and decode the video frames with only partial distortion. In Nguyen and Zakhor [69], the path diversity concept is again used in video communication systems where a packet level FEC is used. The use of FEC enables the receiver to recover the lost packets caused by corrupted paths.

To take advantage of path diversity in a multisource P2P streaming system, a child peer could select parents located in different places so as to receive video packets from different parents and network paths in hope of reducing the impact of path failures. For example, the topology-aware parent selection algorithm used in PROMISE [37] uses connection conditions to select parents. In CoopNet [45] and SplitStream [35], video packets are transmitted on multiple distribution trees to provide path redundancy. In summary, multisource structures provide a way to achieve source level diversity and the topology-aware parent selection algorithms provide methods to achieve path level diversity.

## 1.6 Conclusion

In this chapter, we covered different architectures for delivering multimedia content to users. We pointed out the potential problems in the traditional unicast architecture. After introducing the multicast technologies and how they would be applied to solve the problems in traditional architecture, we proceeded to describing P2P streaming systems and how they can help reduce server load and network bottlenecks. Finally, we analyzed the theoretical peer behavior and estimated the affect of employing a packet level FEC on robustness. Even though P2P streaming systems seem very promising, there are still some open issues.

First, there are still a lot of open QoS issues. Marfia et al., 2007 [6] compares the performance of different P2P schemes and the results show that the startup delay of current P2P live streaming systems are still very high (see Figure 1.38; Table 1.1) [51–53].

Second, current P2P live streaming technologies are most suitable for delivering popular video channels where a newly joining peer can easily find

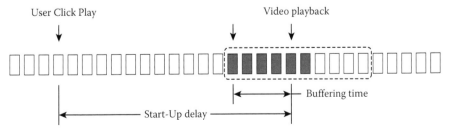

**FIGURE 1.38**
Illustration of startup delay.

**TABLE 1.1**

Performance of Different P2P Schemes

| Scheme | Buffer | Playout Delay | Startup Delay | Data Rate | Definition | Push-Pull | Tree-Mesh |
|---|---|---|---|---|---|---|---|
| PPLive | 2 min | 1 min | 20 sec–2 min | 300–350 kb/s | 320 × 240 | Pull | Mesh |
| CoolStreaming | 2 min | 1 min | 1 min | 300–350 kb/s | 320 × 240 | Pull | Mesh |
| Anysee [43,44] | 40 sec | 20–30 sec | 20 sec | 300–350 kb/s | 320 × 240 | Push | Hybrid |
| SopCast | 1 min | 1 min | 1–5 min | 300–350 kb/s | 320 × 240 | Pull | Mesh |

*Source:* Marfia, G., Sentivelli, A., Tewari, S., Gerla, M., and Kleinrock, L., *IEEE Communications Magazine*, Special Issue on Peer-to-Peer Streaming, June 2007. With permission.

other peers watching the same channel. However for less popular channels, users can experience difficulties getting smooth playback. Therefore, adaptively employing different technologies (e.g., combining P2P with CDN), to get better QoS is becoming a new trend.

Third, the increasing amount of P2P traffic over the whole network causes network congestion. Research about developing mechanisms urging the cooperation between ISPs and P2P streaming applications has been conducted.

Even though there are still these open issues, pure P2P streaming systems are increasingly popular. Much research has been conducted to improve the QoS of P2P streaming. There are also new trends combining different technologies to increase the users' streaming experience, and P2P streaming will continue to play a crucial role in them.

## References

1. F. Kozamernik, "Media Streaming over the Internet: An Overview of Delivery Technologies," *EBU Technical Review,* no. 292, October 2002.
2. S. Alstrup and T. Rauhe, "Introducing Octoshape: A New Technology for Large-Scale Streaming over the Internet," *EBU Technical Review,* no. 303, July 2005.
3. L. Lao, J.-H. Cui, M. Geria, and D. Maggiorini, "A Comparative Study of Multicast Protocols: Top, Bottom, or In the Middle?" Proceedings of 8th IEEE Global internet Symposium (GI'05) in conjunction with IEEE (INFOCOM'05), Miami, FL, March 2005.
4. L. Lao, J.-H. Cui, M. Gerla, and D. Maggiorini, "A Comparative Study of Multicast Protocols: Top, Bottom, or In the Middle?" Technical Report TR040054, Computer Science Department, University of California–Los Angeles, January 2005.
5. S. Tewari and L. Klenirock, "Analytical Model for BitTorrent-Based Video Streaming," in IEEE Workshop on Networking Issues in Multimedia Entertainment, 2007.
6. G. Marfia, A. Sentivelli, S. Tewari, M. Gerla, and L. Kleinrock, "Will IPTV Ride the Peer-to-Peer Stream?" *IEEE Communications Magazine,* Special Issue on Peer-to-Peer Streaming, June 2007.
7. Y. Xiao, X. Du, J. Zhang, F. Hu, and S. Guizani, "Internet Protocol Television (IPTV): The Killer Application for the Next-Generation Internet," *IEEE Communication Magazine,* November 2007.
8. J. Buford, "Survey of ALM, OM, Hybrid Technologies," Presentation Slides, Scalable Adaptive Multicast Research Group, July 2006.
9. W.-P. K. Yiu, J. Xing, and S.-H. G. Chan, "Challenges and Approaches in Large-Scale P2P Media Streaming," *IEEE Multimedia*, vol. 14, no. 2, pp. 50–59, April–June 2007.
10. A. Ganjam and H. Zhang, "Internet Multicast Video Delivery," *Proceedings of the IEEE,* vol. 93, no. 1, pp. 159–70, January 2005.

11. H. Song, D. S. Lee, and H. R. Oh, "Application Layer Multicast Tree for Real-Time Media Delivery," *Computer Communications,* vol. 29, no. 9, pp. 1480–91, May 2006.

12. H. Deshpande, M. Bawa, and H. Garcia-Molina, "Streaming Live Media over a Peer-to-peer Network," Technical report, Stanford University, California, 2001.

13. D. Xu, M. Hefeeda, S. Hambrusch, and B. Bhargava, "On Peer-to-Peer Media Streaming," Proceedings of IEEE (ICDCS'02), Vienna, Austria, July 2002.

14. C. K. Yeo, B. S. Lee, and M. H. Er, "A Survey of Application Level Multicast Techniques," *Computer Communications,* vol. 27, no. 15, pp. 1547–68, September 2004.

15. S. W. Tan, G. Waters, and J. Crawford, "A Survey and Performance Evaluation of Scalable Tree-based Application Layer Multicast Protocols," Technical Report No. 9-03, University of Kent at Canterbury, United Kingdom, July 2003.

16. J. F. Kurose and K. W. Ross, *Computer Networking: A Top-Down Approach Featuring the Internet,* 3rd ed., Addison Wesley, Boston, MA, 2005.

17. R. Wittmann and M. Zitterbart, *Multicast Communication: Protocols, Programming, and Applications,* Morgan Kaufmann Publishers, San Francisco, CA, 2000.

18. A. I. Holub, *Compiler Design in C,* Prentice Hall, Englewood Cliffs, NJ, 1990.

19. J. Jannotti, D. K. Gifford, K. L. Johnson, M. F. Kaashoek, and J. W. O'Toole, Jr., "Overcast: Reliable Multicasting with an Overlay Network," Proceedings of 4th Symposium on Operating System Design and Implementation (USENIX OSDI), San Diego, California, October 2000.

20. Y. Chawathe, "Scattercast: An Adaptable Broadcast Distribution Framework," *Multimedia Systems,* vol. 9, no. 1, pp. 104–18, July 2003.

21. B. Zhang, S. Jamin, and L. Zhang, "Host Multicast: A Framework for Delivering Multicast To End Users," Proceedings of 21st IEEE Computer and Communications Societies (INFOCOM'02), vol. 3, pp. 1366–75, New York City, June 2002.

22. B. Zhang, S. Jamin, and L. Zhang, "Universal IP Multicast Delivery," Proceedings of the Int'l Workshop on Networked Group Communication (NGC), October 2002.

23. B. Zhang, W. Wang, S. Jamin, D. Massey, and L. Zhang, "Universal IP Multicast Delivery," *Computer Networks,* vol. 50, no. 6, pp. 781–806, April 2006.

24. D. G. Andersen, H. Balakrishnan, M. F. Kaashoek, and R. Morris, "Resilient Overlay Networks," Proceedings of 18th ACM Symposium on Operating Systems Principles (SOSP), Banff, Canada, October 2001.

25. A. Rowstron and P. Druschel, "Pastry: Scalable, Distributed Object Location and Routing for Large-Scale Peer-to-Peer Systems," Proceedings of 18th IFIP/ACM International Conference on Distributed Systems Platforms (Middleware 2001), Heidelberg, Germany, November 2001.

26. M. Castro, P. Druschel, A-M. Kermarrec, and A. Rowstron, "One Ring to Rule Them All: Service Discovery and Binding in Structured Peer-to-Peer Overlay Networks," Proceedings of SIGOPS European Workshop, France, September 2002.

27. B. Y. Zhao, J. D. Kubiatowicz, and A. D. Joseph, "Tapestry: An Infrastructure for Fault-Tolerant Wide-Area Location and Routing," Technical Report UCB/USD-01-1141, University of California–Berkeley, April 2000.

28. B. Y. Zhao, L. Huang, J. Stribling, S. C. Rhea, A. D. Joseph, and J. D. Kubiatowicz, "Tapestry: A Resilient Global-Scale Overlay for Service Deployment," IEEE *Journal on Selected Areas in Communications,* vol. 22, no. 1, January 2004.

29. S. Ratnasamy, P. Francis, M. Handley, R. Karp, and S. Shenker, "A Scalable Content-Addressable Network," Proceedings of (SIGCOMM'01), San Diego, California, August 2001.

30. I. Stoica, R. Morris, D. Karger, M. F. Kaashoek, and H. Balakrishnan, "Chord: A Scalable Peer-to-Peer Lookup Service for Internet Applications," Proceedings of ACM S (SIGCOMM'01), San Diego, California, August 2001.

31. Y. H. Chu, S. G. Rao, S. Seshan, and H. Zhang, "A Case for End System Multicast," *IEEE Journal on Selected Areas in Communications,* vol. 20, no. 8, October 2002.

32. D. Pendarakis, S. Shi, D. Verma, and M. Waldvogel, "ALMI: An Application Level Multicast Infrastructure," Proceedings of the 3rd Symposium on Internet Technologies and Systems, San Francisco, California, March 2001.

33. H. Deshpande, M. Bawa, and H. Garcia-Molina, "Streaming Live Media Over a Peer-to-Peer Network," Technical Report, Stanford University, California, 2001.

34. M. Castro, P. Druschel, A-M. Kermarrec, and A. Rowstron, "SCRIBE: A Large-Scale and Decentralized Application-Level Multicast Infrastructure," *IEEE Journal on Selected Areas in Communications,* vol. 20, no. 8, pp. 1489–99, October 2002.

35. M. Castro, P. Druschel, A. M. Kermarrec, A. Nandi, A. Rowstron, and A. Singh, "SplitStream: High-Bandwidth Multicast in Cooperative Environments," Proceedings of the 19th ACM Symposium on Operating Systems Principles, October 19–22, 2003.

36. M. Hefeeda, A. Habib, D. Xu, B. Bhargava, and B. Botev, "CollectCast: A Peer-to-Peer Service for Media Streaming," *ACM Multimedia Systems Journal,* vol. 11, no. 1, pp. 68–81, November 2005.

37. M. Hefeeda, A. Habib, B. Botev, D. Xu, and B. Bhargava, "PROMISE: Peer-to-Peer Media Streaming Using CollectCast," Proceedings of the 11th ACM international conference on Multimedia, Berkeley, California, November 2003.

38. V. Padmanabhan, H. Wang, P. Chou, and K. Sripanidkulchai, "Distributing Streaming Media Content Using Cooperative Networking," Proceedings of 12th ACM International Workshop on Network and Operating System Support for Digital Audio and Video (NOSSDAV'02), Miami Beach, FL, May 2002.

39. V. Padmanabhan, H. Wang, and P. Chou, "Resilient Peer-to-Peer Streaming," Proceedings of 11th IEEE International Conference on Network Protocols (ICNP'03), Atlanta, Georgia, November 2003.

40. S. Banerjee, B. Bhattacharjee, and C. Kommareddy, "Scalable Application Layer Multicast," Technical report, UMIACS TR-2002-53 and CS-TR 4373, Department of Computer Science, University of Maryland–College Park, May 2002.

41. D. A. Tran, K. A. Hua, and T. T. Do, "ZIGZAG: An Efficient Peer-to-Peer Scheme for Media Streaming," Proceedings of 22nd IEEE Computer and Communications Societies (INFOCOM), San Francisco, California, March–April 2003.

42. D. A. Tran, K. A. Hua, and T. T. Do, "A Peer-to-Peer Architecture for Media Streaming," *IEEE Journal on Selected Areas in Communications,* vol. 22, no. 1, pp. 121–33, January 2004.

43. C. Zhang, H. Jin, D. Deng, S. Yang, Q. Yuan, and Z. Yin, "Anysee: Multicast-Based Peer-to-Peer Media Streaming Service System," Proceedings of 2005 Asia-Pacific Conference on Communications, Perth, Western Australia, October 2005.

44. X. Liao, H. Jin, Y. Liu, L. Ni, and D. Deng, "Anysee: Peer-to-Peer Live Streaming," Proceedings of 25th IEEE Computer and Communications Societies (INFOCOM), Barcelona, Catalunya, Spain, April 2006.

45. X. Zhang, J. Liu, B. Li, and T. Yum, "Coolstreaming/DONet: A Data-Driven Overlay Network for Efficient Live Media Streaming," Proceedings of 24th IEEE Computer and Communications Societies (INFOCOM), Miami, Florida, March 2005.

46. X. Hei, C. Liang, J. Liang, Y. Liu, and K. Ross, "A Measurement Study of a Large-Scale P2P IPTV System," *Multimedia, IEEE Transactions*, vol. 9, pp. 1672–87, 2007.

47. X. Hei, C. Liang, J. Liang, Y. Liu, and K. W. Ross, "Insight into PPLive: Measurement Study of A Large Scale P2P IPTV System," Proceedings of WWW'06, Edinburgh, Scotland, May 2006.

48. X. Hei, C. Liang, J. Liang, Y. Liu, and K. W. Ross, "A Measurement Study of a Large-Scale P2P IPTV System," Technical report, Department of Computer and Information Science at Polytechnic University, 2006.

49. M. Zhang, Y. Tang, L. Zhao, J. G. Luo, and S. Q. Yang, "Gridmedia: A Multi-Sender Based Peer-To-Peer Multicast System For Video Streaming," Proceedings of Conference on Multimedia and Expo (ICME), Amsterdam, the Netherlands, July 2005.

50. M. Zhang, Y. Tang, L. Zhao, J. G. Luo, and S. Q. Yang, "Design and Deployment of a Peer-to-Peer Based IPTV System over Global Internet," Proceedings of Communications and Networking (CHINACOM), Beijing, China, October 2006.

51. T. Silverston and O. Fourmaux, "Measuring P2P IPTV Systems," Proceedings of ACM Network and Operating Systems Support for Digital Audio & Video (NOSSDAV'07), Urbana-Champaign, Illinois, June 2007.

52. T. Silverston and O. Fourmaux, "P2P IPTV Measurement: A Comparison Study," arXiv: cs/061033v4.

53. D. Qiu and R. Srikant, "Modeling and Performance Analysis of BitTorrent-Like Peer-to-Peer Networks," Proceedings of (SIGCOMM'04), Portland, Oregon, August–September 2004.

54. X. Jiang, Y. Dong, D. Xu, and B. Bhargava, "GnuStream: A P2P Media Streaming System Prototype." Technical report, Purdue University, Indiana, 2003.

55. F. Wang, Y. Xiong, and J. Liu, "mTreebone: A Hybrid Tree/Mesh Overlay for Application-Layer Live Video Multicast," Proceedings of International Conference on Distributed Computing Systems, Toronto, Ontario, Canada, June 2007.

56. Y. Liu, Y. Guo, and C. Liang, "A Survey on Peer-to-Peer Video Streaming Systems," *Peer-to-Peer Networking and Applications*, vol. 1, no. 1, March 2008.

57. G. de Veciana and X. Yang, "Fairness, Incentives and Performance in Peer-to-Peer Networks," Proceedings of the 41st Annual Allerton Conference on Communication, Control and Computing, Monticello, Illinois, October 2003.

58. P. J. Wu, C. N. Lee, V. Gau, and J. N. Hwang, "Overcoming Burst Packet Loss in Peer-to-Peer Live Streaming Systems," Proceedings of IEEE Symposium on Circuits and Systems (ISCAS), Seattle, Washington, May 2008.

59. S. M. Ross, "Introduction to Probability Models," New York: Academic, 1985.

60. W. T. Tan and A. Zakhor, "Video Multicast Using Layered FEC and Scalable Compression," *IEEE Trans. Circuits and System for Video Technology*, vol. 11, no. 3, pp. 373–86, March 2001.

61. A. Mohr, E. Riskin, and R. Ladner, "Unequal Loss Protection: Graceful Degradation over Packet Erasure Channels Through Forward Error Correction," *IEEE Journal on Selected Areas in Communication*, vol. 18, pp. 819–28, April 2000.

62. P. J. Wu, J. N. Hwang, C. N. Lee, and Y. C. Teng, "Receiver Driven Overlap FEC for Scalable Video Coding Extension of the H.264/AVC," Proceedings IEEE Symposium on Circuits and Systems (ISCAS'09), Seattle, Washington, May 2009.

63. W. T. Tan and A. Zakhor, "Real-Time Internet Video using Error Resilient Scalable Compression and TCP-Friendly Transport Protocol," *IEEE Transactions on Multimedia*, vol. 1, no. 2, June 1999.

64. I. Cidon, A. Khamisy, and M. Sidi, "Analysis of Packet Loss Processes in High-Speed Networks," *IEEE Transactions on Information Theory*, vol. 39, pp. 98–108, January 1993.

65. N. Magharei and R. Rejaie, "Prime: Peer-to-Peer Receiver-Driven Mesh-Based Streaming," Proceedings of IEEE Computer and Communications Societies (INFOCOM), pp. 1415–23, Anchorage, Alaska, May 2007.

66. V. Gau, Y. H. Wang, and J. N. Hwang, "A Hierarchical Push-Pull Scheme for Peer-to-Peer Live Streaming," IEEE Symposium on Circuits and Systems (ISCAS), Seattle, Washington, May 2008.

67. P. J. Wu, J. N. Hwang, C. N. Lee, C. C. Gau, and H. H. Kao, "Eliminating Packet Loss Accumulation in Peer-to-Peer Streaming Systems," to appear in IEEE Transactions on Circuits and Systems for Video Technology.

68. J. G. Apostolopoulos, "Reliable Video Communication over Lossy Packet Networks using Multiple State Encoding and Path Diversity," Proceedings of Visual Communications and Image Processing, pp. 392–409, San Jose, California, January 2001.

69. T. Nguyen and A. Zakhor, "Path Diversity with Forward Error Correction (PDF) System for Packet Switched Networks," Proceedings of IEEE Computer and Communications Societies (INFOCOM), vol. 1, pp. 663–72, San Francisco, California, April 2003.

70. J. Liu, S. Rao, B. Li, and H. Zhang, "Opportunities and Challenges of Peer-to-Peer Internet Video Broadcast," *Proceedings of the IEEE,* vol. 96, pp. 11–24, 2008.

71. X. Hei, C. Liang, J. Liang, Y. Liu, and K. Ross, "A Measurement Study of a Large-Scale P2P IPTV System," *Multimedia, IEEE Transactions*, vol. 9, pp. 1672–87, 2007.

# 2

## The Design of VoIP Systems with High Perceptual Conversational Quality

Benjamin W. Wah and Batu Sat

## CONTENTS

This chapter describes our work on real-time, two-party, and multi-party voice-over-IP (VoIP) systems that can achieve high perceptual conversational quality. It focuses on the fundamental understanding of conversational quality and its trade-offs among the design of speech codecs and strategies for network control, playout scheduling (POS), and loss concealments. We have studied three key aspects that address the limitations of existing work and improve the perceptual quality of VoIP systems. Firstly, we have developed a statistical approach based on a just-noticeable difference (JND) to significantly reduce the large number of subjective tests, as well as a classification method to automatically learn and generalize the results to unseen conditions. Using network and conversational conditions measured at run time, the classifier learned helps adjust the control algorithms in achieving high perceptual conversational quality. Secondly, we have designed a cross-layer speech codec to interface with the loss-concealment (LC) and POS algorithms in the packet-stream layer in order to be more robust and effective against packet losses. Thirdly, we have developed a distributed algorithm for equalizing mutual silences and an overlay network for multi-party VoIP systems. The approach leads to multi-party conversations with high listening-only speech quality (LOSQ) and balanced mutual silences.

---

## 2.1 Introduction

This chapter summarizes the results on real-time, two-party and multi-party VoIP (voice-over-IP) systems that can achieve high perceptual conversational quality. It focuses on the fundamental understanding of conversational quality and its trade-offs among the design of speech codecs and strategies for network control and loss concealments. An important aspect of this research is on the development of new methods for reducing the large number of subjective tests and for automated learning and generalization of the results of subjective evaluations. Since the network delays in VoIP can be long and time varying, its design is different from those for public switched telephone network (PSTN) with short and consistent delays [1].

Figure 2.1 outlines the components in the design of a VoIP system. The first component on *conversational quality* entails the study of human conversational behavior, modeling conversational dynamics, and identifying user-perceptible attributes that affect quality. Its study also includes the design of off-line subjective tests and algorithms for learning the test results. Next, the study of *network* and *conversational environments* entails the identification of objective metrics for characterizing network and conversational conditions and the dissemination of this information at run time. The design of loss-concealment (LC) and playout scheduling (POS) strategies in the *packet-stream layer* involves delay-quality trade-offs that optimize user-perceptible attributes. The *network-control layer* provides support for network transport and admissions control in multi-party VoIP. Lastly, the design of the *speech codec* and its LC and compression capabilities must take into account its interactions with the LC and POS strategies in the packet-stream layer.

**Effects of Delays on Conversations.** In a two-party conversation, each participant takes turns speaking and listening [3,5], and both perceive a silence period (called mutual silence or MS) when the conversation switches from one party to another. Hence, a conversation consists of alternating speech segments and silence periods.

In a face-to-face setting, both participants have a common reality of the conversation: one speech segment is separated from another by a silence period that is identically perceived by both. However, when the same conversation is conducted over the Internet, the participants' perception of the conversation is different due to delays, jitters, and losses incurred on the speech segments during their transmission [4,5].

Richards [6] has identified three factors that influence the quality of service in telephone systems: difficulty in listening to one-way speech, difficulty in

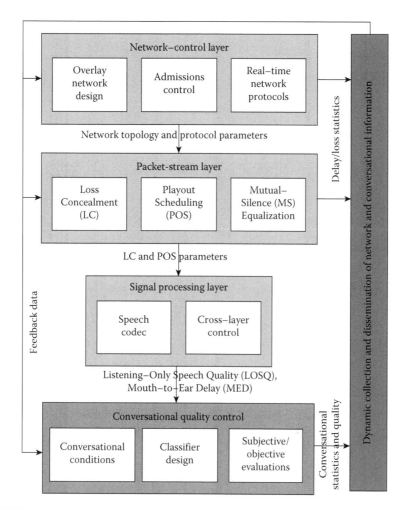

**FIGURE 2.1**
Layers in the architecture of a VoIP system.

talking, and difficulty in conversing during turn taking. Hence, we evaluate the quality of a conversation over a network connection by the quality of the one-way speech segments received (the listening-only speech quality or LOSQ) and that of the interactions [4]; the latter is measured by the delay incurred from the mouth of the speaker to the ear of the listener (the mouth-to-ear delay or MED) [6].

When a connection has delays, the MSs perceived by a participant consist of alternating short and long silence periods between turns [4]. This asymmetry is caused by the fact that after A speaks, the MS experienced by A ($MS_A$ in Figure 2.2) consists of the time for A's ($MED_{A,B}$), the time for B to construct a response (human response delay or $HRD_B$)), and the time for B's response to travel to A ($MEB_{B,A}$). In contrast, after B hear the speech from A, the MS

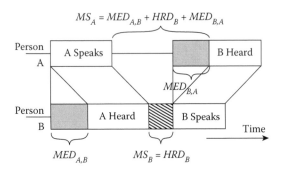

**FIGURE 2.2**
Asymmetric mutual silences.

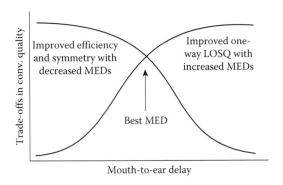

**FIGURE 2.3**
Trade-off considerations.

experienced by B is only governed by his or her HRD ($MS_B = HRD_B$). This asymmetry leads to a perception that each user is responding slowly to the other, and consequently results in degraded efficiency and perceptual quality [4].

Conversational quality cannot be improved by simultaneously improving LOSQ and reducing MED. A longer MED will improve LOSQ because segments will have a higher chance to be received, but will worsen the symmetry of MSs. Figure 2.3 shows the delay-quality trade-off and a suitable MED with the best quality. This trade-off also depends on the turn-switching frequency [4,7] and on changes in network and conversational conditions [8]. It has been shown that long MEDs can cause doubletalk and interruptions even when MED is constant [6,9].

The perceived effect of delays in multi-party VoIP are more complex than those in two-party VoIP because there may be large disparities in network conditions across the participants [10]. In the multi-party case, the conversational quality depends on the LOSQ and the latency of the one-way speech from each speaker, as well as the symmetry of the conversation among the participants. Hence, each listener may have a slightly different perception of the same conversation. Figure 2.4 depicts two conversation units (CU) in a

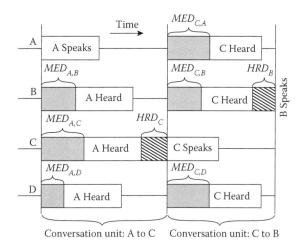

**FIGURE 2.4**
A four-party conversation.

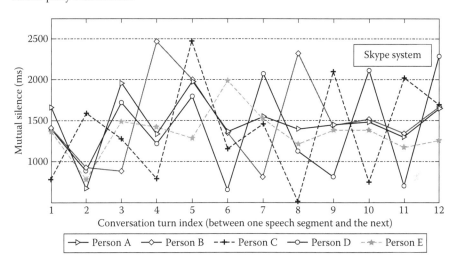

**FIGURE 2.5**
MSs in a simulated Skype conversation.

four-party conversation. Each listener experiences different asymmetric MSs across the speakers: some appear to be more distant than others or some respond slower than others.

Figure 2.5 illustrates the disparities in MSs perceived by five participants in 13 CUs when we simulate a multi-party Skype (Version 3.5.0.214) conversation (with HRD of 750 ms) using UDP traces of five sites (three in North America and two in Asia) collected in PlanetLab [10]. Since all traffic in Skype is routed through a common client, there are large disparities in MSs from one CU to the next.

**Objective Metrics of Interactive VoIP.** We define two objective metrics for characterizing the quality of multi-party VoIP [10]. These metrics can be tailored to a two-party VoIP in a straightforward fashion.

1. The conversational efficiency (CE) measures the extension in time to accomplish a VoIP conversation when there are communication delays (Figure 2.2). Since a conversation over a network may be charged according to its duration, the same conversation will cost more for a network with lower CE.

$$CE = \frac{\text{Speaking Time} + \text{Listening Time}}{\text{Total Time of Call}} \qquad (2.1)$$

   Note that CE is identically perceived by all participants (see Table 2.1).

2. Conversational symmetry (CS). When a participant experiences highly asymmetric response times in a conversation, he or she tends to perceive degradation in the naturalness of the conversation because it does not resemble a face-to-face conversation with uniform delays. One possible effect is, when A perceives B to be responding slowly, then A tends to respond slowly as well. To capture the asymmetry of MSs perceived by A, we define $CS_A$ to be the ratio of the maximum and the minimum MSs experienced by A in a past window when the conversation switches from $i$ to $j$ $(i \rightarrow j)$, and $j$ is the speaker:

$$(A(mp)) = CS_{A(2p)} = \frac{\max_j MS_A^{i \rightarrow j}}{\min_j MS_A^{i \rightarrow j}} (\text{2-party VoIP}). \qquad (2.2)$$

**TABLE 2.1**

Research Issues in Interactive VOIP Systems

| |
|---|
| Section II: Evaluations of conversational quality |
|   –Statistical model of subjective tests |
|   –Trade-offs on objective metrics and JND |
|   –Testbed for evaluating algorithms and systems |
|   –Classifiers for learning and generalization |
| Section III: Cross-layer designs of speech codecs |
|   –Codec design for dynamic encapsulation |
|   –Rate-adaptive generation of enhancement layers |
| Section IV: Packet-stream-layer control algorithms |
|   –Dynamic POS and LC control using classifiers |
|   –Distributed equalization of MSs |
| Section V: Network-layer control algorithms |
|   –Overlay networks and transport protocols |
|   –Admissions control for multi-party VoIP |

$CS_A$ is approximately one in a face-to-face conversation, but increases as the round-trip delay increases. In the two-party case, the minimum MS is always experienced by the speaker and the maximum experienced by the listener. Since there are multiple listeners in the multi-party case and the majority of the clients are passive listeners, it is important to identify the asymmetry for the passive listeners alone. Hence, we choose to eliminate the current speaker when evaluating the minimum MS in the multi-party case. Note that *CS* and *CE* are counteracting: as *CS* improves, *CE* degrades.

In VoIP, a user does not have an absolute perception of MEDs because he or she does not know when the other person will start talking or who will speak next (in multi-party VoIP). However, by perceiving the indirect effects of MED, such as MS and CE, the participants can deduce the existence of MED.

Table 2.1 summarizes the four research issues in the design of interactive VoIP systems. These are related to the classification and generalization of subjective test results and the design of network control and coding algorithms. For each of these issues, we present some existing work and our approaches in Sections 2.2–2.5. Lastly, Section 2.6 concludes the chapter.

## 2.2 Evaluating Conversational Quality

In this section, we first survey existing metrics for measuring conversational quality. We then present results on evaluating subjective quality of VoIP systems and methods for learning the mapping from objective metrics to control algorithms that optimize subjective quality.

### 2.2.1 Previous Work

**Effects of MED on Conversational Quality.** Subjective tests by Brady [9] and Richards [6] in the 1970s have led to the conclusion that MED affects the user perception of conversational quality, and that longer MEDs increase the dissatisfaction rate. However, their conclusions are limited when used for evaluating VoIP systems, since only a few constant delays were experimented. Subjective tests by Kiatawaki and Itoh [7] at NTT show that one-way delays are detectable, with a detectability threshold of 100–700 ms for a trained crew and of 350–1,100 ms for untrained subjects. The International Telecommunication Union (ITU) G.114 [11] prescribes that a one-way delay of less than 150 ms is desirable in voice communication, and that a delay of more than 400 ms is unacceptable. Without specifying the trade-offs with LOSQ, MED alone is not adequate for evaluating VoIP.

**Objective Measures on Conversational Quality.** The ITU has several recommendations for the objective and subjective evaluations of the end-to-end quality of a voice transmission system. Table 2.2 shows the naming standard

**TABLE 2.2**

ITU P.800.1 Terminology on Telephone Transmission Quality

| Methodology | Listening-Only Conditions Tested | Conversational Conditions Tested |
|---|---|---|
| Subjective | $MOS_{LQS}$: P.800 Listening-only tests | $MOS_{CQS}$: P.800 Conversational tests |
| Objective | $MOS_{LQO}$: P.862 PESQ | $MOS_{CQO}$: P.562 for PSTN, not defined for VoIP |
| Estimated | $MOS_{LQE}$: Not defined | $MOS_{CQE}$: G.107 E-model |

established by ITU for the evaluation of the telephone transmission quality [12]. There are several recommendations for evaluating the objective conversational quality of a system in absolute category rating (ACR).

1. *PESQ* (ITU P.862) is an objective measure for evaluating speech quality based on the original and the degraded waveforms. It has been shown to have high correlations to subjective mean-opinion-score (MOS) results for a variety of landline, mobile, and VoIP applications. Since it only assesses the LOSQ and not the effects of delay, it must be used in conjunction with other metrics when evaluating conversational quality.

2. The *E-Model* (ITU G.107) was designed for estimating conversational quality in network planning. It considers the effects of the codec, packet losses, one-way delay, and echo. It is oversimplifying because it assumes the independence and additivity of degradations due to LOSQ and delay. Despite a number of extensions [13–17] that try to address its limitations, it is difficult to extend its role beyond network planning and use it for evaluating conversational quality in actual systems.

3. The *Call Clarity Index* (ITU P.561 and P.562) was developed for estimating the customer opinion of a voice communication system in a way similar to the E-Model. Although it provides models for PSTN systems, it does not have a user opinion model for packet-switched networks with long delays and with nonlinear and time variant signal processing devices, such as echo control and speech compression.

At this time, there is no single objective metric that can adequately capture the trade-offs among the factors that affect subjective conversational quality under all network and conversational conditions.

**Subjective Measures on Conversational Quality.** A user's perception of a speech segment mainly depends on the intelligibility of the speech heard because the user lacks a reference to the original segment. To assess subjective conversational quality, formal MOS tests (ITU P.800) [12] are usually conducted. The method asks two subjects to complete a specific task over a communication system, ranks the quality using an ACR, and averages the opinion of multiple subjects.

There are several shortcomings of this approach for evaluating VoIP. Firstly, when completing a task and evaluating the quality of a conversation simultaneously, the cognitive attention required for both may interfere with each other. Secondly, the type and complexity of the task affects the quality perception. Tasks requiring faster turn taking can be more adversely affected by transmission delays than others. Thirdly, there is no reference in subjective evaluations, and ACR highly depends on the expertise of the subjects. Lastly, the results are hard to repeat, even for the same subjects and the same task.

In the Nippon Telegraph and Telephone Corp. (NTT) study [7] discussed earlier, subjective conversational experiments were conducted between two parties using a voice system with adjustable delays. Since the study did not consider the effect of losses and variations in delay, it is not applicable for VoIP systems.

ITU-T Study Group [10] has realized a lack of methods for evaluating conversational speech quality in networks and is currently conducting a study. However, it is not clear if the study will lead to an objective or a subjective methodology and whether the results can help design better VoIP systems.

### 2.2.2 Evaluations-Generalization of Conversational Quality

**Testbed for Evaluating VoIP Systems.** We have developed a testbed for emulating two-party [4] and multi-party [10] VoIP. This entails the collection of Internet packet traces and multi-party interactive conversations and the design of a system to replay these traces and conversations. The prototype allows subjective tests to be repeated for different VoIP systems under identical network and conversational conditions [18].

The prototype consists of multiple computers, each running the VoIP client software, and a Linux router for emulating the real-time network traffic [4,10]. We have modified the kernel of the router in order to intercept all UDP packets carrying encoded speech packets between any two clients. The router runs a troll program that drops or delays intercepted packets in each direction according to packet traces collected in the PlanetLab. We have also developed a human-response-simulator (HRS) that runs on each end-client. The HRSs simulate a conversation with prerecorded speech segments by taking turns speaking their respective segments. We use a software interface to digitally transfer the waveforms to and from the clients without quality loss.

**Subjective Evaluations of Four VoIP Systems.** We have compared four, two-party VoIP clients: Skype (3.6), Google-Talk (beta), Windows Live Messenger (8.1), and Yahoo Messenger (8.1) [18]. Using conversations recorded by our testbed under some network and conversational conditions, human subjects were asked to comparatively evaluate two conversations by the CCR scale. The tests were conducted using six Internet traces under different network conditions and an additional trace representing an ideal condition with no loss and delay. We use three distinct conversations of different single-talk durations, HRD, and switching frequencies. The subjective test results in

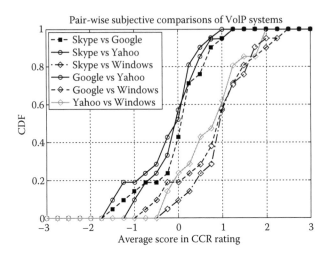

**FIGURE 2.6**
Distribution of pair-wise subjective scores of four VoIP systems.

Figure 2.6 illustrate that Windows Live is preferred over the others. These are consistent with the objective metrics in terms of PESQ, CS, and CE (not shown). Similar tests have also been conducted to compare the multi-party version of Skype and our proposed system [10].

**Statistical Off-Line Subjective Tests.** We have studied the statistical scheduling of off-line subjective tests for evaluating alternative control schemes in real-time multimedia applications. These applications are characterized by multiple counteracting objective quality metrics (such as delay and signal quality) that can be affected by various control schemes. However, the trade-offs among these metrics with respect to the subjective preferences of users are not defined. As a result, it is difficult to select the proper control parameter value(s) that leads to the best subjective quality at run time. Since subjective tests are expensive to conduct and the number of possible control values and run-time conditions is prohibitively large, it is important that a minimum number of such tests be conducted off-line, and that the results learned can be generalized to unseen conditions with statistical confidence. To this end, we have developed efficient algorithms for scheduling a sequence of subjective tests under given conditions. Our goal is to minimize the number of subjective tests needed in order to determine the best point for operating the multimedia system to within some prescribed level of statistical confidence. A secondary goal is to efficiently schedule subjective tests under a multitude of operating conditions. Its success is based on the fact that humans can differentiate two such conversations when they are beyond the JND, aka difference limen [19]. Here JND is a difference in the physical sensory input that results in the detection of the change 50 percent of the time.

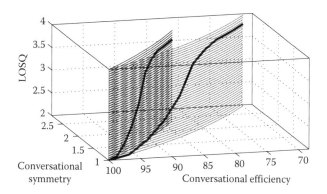

**FIGURE 2.7**
The points on each operating curve are a function of the MED imposed. Each plane represents some network and conversational conditions.

For a two-party conversation, we use the following triplet in 3-D space to denote the operating point under a given codec and some network and conversational conditions:

$$CQ_{2\text{-party}} = \{LOSQ(MED, R), CE(MED), CS(MED)\},$$

where each axis represents an objective metric measured over a past window, and $R$ is the redundancy degree. Figure 2.7 depicts the trade-offs as a function of MED and $R$ for two conversations of different HRDs and switching frequencies [4]. For a conversation under some given conditions, the trade-offs between $CE$ and $CS$ are shown as a plane parallel to the LOSQ axis. Under these conditions, the possible LOSQs as a function of MED are shown by an operating curve on this plane.

The trade-offs shown by each operating curve are very complex and cannot be represented in closed forms because they involve some network and conversational conditions that cannot be modeled. Finding the *most probable* operating point on each curve with the best subjective quality (by selecting a proper MED) proves to be difficult because there are infinitely many operating points and each involves subjective tests. Also, the operating points do not have a total order because it may not always be possible to compare two conversations, one with high LOSQ but low $CS$ and another with high $CS$ but low LOSQ. To this end, we use JND as a vehicle to prune operating points with slightly different conditions that cannot be distinguished.

We have developed a method that uses the JND framework to discretize an operating curve in Figure 2.7 into a finite and manageable set [4]. Figure 2.8 illustrates the pruning of points on an operating curve that do not need to be compared against point A. Using subjective tests based on comparison MOS, which gives a relative subjective comparison of two conversations (similar to the Comparison Category Rating–CCR–method in ITU P.800 Annex E [12]),

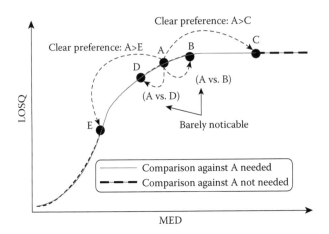

**FIGURE 2.8**
JND helps reduce the number of subjective evaluations.

we have carried out simulated conversations and have conducted pair-wise subjective tests to capture the relative user preferences.

It is difficult to find the best operating point on an operating curve by comparing MOS scores because there are infinitely many such scores to be evaluated. Moreover, some operating points do not have to be assessed to the same confidence level when they are obviously inferior or unnecessary. To this end, we have developed a statistical method that minimizes the total number of tests for finding an operating point with the best subjective conversational quality to within a prescribed confidence level [20]. This is possible because statistical tests for comparing two conversations by human subjects follow a multinomial distribution.

When comparing two points on an operating curve, we have developed axioms on reflexivity, independence, identical statistical distribution, symmetry, indistinguishability, incomparability, and subjective preference. By using the axioms, we have constructed a general model for comparing two points on an operating curve that allows us to determine a likely direction on the location of the local optimum of subjective preference. However, the nonparametric nature of the model makes it difficult to combine the result of a test with the prior information obtained. Hence, we have also developed a parametric model of subjective comparisons after simplifying the general model. The simple model allows a probabilistic representation of our knowledge on the location of the local optimum and a way to statistically combine the deductions from multiple comparisons [20]. It also allows us to develop an adaptive search algorithm that significantly reduces the number of comparisons needed for identifying the local optimum. In addition, an estimate on the confidence of the result provides a consistent stopping condition for our algorithm.

Our results show that sequential evaluations of a single operating curve are the most effective in terms of minimizing the number of tests performed for that curve when identifying a local optimum to within some statistical confidence. Our simulation results show a substantial reduction in the number of comparisons by using a stopping criterion based on a lower confidence level, while incurring a negligible error in the estimation.

Based on this algorithm, the optimal strategy to minimize the total number of subjective tests for a set of operating curves is to test each curve sequentially and all the curves in parallel. In this approach, each subject is presented with a set of operating points to be compared, one from each operating curve to be tested. The tests in each set can be performed in any order and independent of other subjects because the result of the comparisons from one operating curve does not depend on that of another curve. At the end of the tests, the results from all the subjects are combined in order to generate a local optimum estimate and identify the next pair of operating points to be compared for each of the operating curves. As the number of operating curves to be tested is large, this approach allows subjects to independently carry out a batch of independent tests, without having to synchronize their results in a locked-step fashion with other subjects. The number of iterations is bound by the typically small number of iterations to identify a local optimum candidate of an operating curve.

**Classifiers for Learning Evaluation Results.** To address the issue that there are infinitely many possible network and conversational conditions, we propose to develop a classifier [21] that learns from training examples generated under limited conditions and that generalizes to unseen conditions.

Based on the pair-wise comparisons of the conversations recorded on the four VoIP systems discussed last, we have generated training patterns, each consisting of 20 objective measures and a subjective measure. We have then learned these mappings by a classifier implemented as a support vector machine (SVM) [22] with a radial basis kernel function [18]. To simplify learning, we map the average of the user CCR opinions of A against B into three classes: A better than B, B better than A, and about the same. To verify that the classifier can generalize to unseen network and conversational conditions, we use cross-validation techniques commonly employed in statistics. Our results show that we can predict 90 percent of the samples successfully in our training set and 70 percent of the cases when using 10-fold cross validation. We have further used the subjective test results in the design of control algorithms that work well under a variety of conditions observed at run-time. The idea is to collect training data on subjective conversational quality off-line and to design a classifier that learns the mappings between objective metrics measured in a past window and the control parameter value that leads to an operating point with the best subjective quality [23]. Based on a comprehensive set of network and conversational conditions, the training data is obtained by simulating two-way and multi-way VoIP conversations using our testbed. The simulations are carried out under each of the given conditions and values of the system controllable parameters of the POS and LC algorithms, such as MED, redundancy degree, and level of MS equalization. Each element of the training set, therefore, consists of a mapping from the system-controlled parameter values and the objective metrics of the simulated conditions on a pair of conversations to their subjective preference. This method ensures that the conversations compared only differ by one parameter value and that their subjective preference can be attributed to the system-controlled value that leads to that opinion. We then learn a SVM classifier using training data based on the results of the subjective tests and the conditions under which the tests are conducted.

At run-time, the parameters representing the current conditions are estimated and input to the SVM. For example, in the design of the POS algorithm for two-party VoIP, loss, delay, and jitter parameters are used to represent network conditions, and switching frequency and singe-talk duration parameters represent conversational conditions. The SVM learned outputs the subjective preference for a given pair of points on the operating curve that corresponds to the network and conversational conditions observed. Its predictions on the subjective preference between multiple pairs of points on the same operating curve are combined using the statistical method described earlier in order to identify the optimal MED value, which is then used by the POS algorithm to adjust the jitter-buffer delay in order to achieve the operating point with the highest subjective quality.

## 2.3 Cross-Layer Speech Codecs for VoIP

Traditional codecs developed for cellular communications and PSTN calls are not suitable for VoIP because they have been designed for circuit switching under low bandwidth, fixed bit rates, and random bit errors. These codecs are not effective in packet-switched networks, whose loss rates and delay jitters are dynamic. Some recent codecs have been developed for VoIP applications. They can encode wide-band speech and exploit trade-offs between bit rate and delay in order to be more robust against bursty losses. However, they have been designed without due consideration of LC strategies in other layers of the protocol stack. Without such considerations, the LC strategies in these codecs can be inadequate and give subpar performance, or redundant and unnecessary.

In this section, we first briefly survey speech codecs designed for VoIP. We then present the design of cross-layer speech codecs that are done in conjunction with LC strategies in the packet-stream layer.

### 2.3.1 Previous Work on Speech Codecs

Speech codecs were traditionally designed for applications in cellular and PSTN communications. With the proliferation of IP networks, they have been increasingly used in VoIP. They can be classified based on their coding techniques. *Waveform* codecs, such as ITU G.711 and G.726 [11], were designed to reconstruct a sample-wise waveform as closely as possible. *Parametric* codecs, such as G.722.2, G.723.1, G.728, and G.729A [11], model the production of speech in order to reconstruct a waveform that perceptually resembles the original speech. *Hybrid* codecs, such as G.729.1 [11], iLBC [44], and iSAC [43], combine techniques from both. Under no-loss conditions, the perceptual quality of a codec is a function of its coding technique and bit rate. However, it is difficult to compare codecs under loss conditions.

Parametric and hybrid codecs are popular in VoIP because they have lower bit rates and better perceptual quality. By controlling the frame size and the frame period, their design involves trade-offs among robustness, quality, and algorithmic delay [44]. Frame size generally varies between 10 bytes (G.729A) and 80 bytes (G.729.1, 32-kbps wideband mode), with multimode codecs having a wide range. (For example, G.722.2 has frames between 17 and 60 bytes.) Frame period varies between 10 ms (G.729A) and 60 ms (iSAC with 30–60 ms adaptive size). The most common periods are 20 ms (such as iLBC 15.2-kbps mode, G.729.1 and G722.2) and 30ms (iLBC 13.3-kbps mode and G723.1). In general, a larger frame with a shorter period achieves higher quality within the multiple modes of a codec or within a family of codecs with similar coding techniques. However, a longer frame and look-ahead window may incur more algorithmic delay.

Figure 2.9a summarizes some of the LC techniques employed in speech codecs. Receiver-based schemes that do not use redundant information can be classified into sample-based and model-based schemes. In early systems, silence or comfort-noise substitution [24] or repetition of the previous frame [26] was proposed in place of a lost frame. Also proposed was the transmission of even and odd samples in separate packets and using sample-based interpolation when a packet was lost [25]. Early model-based schemes simply repeat the codec parameters of the last successfully received frame [28]. Later, interpolation of codec parameters from the previous and the next frame was proposed [27]. Other schemes utilized the information about the voiced–unvoiced properties of a speech frame to apply specialized LC for reducing the perception of degradation [29,30]. Schemes that require the cooperation of the sender and the receiver utilize partial redundant information [31–33] that is made available by the packet-stream-level LC (see Section 2.4).

The trade-offs between frame size and frame period are different from those between packet size and packet period in the packet-stream layer. To avoid excessive losses in the Internet, it is important to choose an appropriate packet period, as long as packets are smaller than an MTU of 576 bytes and can be sent without fragmentation [45]. (In practice, an MTU of 1,500 bytes will not cause fragmentation.) For VoIP using IPv4, a packet period between 30 ms and 60 ms generally works well. When the frame period is shorter than the packet period, multiple frames have to be encapsulated in a packet before they are sent. For some codecs, the loss of a single packet can cause a misalignment of its internal states and degrade its decoded output. For

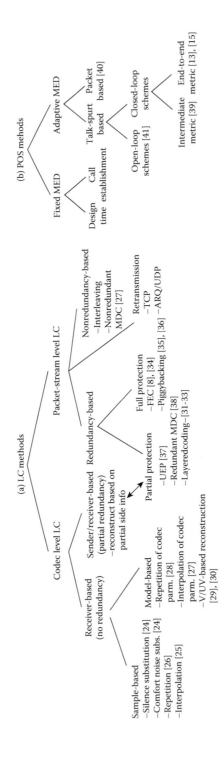

**FIGURE 2.9**

A classification of existing LC and POS methods at (a) the codec and (b) the packet-stream layers.

example, a single lost frame in G.729A [46] can lead to perceptible distortions in the reconstructed speech for up to 20 frames [31,32].

Although the large MTU in packets relative to the frame size provides new opportunities for LC in the packet-stream layer (see Section 2.4), the LC mechanisms in the packet-stream and the codec layers may not work well together. This happens because techniques for recovering lost frames [47] in codecs often exploit information in the coded speech [48] that may not be properly encapsulated into packets. For instance, G.729.1 [33], the wide-band version of G.729A, recovers a single lost frame using multiple layers of information from the previous and the next frames received. This LC technique will be not useful when multiple adjacent frames are encapsulated into a single packet that is lost. For this LC technique to be effective, the information for reconstructing a lost frame must be encapsulated in different packets.

The Internet Low Bit-rate Codec (iLBC) [42] used in early versions of Skype and its extensions [49,50] address this issue by encoding frames into self-decodable units using a modified CELP coder with increased bit rate. Although this approach avoids the propagation of internal-state errors after a loss, distortions are still perceptible unless additional LC mechanisms are implemented in the packet-stream layer.

Recently, Global IP Sound (GIPS) released the second version of its proprietary iSAC wideband speech codec for use in Skype. Its white paper [43] indicates that iSAC uses an adaptive packet period of 30–60 ms, with an adaptive bit rate of 10–32 kbps and a separate low complexity mode. Although the white paper claims that the codec achieves better performance than G.722.2 for comparable bit rates, there is no independent validation of the claim and no information on its LC capability.

In evaluating a speech codec designed, its performance is commonly evaluated by comparing the quality of its reconstructed waveforms under ideal and nonideal network conditions. One common method of generating nonideal network conditions is to use stochastic models.

Sun and Ifeachor proposed a simple Bernoulli model with independent packet losses for modeling the loss behavior in VoIP [13]. The model is highly approximate because Internet packet losses exhibit temporal dependencies [51], especially for periodic transmissions. Further, speech quality can vary significantly across different loss patterns with the same average rate [8]. A second approach based on the Gilbert model has been used for modeling the loss behavior of Internet traces in IP telephony [8] and multimedia [51] and for evaluating speech codecs [49,50]. The model is approximate because it assumes that a packet loss only depends on the loss of the previous packet. Extended models, such as the $n$-state Markov chain and the extended Gilbert model [51], use additional parameters to model the dependency of losses. The main deficiency of these models is that they do not consider the LC algorithm in the packet-stream layer, such as redundant piggybacking [32] and multidescription coding (MDC) [47,52,53]. There are a number of recent studies on cross-layer designs of codecs [54–58], but none has focused on designs for real-time VoIP. To provide efficient use of resources and the best conversational quality, the LC mechanisms in the packet-stream and the codec layers must be developed in a coupled fashion.

To optimize perceptual quality, our survey identifies the need to design the LC mechanism in codecs with that of the packet-stream layer. This means that the encoding of speech into frames must dynamically adapt to the packet rate, which in turn adapts to network congestion. In the next section, we describe our approach to improve the design of codecs for VoIP.

### 2.3.2 Cross-Layer Designs of Speech Codecs

**Codecs with Self-Decodable Units.** To avoid the propagation of errors in internal states across packet boundaries and to maximize coding efficiency within a packet, we have designed codecs that encode frames in such a way that are self-decodable but may have dependent internal states when encapsulated into a packet. This is similar to what was done in iLBC [42]. In addition, these codecs can operate in multiple modes, in terms of frame size and packet period selected by the SVM learned in the packet-stream layer for optimizing conversational quality.

Based on the ITU G.729 CS-ACELP speech codec operating at 8 kbps, we have developed cross-layer designs with redundancy-based LC. In our first design [31], we have increased the frame length in order to reserve space for redundancies at the packet level, without increasing the bit rate. It uses MDC to conceal losses at the packet level and adapts to dynamic loss conditions, while maintaining an acceptable end-to-end delay. By protecting only the most important excitation parameters of each frame according to its speech type, our approach enables more efficient usage of the bit budget. In our second design [32], we have developed a variable bit-rate layered coding scheme that dynamically adapts to the characteristics of the speech encoded and the network loss conditions. To cope with bursty losses while maintaining an acceptable end-to-end delay, our scheme employs layered coding with redundant piggybacking of perceptually important parameters in the base layer, with a degree of redundancy adjusted according to feedbacks from receivers. Under various delay constraints, we study trade-offs between the additional bit rate required for redundant piggybacking and the protection of perceptually important parameters. Although these cross-layer designs incorporate LC information in the packet-stream layer, G.729 is not the perfect codec because it suffers from the propagation of errors in internal states across packet boundaries. We have also applied a similar approach to the design of the G.722.2 and G.729.1 wide-band speech codecs that have self-decodable frames.

**Cross-Layer Designs of Speech codecs.** To facilitate the generation of information for effective LC, the encoder needs to know the amount of payload in each packet available for carrying the LC information. This is important in multi-party VoIP when multiple voice streams have to be encapsulated in the same packet and the payload for carrying redundant information is limited.

In this case, the codec needs to generate enhancement layers, and the piggy-backing of enhancement layers depends on the payload available.

**Codec Design for Dynamic Encapsulation.** We have developed a codec that allows the flexible encapsulation of frames into packets with self-contained internal states and that avoids the propagation of internal-state errors across packet boundaries. We have modified the G.722.2 codec that can operate in a wide range of frame periods (10–60 ms) and bit-rates (8–40 kbps). Based on the information provided by the packet-stream layer of the VoIP client, the coder can switch among different modes during a conversation in order to adapt to changes in the network and conversational conditions. The packet period used is propagated to the coder to ensure that it removes any redundancies among the multiple frames encapsulated in a packet. A similar adaptation has been observed in Skype, although its mechanism cannot be conclusively deduced to changes in network conditions, as its source code is unavailable.

**Off-Line Learning of Codec's Robustness to LC.** We have extended the learning of classifiers in the packet-stream layer to accurately predict LOSQ at run time, using unconcealed frame patterns available at the receiver [4]. The advantage of this approach over PESQ is that it does not require the original speech frames in the computation (which are needed for the calculation of PESQ). We have evaluated the robustness of our codec against a variety of loss conditions using off-line experiments. Network traces collected were processed using several different LC and POS schemes, and their unconcealed frame patterns were generated. Speech segments would then be encoded and decoded using various unconcealed frame patterns, and the output speech was objectively evaluated by PESQ. Finally, a classifier was trained using the unconcealed frame patterns and the corresponding PESQs.

## 2.4 Loss Concealments and Play-Out Scheduling

The packet-stream layer mitigates delays, losses, and jitters at the packet level through its LC and POS algorithms. As is discussed in the last section, these algorithms should be designed in conjunction with the speech codec and with an understanding of conversational quality. In this section, we identify the limitations of existing LC and POS algorithms. We then discuss our approaches to address the issues in their design.

### 2.4.1 Previous Work

Figure 2.9 summarizes some existing LC techniques at the packet-stream layer. They aim to either reduce the amount of unconcealable frames experienced by the decoder or provide partial redundancy for helping the decoder reduce perceptual degradations due to losses.

Retransmissions of speech frames after the detection of a network loss is infeasible in real-time VoIP, due to the excessive delays involved and their effects on MED.

Nonredundant LC schemes are generally based on the interleaving of frames during packetization [59]. One way is to exploit the fact that shorter distortions are less likely to be perceived, and to break an otherwise long segment into several shorter segments that are close by but not consecutive. This is not strictly an LC technique because it does not actually recover losses. Another way is MDC [47,52,53] that generates multiple descriptions with correlated information from the original speech data. This may be hard in low bit-rate streams whose correlated information has been largely removed during coding [47]. Another disadvantage is that the receiver will incur a longer MED when waiting for all the descriptions to arrive before declaring a description is lost.

Redundant LC schemes exploit trade-offs among the redundancy level, the delay for recovering losses from the redundant information, and the quality of the reconstructed speech. They work on the Internet because increases in packet size, as long as they are less than the MTU [45], do not lead to noticeable increases in the loss rate [36]. They consist of schemes that use partial and full redundancies. Examples employing partial redundancies include layered coding [31–33], unequal error protection (UEP) [37], and redundant MDC [38]. Examples employing full redundancies include FEC (forward error correction) [9,34] and redundant piggybacking [35,36]. An FEC-based LC scheme [15] for VoIP incorporates into its optimization metric the additional delay incurred due to redundancy. In our previous work, we have used piggybacking as a simple yet effective technique for sending copies of previously sent frames together with new frames in the same packet, without increasing the packet rate [4,10,36].

The main difficulty of using redundant LC schemes is that it is hard to know a suitable redundancy level. Its dynamic adaption to network conditions may either be too slow, as in Skype [36], or too conservative [4]. Another consideration is that the redundancy level is application-dependent. Fully redundant piggybacking is suitable in two-party VoIP, but partial redundancy may need to be used in multi-party VoIP when speech frames from multiple clients are encapsulated in the same packet.

Figure 2.9b also summarizes the various POS methods. Due to nonstationary and path-dependent delays and losses, simple schemes with fixed MEDs either hardcoded at design time or during call establishment do not provide consistent protection against late losses. Adaptive POS schemes that adjust the playout schedule at the talk spurt or the packet level are more prevalent.

At the talk-spurt level, silence segments can be added or omitted at the beginning of a talk spurt in order to make the changes virtually imperceptible to the listener. Adjustments can also be made for each frame using time-scale modification [40] that stretches or compresses frames without changing its pitch period. However, it requires additional computational resources, has small effects on MEDs, and is generally perceptible.

At the packet level, there have been several studies that aim to balance between the number of packets late for playout and the jitter-buffer delay that packets wait before their scheduled playout times. Open-loop schemes use heuristics for picking some system-controllable metrics (such as MED), based on network statistics available [41]. They are less robust because they do not explicitly optimize a target objective. Moreover, they do not consider the effects of the codec on speech quality, although their performance depends on the codec used. Closed-loop schemes with intermediate quality metrics [39] control an intermediate metric based on the late-loss rate collected in a window. Their difficulty lies in choosing a good intermediate metric. Closed-loop schemes with end-to-end quality metrics generally use the E-model [1] for estimating conversational quality as a function of some objective metrics. One study uses this estimate in a closed-loop framework to jointly optimize the POS and FEC-based LC [15]. Another study [13] proposes to use the E-model but separately trains a regression model for modeling the effects of the loss rate and the codec on PESQ. These models are limited because, without a redundancy-based LC scheme, lost frames cannot be recovered by adjusting the playout delays alone.

Existing VoIP systems usually employ redundancy-based LC algorithms for recovering losses when using UDP. However, none of these approaches considers delay-quality trade-offs for delivering VoIP of high perceptual quality to users. Previous LC algorithms based on analytic loss models [39,41] do not always perform well, as these models may not fully capture the dynamic network behavior and do not take into account the LC strategies in codecs. Existing POS algorithms based on open-loop heuristic functions [41] may not be robust under all conditions, whereas closed-loop approaches [39] are difficult to optimize without a good intermediate metric. Some recent approaches [13,15] have employed an end-to-end objective metric, such as the E-model, as their intermediate metric. There is also very little reported results on POS algorithms for multi-party VoIP [60].

We present in the next section new LC and POS control algorithms that address the trade-offs related to conversational quality. Using the classifier learned, we use run-time network and conversational conditions to select the best operating point of these control algorithms. A related problem studied is the equalization of MSs for improving perceptual quality in multi-party VoIP. We also consider the design of these algorithms with the design of the codec and the network-control algorithms in multi-party VoIP.

### 2.4.2  Packet-Stream Layer LC and POS Algorithms

**Two-Party LC and POS Schemes.** We have developed new POS/LC schemes for dynamically selecting a playout schedule for each talk spurt [4] and an appropriate redundancy degree for each packet. Using the loss information of the 100 most recent packets ($\approx$3 sec) from the receiver, the sender selects a redundant piggybacking degree in order to achieve 2 percent target unconcealed packet loss rate. Our POS scheme at the receiver uses delay information, redundant piggybacking degree, and predictions of objective metrics (LOSQ, CS, and CE) for the upcoming talk spurt in order to select a suitable playout schedule for that talk spurt. Since the conversational condition changes slowly during a conversation, CS and CE can be accurately estimated by monitoring the silence and the voiced durations at the receiver. In estimating the LOSQ curve, we first conduct off-line experiments to learn a classifier that maps network conditions to the corresponding PESQs. By considering bursty loss patterns in real traces, our approach leads to significantly more accurate estimates of PESQ when compared to those in the previous work [13] with IID loss patterns. Our classifier is then used at run time to estimate the relation between MED and LOSQ.

Figure 2.10 depicts the decisions made by our control schemes for two connections with large jitters and medium losses. It shows that our schemes

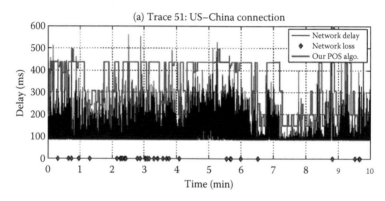

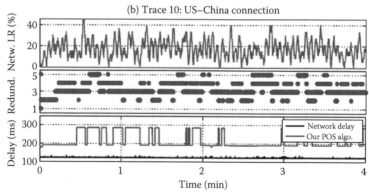

**FIGURE 2.10**
Network delays and POS/LC control decisions made by our JND-based POS and LC schemes for two connections: (a) connection with high and time-varying delay spikes; and (b) connection with medium packet losses.

can closely track the changing network conditions, while making discrete adjustments when needed in order to keep the conversational quality in a user preferred state. Our system also performs better than the *p*-optimum algorithm [13], which estimates conversational quality by a hybrid E-Model and PESQ [4].

Lastly, we have integrated the classifiers learned into the design of a prototype VoIP system. The classifiers enable the systematic tuning of the control algorithms in our prototype, without the need to carry out expensive subjective tests.

**Multi-Party LC and POS Schemes.** We observe that, from a client's (say A) perspective, the decisions made by the POS of other clients in equalizing MSs in the current turn do not affect the MS observed by A. Hence, A's decision can be assumed to be independent of the concurrent decisions made by the other clients. Figure 2.11 extends the trade-off curve for the two-party VoIP in Figure 2.7 to the multi-party case. It depicts two trade-off curves: the curve connecting A, B, and C corresponds to a network condition with high disparities in delays among the connections; and the curve connecting A, B′ and C′ corresponds to a condition with similar average delay but considerably less disparities. The control from A to B (*resp.*, A to B′) is similar to the two-party case: increasing the MED toward B (*resp.*, B′) conceals more packets and improves LOSQ but degrades CS and CE. In the multi-party case, further increasing the MED from B to C (*resp.*, B′ to C′) to achieve full equalization will lead to a high LOSQ with improved CS but degraded CE. Hence, operating at C will result in a highly inefficient conversation with low CE; whereas operating at C′ is relatively more efficient than at C.

We have developed control schemes that operate in conjunction with an overlay network (described in the next section) in multi-party VoIP [10]. To reduce the overhead and to improve loss adaptations, our LC scheme operates

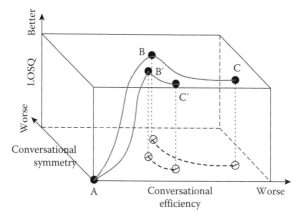

**FIGURE 2.11**
Trade-offs under different multi-party conditions.

on a link-by-link basis to select the redundant piggybacking degree. When multiple speakers are talking, we combine their packets destined to the same client into one (without exceeding the MTU). Although our scheme requires each node to maintain retransmission buffers for storing recently received frames, the overhead is manageable when the piggybacking degree is four or less.

Our POS scheme is different than the two-party counterpart. In the multi-party case, the order of the speakers is unknown, and the network conditions among the participants may have large disparities. We address this problem by equalizing the MSs observed by different listeners in the same turn. For listeners whose MED affects the efficiency of the whole conversation (bottleneck clients), we use stricter MED values that closely hugs the delay curve, similar to that in Figure 2.10a. For other listeners, we use less strict MEDs. In comparison to the MSs in Skype in Figure 2.5, Figure 2.12 shows that our system has a lower average MS as well as less variations. We are in the process of developing a general equalization algorithm with parameters that operate in continuous equalization levels [10]. Using simulated conversations generated under various control parameters and network and conversational conditions, we plan to conduct subjective JND tests to determine the perceptual sensitivity of humans in conversational quality as a function of the level of equalization. We will develop a statistical procedure similar to that described in Section 2.2 and experiment on some limited cases to within a prescribed confidence level. Finally, we plan to train a classifier off-line and apply it at run time to determine the control parameters that will lead to the most preferred conversational quality. We expect our approach to be general and lead to better equalization than that in Figure 2.12.

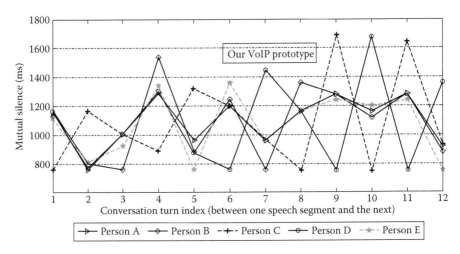

**FIGURE 2.12**
MSs experienced in our VoIP prototype.

## 2.5  Network-Layer Support for VoIP

Several standardized and proposed transport protocols, such as TCP, UDP, and RTP, were designed with different trade-offs between reliable delivery of packets and application-layer end-to-end delay. By providing hooks for synchronization, reliability, QoS feedback, and flow control, RTP can support real-time multimedia applications [61–63]. Although RTP does not guarantee the real-time delivery of continuous media, it relies on resource reservations protocols, such as RSVP [64], for resource scheduling. RTP can be used in conjunction with the mechanisms described in this chapter to ensure high conversational quality. Another approach is to design end-to-end protocols that are more TCP friendly [65,66]. These protocols will need to be extended in order to address the trade-offs in conversational quality.

In multi-party VoIP, the current speaker(s) needs to convey his or her speech information to all the listeners. A good design should accommodate dynamic and diverse network conditions among the clients. The protocol and the connection topology used are generally dictated by where audio mixing is done [67]. When a VoIP client (as in Skype [68]) or a bridge (as in QQTalk [69]) is responsible for decoding, mixing, and encoding the signals from the clients, it is natural for all the clients to send their packets to the centralized site to be mixed and forwarded. The approach may not be scalable because it can create a bottleneck near that site. Moreover, the maximum end-to-end delay (ME2ED) between any speaker-listener pair can be large when the clients are geographically distributed [10].

On the other hand, a distributed approach asks each client to independently manage its transmissions. One way is for each speaker to multicast the speech information to all listeners. Although multicasts are available on the Internet [70], the support of reliable real-time multicasts for receivers of different loss and delay behavior is very preliminary [71]. The focus in the IETF working group on a NACK-based asynchronous-layered coding protocol with FEC [72] is inadequate for multi-party VoIP.

A hybrid approach is to have an overlay network [73] that uses a subset of the clients to manage the mixing and forwarding of unicast packets. The approach achieves a shorter ME2ED than a centralized approach and a smaller number of unicast messages than a fully distributed approach. One issue, however, is that it is complex for the overlay clients to coordinate the decoding, mixing, and encoding of the speech signals. Alternatively, we have taken the approach for the overlay clients to simply encapsulate the speech frames from the multiple clients into a single packet before forwarding the frames [10]. This is not an issue as long as the number of streams to be encapsulated fits within the MTU of each packet.

To design a topology with proper trade-offs between ME2ED among the clients and the maximum number of packets relayed in a single packet period by any client, we have studied a commonly used overlay topology constructed from a subset of the clients (called parent nodes) [10]. The topology assumes that all the parent nodes are fully connected, and that each remaining node (called child node) is connected to only one parent node. When a call is set up, the client that initiated the call collects delay and loss information among the clients. Due to the prohibitive nature of enumerating all topologies, we use a greedy algorithm to find a good topology. The heuristic first determines the client pair with ME2ED (called bottleneck pair) in a fully connected topology. It then finds a single-parent topology that minimizes ME2ED. If the improvement in ME2ED is small (say less than 50 ms), then it uses the best single-parent topology as the overlay network; otherwise, it adds a second parent node. It iteratively increases the number of parents until either the difference between the ME2EDs of the current topology and the fully connected topology is small or the bottleneck pair in the current topology is directly connected. The process can be repeated whenever there is a significant change in the network conditions.

We are in the process of developing admissions-control algorithms for deciding whether a new client can be added without adversely affecting the conversational quality of all the listeners, and the dynamic dissemination of network information for operating the control algorithms. We plan to initially use UDP as the transport protocol but will consider RTP and other protocols at a later stage.

## 2.6 Conclusions

In this chapter, we have presented some solutions on the study of real-time, two-party, and multi-party VoIP systems that can achieve high perceptual conversational quality. Our solutions focus on the fundamental understanding of conversational quality and its trade-offs among the design of speech codecs and strategies for network control, POS, and loss concealments.

The degradation in the perceptual quality of an interactive conversation over a network connection is caused by a combination of the decrease in speech quality when packets are lost or delayed, and the asymmetry in silence periods when the conversation switches from one speaker to another. Its study is largely unexplored because there is no single objective metric for assessing the quality of a VoIP conversation whose results match well with subjective results. Indiscriminate subjective testing is not feasible because it is prohibitively expensive to carry out many such tests under various conversational and network conditions. Moreover, there is no systematic method to generalize the subjective test results to unseen conditions. To this end, there are three issues studied in this chapter that address the limitations of existing work and that improve the perceptual quality of VoIP systems.

1. We have presented a statistical approach based on a JND to significantly reduce the large number of subjective tests, as well as a classification method to automatically learn and generalize the results to unseen conditions. Using network and conversational conditions measured at run time, the classifier learned helps adjust the control algorithms in achieving high perceptual conversational quality.

2. We have described the concept of a cross-layer speech codec to interface with the LC and POS algorithms in the packet-stream layer in order to be more robust and effective against packet losses.

3. We have presented a distributed algorithm for equalizing mutual silences and an overlay network for multi-party VoIP systems. The approach leads to multi-party conversations with high listening-only speech quality and balanced mutual silences.

## Acknowledgment

Research supported by National Science Foundation Grant CNS 08-41336.

## References

1. ITU-G.107, "The E-model, a computational model for use in transmission planning," http://www.itu.int/rec/T-REC-G.107/en.
2. K. Weilhammer and S. Rabold, "Durational aspects in turn taking," in *Proc. Int'l Conf. on Phonetic Sciences*, 2003.
3. K. Weilhammer and S. Rabold, "Durational aspects in turn taking," in *Proc. Int'l Conf. on Phonetic Sciences*, 2003. [Online]. Available: http://novel.crhc.illinois.edu/ papers.db/k/KWSR03.pdf
4. B. Sat and B. W. Wah, "Playout scheduling and lossconcealments in VoIP for optimizing conversational voice communication quality," in *Proc. ACM Multimedia*, Augsburg, Germany, September 2007, pp. 137–146.
5. B. Sat and B. W. Wah, "Evaluation of conversational voice quality of the Skype, Google-Talk, Windows Live, and Yahoo Messenger VoIP systems," in *IEEE Int'l Workshop on Multimedia Signal Processing*, October 2007.
6. D. L. Richards, *Telecommunication by Speech*. London, UK: Butterworths, 1973.
7. N. Kiatawaki and K. Itoh, "Pure delay effect on speech quality in telecommunications," *IEEE Journal on Selected Areas of Communication*, vol. 9, no. 4, pp. 586–593, May 1991.
8. J.-C. Bolot, S. Fosse-Parisis, and D. Towsley, "Adaptive FEC-based error control for Internet telephony," in *Proc. IEEE INFOCOM*, vol. 3, 1999, pp. 1453–1460.

9. P. T. Brady, "Effects of transmission delay on conversational behaviour on echo-free telephone circuits," *Bell System Technical Journal*, vol. 50, no. 1, pp. 115–134, January 1971.

10. B. Sat, Z. X. Huang, and B. W. Wah, "The design of a multi-party VoIP conferencing system over the Internet," in *Proc. IEEE Int'l Symposium on Multimedia*, Taichung, Taiwan, December 2007, pp. 3–10.

11. International Telecommunication Union, "ITU-T G-Series recommendations," http://www.itu.int/rec/T-REC-G/en

12. International Telecommunication Union, "ITU-T P-Series recommendations," http://www.itu.int/rec/T-REC-P/en

13. L. Sun and E. Ifeachor, "New models for perceived voice quality prediction and their applications in playout buffer optimization for VoIP networks," in *Proc. IEEE Communications*, vol. 3, 2004, pp. 1478–1483.

14. L. Sun and E. Ifeachor, "Voice quality prediction models and their applications in VoIP networks," *IEEE Trans. on Multimedia*, vol. 9, no. 4, pp. 809–820, 2006.

15. C. Boutremans and J.-Y. L. Boudec, "Adaptive joint playout buffer and FEC adjustment for Internet telephony," in *Proc. IEEE INFOCOM*, vol. 1, 2003, pp. 652–662.

16. A. Takahashi, H. Yoshino, and N. Kitawaki, "Perceptual QoS assessment technologies for VoIP," *IEEE Communications Magazine*, pp. 28–34, July 2004.

17. A. P. Markopoulou, F. A. Tobagi, and M. J. Karam, "Assessing the quality of voice communications over Internet backbones," *IEEE/ACM Trans. on Networking*, vol. 11, no. 5, pp. 747–760, 2003.

18. B. Sat and B. W. Wah, "Evaluating the conversational voice quality of the Skype, Google-Talk, Windows Live, and Yahoo Messenger VoIP systems," *IEEE Multimedia*, pp. 46–58, January–March 2009.

19. J. L. Flanagan, *Speech Analysis, Synthesis and Perception.* New York: Springer, 1972.

20. B. Sat and B. W. Wah, "Statistical testing of off-line comparative subjective evaluations for optimizing perceptual conversational quality in voip," in *Proc. IEEE Int'l Symposium on Multimedia*, pp. 424–431, December 2008.

21. P. L. Lanzi, "Learning classifier systems: Then and now," *Evolutionary Intelligence*, vol. 1, pp. 63–82, 2008.

22. C.-C. Chang and C.-J. Lin, "A library for support vector machines," http://www.csie.ntu.edu.tw/~cjlin/libsvm/.

23. Z. X. Huang, B. Sat, and B. W. Wah, "Automated learning of play-out scheduling algorithms for improving the perceptual conversational quality in multi-party VoIP," in *Proc. IEEE Int'l Conf. on Multimedia and Expo*, Hannorer, Germany, July 2008, pp. 493–496.

24. J. Suzuki and M. Taka, "Missing packet recovery techniques for low-bit-rate coded speech," *IEEE Journal on Selected Areas in Communications*, vol. 7, no. 5, pp. 707–717, June 1989.

25. N. S. Jayant and S. W. Christensen, "Effects of packet losses in waveform coded speech and improvements due to odd-even sample-interpolation procedure," *IEEE Trans. on Communications*, vol. 29, no. 2, pp. 101–110, February 1981.

26. R. C. F. Tucker and J. E. Flood, "Optimizing the performance of packet-switch speech," in *IEEE Conf. on Digital Processing of Signals in Communications*, Loughborough University, April 1985, pp. 227–234.

27. D. Lin, *Loss Concealments for Low Bit-Rate Packet Voice*.1em plus 0.5em minus 0.4em Urbana, IL: Ph.D. Thesis, Dept. of Electrical and Computer Engineering, Univ. of Illinois, August 2002.

28. S. Atungsiri, A. Kondoz, and B. Evans, "Error control for low-bit-rate speech communication systems," *IEEE Proc. I: Communications, Speech and Vision*, vol. 140, no. 2, pp. 97–103, April 1993.

29. J. Wang and J. D. Gibson, "Parameter interpolation to enhance the frame erasure robustness of CELP coders in packet networks," in *IEEE Int'l Conf. on Acoustics, Speech, and Signal Processing*, vol. 2, 2001, pp. 745–748.

30. J. F. Wang, J. C. Wang, J. Yang, and J. J. Wang, "A voicing-driven packet loss recovery algorithm for analysis-by-synthesis predictive speech coders over Internet," *IEEE Trans. on Multimedia*, vol. 3, no. 1, pp. 98–107, March 2001.

31. B. Sat and B. W. Wah, "Speech-adaptive layered G.729 coder for loss conceal-ments of real-time voice over IP," in *Proc. IEEE Int'l Conf. on Multimedia and Expo*, Amsterdam, The Netherlands, July 2005.

32. ------, "Speech- and network-adaptive layered G.729 coder for loss concealments of real-time voice over IP," in *IEEE Int'l Workshop on Multimedia Signal Processing*, Oct. 2005.

33. ITU-G.729.1, "G.729-based embedded variable bit-rate coder: An 8-32 kbit/s scalable wideband coder bitstream interoperable with G.729," http://www.itu.int/rec/T-REC-G.729.1/en.

34. N. Shacham and P. McKenney, "Packet recovery in high-speed networks using coding and buffer management," in *Proc. of IEEE INFOCOM*, San Francisco, CA, May 1990, pp. 124–131.

35. T. J. Kostas, M. S. Borella, I. Sidhu, G. M. Schuster, J. Grabiec, and J. Mahler, "Real-time voice over packet-switched networks," *IEEE Network*, vol. 12, no. 1, pp. 18–27, January–February 1998.

36. B. Sat and B. W. Wah, "Analysis and evaluation of the Skype and Google-Talk VoIP systems," in *Proc. IEEE Int'l Conf. on Multimedia and Expo*, Toronto, Ontario, Canada, July 2006.

37. M. Chen and M. N. Murthi, "Optimized unequal error protection for voice over IP," in *ICASSP*, 2004, pp. 865–868.

38. W. Jiang and A. Ortega, "Multiple description coding via polyphase transform and selective quantization," in *Proc. Visual Communications and Image Processing*, vol. 3653, December 1998, pp. 998–1008.

39. S. B. Moon, J. Kurose, and D. Towsley, "Packet audio playout delay adjustment: performance bounds and algorithms," *Multimedia Systems*, vol. 6, no. 1, pp. 17–28, January 1998.

40. Y. J. Liang, N. Faber, and B. Girod, "Adaptive playout scheduling and loss con-cealment for voice communication over IP networks," *IEEE Trans. on Multimedia*, vol. 5, no. 4, pp. 532–543, December 2003.

41. R. Ramjee, J. Kurose, D. Towsley, and H. Schulzrinne, "Adaptive playout mechanisms for packetized audio applications in wide-area networks," in *Proc. 13th Annual Joint Conf. IEEE Computer and Communications Societies on Networking for Global Commmunication*, Toronto, Ontario, Canada, vol. 2, 1994, pp. 680–688.

42. S. Andersen, A. Duric, H. Astrom, R. Hagen, W. Kleijn, and J. Linden, "Internet low bit rate codec (iLBC)," December 2004, http://www.ietf.org/rfc/rfc3951.txt.

43. Global IP Sound, "Datasheet of GIPS Internet Speech Audio Codec," 2007, http://www.gipscorp.com/files/english/datasheets/iSAC.pdf.

44. W. C. Chu, *Speech Coding Algorithms*. Wiley-Interscience, A John Wiley & Sons, Inc., Publication, 2003.

45. IETF, "RFC 791, Internet Protocol: DARPA Internet program protocol specification," September 1981, http://www.ietf.org/rfc/rfc791.txt

46. ITU-G.729, "Coding of speech at 8 kbit/s using conjugate-structure algebraic-code-excited linear prediction (CS-ACELP)," http://www.itu.int/rec/T-REC-G.729/en.

47. D. Lin and B. W. Wah, "LSP-based multiple-description coding for real-time low bit-rate voice over IP," *IEEE Trans. on Multimedia*, vol. 7, no. 1, pp. 167–178, February 2005.

48. T. Chua and D. C. Pheanis, "QoS evaluation of sender-based loss-recovery techniques for VoIP," *IEEE Network*, pp. 14–22, December 2006.

49. C. M. Garrido, M. N. Murthi, and S. V. Andersen, "Towards iLBC speech coding at lower rates through a new formulation of the start state search," in *Proc. ICASSP*, vol. 1, 2005, pp. 769–772.

50. C. M. Garrido, M. N. Murthi, and S. V. Andersen, "On variable rate frame independent predictive speech coding: Re-engineering iLBC," in *Proc. ICASSP*, vol. 1, 2006, pp. 717–720.

51. W. Jiang and H. Schulzrinne, "Modelling of packet loss and delay and their effect on real-time multimedia service quality," in *Proc. Int'l Workshop on Network and Operating System Support for Digital Audio and Video (NOSSDAV)*, Chapel Hill, North Carolina, USA, 2000.

52. A. A. E. Gamal and T. M. Cover, "Achievable rates for multiple descriptions," *IEEE Trans. on Information Theory*, vol. 28, pp. 851–857, November 1982.

53. V. K. Goyal, "Multiple description coding: Compression meets the network," *IEEE Signal Processing Magazine*, pp. 74–93, September 2001.

54. D. Triantafyllopoulou, N. Passas, A. K. Salkintzis, and A. Kaloxylos, "A heuristic cross-layer mechanism for real-time traffic over IEEE 802.16 networks," *International Journal of Network Management*, vol. 17, no. 5, pp. 347–361, 2007.

55. G. Fairhurst, M. Berioli, and G. Renker, "Cross-layer control of adaptive coding and modulation for satellite Internet multimedia," *International Journal of Satellite Communications and Networking*, vol. 24, no. 6, pp. 471–491, 2006.

56. I. Haratcherev, J. Taal, K. Langendoen, R. Lagendijk, and H. Sips, "Optimized video-streaming over 802.11 by cross-layer signaling," *IEEE Communications Magazine*, pp. 115–121, 2006.

57. J. Makinen, P. Ojala, and H. Toukomaa, "Performance comparison of source controlled GSM AMR and SMV vocoders," *Proc. Int'l Symp. on Intelligent Signal Processing and Communication Systems*, Seoul, Korea, pp. 151–154, 2004.

58. T. Kawata and H. Yamada, "Adaptive multi-rate VoIP for IEEE 802.11 wireless networks with link adaptation function," *Proc. IEEE GlobeCom*, pp. 357–361, 2006.

59. C. Perkins, O. Hodson, and V. Hardman, "A survey of packet loss recovery techniques for streaming audio," *IEEE Network*, pp. 40–48, September–October 1998.

60. T. Z. J. Fu, D. M. Chiu, and J. C. S. Lui, "Performance metrics and configuration strategies for group network communication," in *IEEE International Workshop on Quality of Service*, 2007, pp. 173–181.

61. D. Minoli and E. Minoli, *Delivering Voice over IP Networks*. New York: Wiley Computer Pub., 1998.

62. H. Schulzrinne, S. Casner, R. Frederick, and V. Jacobson, "RFC 1889: RTP: A transport protocol for real time applications," January 1996, http://www.ietf.org/rfc/rfc3951.txt

63. H. Schulzrinne, "Real Time Protocol," 2008, http://www.cs.columbia.edu/~hgs/rtp/. http://www.cs.columbia.edu/~hgs/rtp.

64. R. Braden, L. Zhang, S. Berson, S. Herzog, and S. Jamin, "RFC 2205: Resource ReSerVation protocol (RSVP): Version 1 functional specification," September 1997, http://www.ietf.org/rfc/rfc2205.txt

65. P. S. Center, "Recent work on congestion control algorithms for non-TCP based applications," 2008, http://www.psc.edu/networking/projects/tcpfriendly/

66. S. Liang and D. Cheriton, "TCP-RTM: Using TCP for real-time multimedia applications," Dept. of Electrical Engineering, Stanford University, Stanford, CA, Tech. Rep., 2003.

67. P. J. Smith, P. Kabal, M. L. Blostein, and R. Rabipour, "Tandem-free VoIP conferencing: A bridge to next-generation networks," *IEEE Communications Magazine*, pp. 136–145, May 2003.

68. Skype, "The Skype VoIP system," 2008, http://www.skype.com. [Online].

69. M. Handley and J. Crowcroft, "Internet multicast today," *The Internet Protocol Journal*, vol. 2, no. 4, 1999.

70. I. R. W. Group, "Reliable Multicast Transport (RMT)," 2008, http://www.ietf.org/html.charters/rmt-charter.html. [Online].

71. D. Doval and D. O'Mahony, "Overlay networks: A scalable alternative for P2P," *IEEE Internet Computing*, pp. 1–5, 2003.

# 3

## A Hybrid Approach to Communicate between WSN and UMTS with Video Quality Measurements and Predictions

Yufei Du, Fung Po Tso, and Weijia Jia

## CONTENTS

The 3G-324M is an umbrella standard of the Third Generation Partnership Project (3GPP) for wireless video communications, which was developed to satisfy the stringent requirements of real-time interactive video and audio services. In practice, 3G-324M has been employed in 3G networks today to enable the multimedia services with messaging and streaming. However, the design of the supporting architecture for the unification of the diverse streams with 3G-324M poses lots of challenges. This chapter introduces a new supporting architecture for transmitting the video streams from a sensor to the UMTS user. This architecture integrates mobile network architecture with sensor ad hoc networks, which considerably reduces the implementation cost and the power consumption. It also suits the deployment of large-scale sensor networks. Video sensor can provide important visual information in a number of applications such as health care, environmental monitoring, emergency response, and secure surveillance. However, when transmitted to a mobile user, it is difficult to predict and estimate the quality of a video, since the encoding and decoding and transmission lose will dramatically decrease the original video quality. This chapter describes an algorithm for approximating the received video quality by combining the information of the original video type with channel status. The video type information consists of encoded video quality, which is decided by the original video motion frequency and frame lost rate, which is decided by the original video bit rate. The channel status information indicates the packet lost rate. Predicting the video quality has many uses, such as choosing the best channel to transmit the video.

## 3.1 Introduction

With the development growth of third generation telephony (video telephony), there is a tremendous demand on the third generation Circuit Switched (3G-CS) video service. The 3G wireless network, foreseen to be the enabling technology for multimedia services with up to 64 kbps for CS network and 2 Mbps for the Packet Switched (PS) Network, makes it feasible for visual communication over the wireless link. The transmission rate is usually very unstable in wireless channel, which is caused by multipath fading, intersymbol interference, and noise disturbances. The transmission rate varies with the changing external environment, resulting in devastating effect on multimedia transmission. To cope with this on a wireless channel, accurate network-condition estimation and effective flow control are essential for robust video transmission. It is known that video transmission is delay-sensitive but may be tolerable to some kinds of errors. Moreover, different portions of video bitstream have different importance to the reconstructed video quality; thereby giving rise to different network quality of service (QOS) requirements, such as transmission latency, bit error rate, and so on. For instance, it is intuitive that lower layers of a layered scalable video codec have higher network QOS requirements than those of higher layers. Therefore, adopting different multimedia transmission rate schemes for each portion is more appropriate for such a compressed video bit stream. However, channel coding introduced by dynamic multimedia transmission rate would generate increased computational complexity. Considering the limited bandwidth in the wireless network of the mobile device, the resource, such as bandwidth, should be allocated appropriately for source and channel coding.

To date, the wireless multimedia study has focused on robust video transmission over general PS wireless channels. There is, however, a scarcity of work performed on delivering data from a 3G CS wireless network or system to other PS networks. The key issue for data delivery over a 3G network, which consists of several layers, is that it is not obvious how to achieve end-to-end optimality for data delivery although a single-layer performance can reach optimum. Aiming to solve such a problem, we present a new architecture, which leads to interoperability between the sensor and UMTS networks to move the video from a sensor to a user at a lower cost and lower power.

As pointed out above, this chapter works on UMTS networks and PSN networks from the following three aspects:

1. For the UMTS networks, we will present an efficient implementation of 3G-324M protocol stack for low bit rate multimedia communication, control and data multiplexing protocols H.245 and H.223, targeted for a multipoint video user in both 3G and wireless local area network (WLAN) environments. We will present the design and implementation details of an efficient and robust multimedia gateway that enables ubiquitous multimedia wireless communication. To achieve the performance improvement of the gateway, we focus on efficiently implementing the H.245, which is the main call handling and signaling protocol. In our method, sets of H.245 messages are compiled, used, and stored in a table so as to be reused for further incoming calls. After implementing this procedure, we have successfully tested our 3G-IP gateway, which is robust enough for smoothly handling up to one million concurrent calls.

2. For the PSN networks, we contribute to the effort by introducing a new architecture for video transmission from a sensor to the UMTS terminal at a low cost and power for the purpose that the video sensors can be within reach of UMTS for best utilization of UMTS coverage. The integration of the mobile network infrastructure and sensor ad hoc networks can reduce the cost of building new infrastructure and enable the large scale deployment of sensor networks. Furthermore, we build up the platform and implement the video gateway. The advantages are that communicating between these two systems dynamically and intelligently can reduce the cost and increase the lifetime of sensor networks. It was also shown that this approach is suitable for all organizations and for gathering data on demand. The feasibility and viability of the proposed method has been proven through initial experimental work.

3. For video quality measurements and predictions, we present the video quality measurements and predictions for low bit rate video during our real implementation work. Toward this, the Perceptual Evaluation of Video Quality (PEVQ) is introduced and exploited

for quantifying purposes. The objective of the thesis is to research the relationship of the spatiotemporal dynamics of the content and network status to the deduced perceptual quality of a H.263-based encoded video. Once this relationship has been established, it is shown how it can be further exploited toward proposing a technique for video quality prediction at a pre-encoding state.

## 3.2 Research on Circuit-Switched Mobile Video

3G-324M [15,16] is an umbrella standard that enables real-time multimedia services over CS wireless networks. It includes protocol elements for multiplexing and demultiplexing of speech, video, user, and control data (H.223) [7–11] in a single 64 Kb/s circuit, call control (H.245). video (H.263 [13] and MPEG-4) and audio codecs [19] (AMR-NB [20–23] and G.723.1 [12]). A block diagram of 3G-324M and the relevant 3GPP specifications is presented in Figure 3.1. Note that the 3GPP [17] call setup requirements are spread in several documents. The 3G-324M is derived from H.324 [14], which is a standard made by ITU-T for low bit rate multimedia communication, while H.245 and H.223 are two main parts under H.324 and have given specific descriptions about the procedures of message transformation and data transmission multiplexing. H.324 and its annex C are referred to as H.324M for mobile terminals.

### 3.2.1 Summarization of the H.324

H.324 describes terminals for low bit rate multimedia communication, utilizing V.34 modems operating over the General Switched Telephone Network (GSTN). H.324 terminals may carry real-time voice, data and video, or any combination, including video telephony.

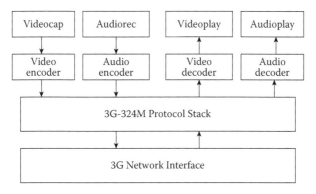

**FIGURE 3.1**
3G324M General structure.

The H.324 terminals may be integrated into personal computers or implemented in the stand-alone devices such as video telephones. Support for each media type (voice, data, and video) is optional, but if supported, the ability to use a specified common mode of operation is required, so that all terminals supporting that media type can interwork. H.324 allows more than one channel of each type to be in use. Other ITU-T recommendations in the H.324 series include the H.223 multiplex, H.245 control, H.263 video codec, and G.723.1 audio codec.

H.324 makes use of the logical channel signaling procedures of ITU-T Recommendations H.245 in which the content of each logical channel is described when the channel is opened. Procedures are provided for expression of receiver and transmitter capabilities, so transmissions are limited to what receivers can decode, and so that receivers may request a particular desired mode from transmitters.

H.324 terminals may be used in multipoint configurations through MCUs, and may interwork with H.320 terminals on the Integrated Services Digital Network (ISDN), as well as with terminals on wireless networks. Annex A defines the data protocol stack for use with the H.324 control channel. Annex B defines High-level Data Link Controller (HDLC) frame structure transparency for asynchronous transmission. Annex C defines the use of H.324 terminals in error-prone transmission environments.

### 3.2.1.1 Functional Elements Covered by 3G324M

The Video codec (H.263 or H.261) carries out redundancy reduction coding and decoding for video streams.

The Audio codec (G.723.1) encodes the audio signal from the microphone for transmission, and decodes the audio code that is output to the speaker. Optional delay in the receiving audio path compensates for the video delay, so as to maintain audio and video synchronization.

The Data protocols support data applications such as electronic whiteboards, still image transfer, file exchange, database access, audio graphics conferencing, remote device control, network protocols, and so on. Other applications and protocols may also be used via H.245 negotiation.

The Control protocol (H.245) provides end-to-end signaling for proper operation of the H.324 terminal, and signals all other end-to-end system functions including reversion to analog speech-only telephony mode. It provides for capability exchange, signaling of commands and indications, and messages to open and fully describe the content of logical channels.

The Multiplex protocol (H.223) multiplexes transmitted video, audio, data, and control streams into a single bit stream, and demultiplexes a received bit stream into various multimedia streams. In addition, it performs logical framing, sequence numbering, error detection, and error correction by means of retransmission, as appropriate to each media type.

The Modem (V.34) converts the H.223 synchronous multiplexed bit stream into an analog signal that can be transmitted over the GSTN, and converts the received analog signal into a synchronous bit stream that is sent to the Multiplex-Demultiplex protocol unit.

### 3.2.1.2 Multimedia Data Streams of 3G324M

At the low level of mobile terminals, multimedia data streams are classified as video streams, audio streams, data streams, and control streams.

- Video streams are continuous traffic, carrying moving color pictures. When used, the bit rate available for video streams may vary according to the needs of the audio and data channels.
- Audio streams are real-time, but may optionally be delayed in the receiver processing path to maintain synchronization with the video streams. In order to reduce the average bit rate of audio streams, voice activation may be provided.
- Data streams may represent still pictures, facsimile, documents, computer files, computer application data, undefined user data, and other data streams.
- Control streams pass control commands and indications between remote counterparts.

### 3.2.1.3 Framework of 3G-324M

A 3G video call communication, covered by 3G-324M, is shown in the following steps: call-setup of voiceband channel and establishment of digital communication, initialization (H.245), communication (video and audio transmission), end of session and supplementary services and call clearing.

#### 3.2.1.3.1 Call-Setup of Voice Band Channel and Establishment of Digital Communication

The calling terminal shall request the connection according to procedures for analog telephony, according to national standards. Upon successful completion of call-setup, the H.324 terminal shall follow the call start-up procedure described in ITU-T Rec. V.8. If the V.8 start-up procedure detects a V.34 modem, the start-up procedure for that modem shall be followed. Upon completion of the modem start-up procedure and establishment of digital communication, the terminal shall proceed to phase D, initialization.

If the V.8 procedure fails to detect a V.34 modem or the handshake and the establishment of the digital connection is not successful after a suitable period the calling terminal may, depending on predetermined configuration, go to telephony mode, disconnect the line, or go to another operating mode more suitable for the detected modem. Such other modes are outside the scope of this recommendation.

This phase is out of the range of 3G324M protocol stack (H.324, H.245, and H.223) but done by modem. A terminal should be able to work under both an active mode, dialing a call-setup, and a passive mode, answering a call-setup request. In our implementation, (Hayes At Command) AT Commands are used to make call-setup with the remote terminal, the procedure is presented in the following codes, including the active mode and passive mode.

### 3.2.1.3.2 *Initialization (H.245)*

After digital communication has been established, a minimum of 16 HDLC flags shall be transmitted in order to ensure synchronization. Following this, system-to-system communication shall be initiated using the H.245 control channel. Since no multiplex table entries have yet been sent to the receiver, initial control messages shall be sent using multiplex table entry 0.

Terminal system capabilities are exchanged by transmission of the H.245 TerminalCapabilitySet message. This capability protocol data unit (PDU) shall be the first message sent. The H.245 MasterSlaveDetermination message shall also be sent at this time in which the terminals exchange random numbers, according to the procedure in ITU-T Rec. H.245, to determine the master and slave terminals. H.324 terminals shall be capable of operating in both master and slave modes, and shall set terminalType to 128 and set StatusDeterminationNumber to a random number in the range 0 to 224. Only one random number shall be chosen by the terminal for each call, except in the case of identical random numbers, as described in ITU-T Rec. H.245.

If the initial capability exchange or master/slave determination procedures fail, these should be retried at least two additional times before the terminal abandons the connection attempt and proceeds to phase D.

The range of terminalTypes from 0 to 127 is reserved for possible use by multipoint control units (MCUs) or other nonterminal devices that may need to be a slave at all times, and the range 129 to 255 is reserved for possible use by MCUs or other nonterminal devices that may need to be master at all the times.

After these procedures are completed, and the far-end capabilities have been received, the procedures of H.245 may then be used to open logical channels for various information streams. Multiplex table entries may be sent before or after logical channels have been opened, but information shall not be transmitted over a logical channel until the channel is opened, and an appropriate H.223 multiplex table entry has been defined.

- **Exchange of Video by Mutual Agreement**
  The indication videoIndicateReadyToActivate, "Video Indicate Ready-To-Activate," is defined in H.245. Its use is optional, but when used the procedure shall be as follows.
  Terminal X has been set so that video is not transmitted unless, and until, the remote terminal has also indicated readiness to transmit

video. Terminal X shall send the indication videoIndicateReadyTo-Activate when the initial capability exchange has been completed, but shall not transmit a video signal until it has received either videoIndicateReadyToActivate or incoming video.

A terminal, which has not been set in this optional way, is not obliged to wait until receipt of videoIndicateReadyToActivate or video before initiating its video transmission.

### 3.2.1.3.3 Communication

During a session, the procedures for changing logical channel attributes, capability, receive mode, for example, shall be carried out as defined in H.245.

- **Rate changes and retrains**
  During phase C communication, the modem may retrain or alter its rate of data transmission, with or without momentary disruption of data transmission and loss of data. Upon any such momentary disruption of data transfer, the terminal shall not restart phase B, but shall remain in phase C and execute the normal H.324 error recovery procedures according to H.223.

- **Involuntary disconnection**
  If the terminal detects involuntary, unrecoverable loss of modem communication, the terminal shall immediately proceed to phase E, analog telephony mode or line disconnection, bypassing phase D.

### 3.2.1.3.4 End of Session

Either terminal may initiate the end of the session. The initiating terminal shall use the following procedure.

For each logical channel carrying video, it shall stop sending video at the end of a complete picture and then close the logical channel.

- It shall close all outgoing logical channels carrying data and audio.
- It shall transmit the H.245 message EndSessionCommand, and then discontinue all H.245 message transmissions. This message shall contain an indication to the far end regarding the mode the terminal will enter after the end of the session (disconnect line, analog telephony, or other mode).
- On subsequent receipt of EndSessionCommand from the remote end, it shall proceed to phase E, except that if the initiating terminal indicated an intention to disconnect the line after the end of session, the terminal shall not wait for receipt of EndSessionCommand from the remote end, but shall proceed directly to phase D.

A terminal receiving EndSessionCommand without first having transmitted it shall

- If the initiating terminal's EndSessionCommand message indicated "disconnect line," optionally follow process above, then proceed to phase E.
- Otherwise, follow process above, then proceed to phase E. If possible, the responding terminal should proceed to the new mode indicated in the initiating terminal's EndSessionCommand message.

### 3.2.1.3.5 *Supplementary Services and Call Clearing*

If the terminal arrived at supplementary services and call clearing by involuntary disconnection, it shall disconnect or revert to analog telephony, depending on a predetermined configuration.

A terminal wishing to terminate a call shall first initiate the session end procedure described in phase endsession.

In supplementary services and call clearing, the terminal should proceed as it indicated in the EndSessionCommand message. If it indicates a change to another digital communication mode, it shall begin the new mode at the equivalent of Initialization; otherwise, it shall initiate the cleardown procedures defined in V.34.

### 3.2.2 Introduction of H.245

H.245 provides a number of different services, some of which are expected to be applicable to all terminals that use it and some are more specific to particular ones. In H.245, these services are named as Signaling Entities (SE). Each SE defines a series of procedures to handle with corresponding messages. Procedures are defined to allow the exchange of audiovisual and data capabilities; to request the transmission of a particular audiovisual and data mode; to manage the logical channels to transport audiovisual and data information; to establish what terminal is the master terminal and what is the slave terminal for the purposes of managing logical channels; to carry various control and indication signals; to control the bit rate of individual logical channels and the whole multiplex; and to measure the round trip delay, from one terminal to the other and back. These procedures are explained in more details below.

### 3.2.2.1 *Master-Slave Determination Signaling Entity*

Conflicts may arise when two terminals involved in a call initiate similar events simultaneously and only one such event is possible or desired. For example, conflicts may arise when resources are available for only one occurrence of the event. To resolve the conflict, one terminal shall act as the master and the other terminal shall act as a slave terminal. Rules specify how the master and slave terminal shall respond at times of conflict.

The master-slave determination procedure allows terminals in a call to determine what terminal is the master and what is the slave. The terminal status may be redetermined at any time during a call; however, a terminal may only initiate the master-slave determination process if no procedure that depends upon its result is locally active.

### 3.2.2.2 Capability Exchange Signaling Entity

The capability exchange procedures are intended to ensure that the only multimedia signals to be transmitted are those that can be received and treated appropriately by the receiving terminal. This requires that the capabilities of each terminal to receive and decode be known to the other terminal. It is not necessary that a terminal understands or stores all incoming capabilities; those that are not understood, or can not be used shall be ignored, and no fault shall be considered to have occurred. When a capability is received that contains extensions not understood by the terminal, the capability shall be accepted as if it did not contain the extensions.

The total capability of a terminal to receive and decode various signals is made known to the other terminal by transmission of its capability set.

Receiving capabilities describe the terminal's ability to receive and process incoming information streams. Transmitters shall limit the content of their transmitted information to which the receiver has indicated it is capable of receiving. The absence of a receive capability indicates that the terminal cannot receive (is a transmitter only).

Transmit capabilities describe the terminal's ability to transmit information streams. Transmit capabilities serve to offer receivers a choice of possible modes of operation, so that the receiver may request the mode that it prefers to receive. The absence of a transmit capability indicates that the terminal is not offering a choice of preferred modes to the receiver (but it may still transmit anything within the capability of the receiver).

These capability sets provide for more than one stream of a given medium type to be sent simultaneously. For example, a terminal may declare its ability to receive (or send) two independent H.263 video streams and two independent G.731.1 audio streams at the same time. Capability messages have been defined to allow a terminal to indicate that it does not have fixed capabilities, but that they depend on other modes that are being used simultaneously. For example, it is possible to indicate that higher resolution video can be decoded when a simpler audio algorithm is used; or that either two low resolution video sequences can be decoded or a single high resolution one. It is also possible to indicate trade-offs between the capability to transmit and the capability to receive.

Nonstandard capabilities and control messages may be issued using the NonstandardParameter structure. Note that while the meaning of nonstandard messages is defined by individual organizations, equipment built by any manufacturer may signal any nonstandard message, if the meaning is known.

### 3.2.2.3 Logical Channel Signaling Entity

An acknowledged protocol is defined for the opening and closing of logical channels that carry the audiovisual and data information. The aim of these procedures is to ensure that a terminal is capable of receiving and decoding the data that will be transmitted on a logical channel at the time the logical channel is opened rather than at the time the first data is transmitted on it; and to ensure that the receiving terminal is ready to receive and decode the data that will be transmitted on the logical channel before that transmission starts. The OpenLogicalChannel message includes a description of the data to be transported, for example, H.263 at 6 Mbit/s. Logical channels should only be opened when there is sufficient capability to receive data on all open logical channels simultaneously.

A part of this protocol is concerned with the opening of bidirectional channels. To avoid conflicts that may arise when two terminals initiate similar events simultaneously, one terminal is defined as the master terminal, and the other as the slave terminal. A protocol is defined to establish the terminal that is the master and the one that is the slave. However, systems that use this recommendation may specify the procedure specified in this recommendation or another means of determining the master and the slave.

### 3.2.2.4 Close Logical Channel Request Signaling Entity

A logical channel is opened and closed from the transmitter side. A mechanism is defined that allows a receiving terminal to request the closure of an incoming logical channel. The transmit terminal may accept or reject the logical channel closure request. A terminal may, for example, use these procedures to request the closure of an incoming logical channel, which, for whatever reason, cannot be decoded. These procedures may also be used to request the closure of a bidirectional logical channel by the terminal that did not open the channel.

### 3.2.2.5 H.223 Multiplex Table Entry Modification Signaling Entity

The H.223 multiplex table associates each octet within an H.223 MUX message with a particular logical channel number. The H.223 multiplex table may have up to 15 entries. A mechanism is provided that allows the transmit terminal to specify and inform the receiver of new H.223 multiplex table entries. A receiving terminal may also request the retransmission of a multiplex table entry.

### 3.2.2.6 Audiovisual and Data Mode Request Signaling Entity

When the capability exchange protocol has been completed, both terminals will be aware of each other's capability to transmit and receive as specified in the capability descriptors that have been exchanged. It is not mandatory for a terminal to declare all its capabilities; it need only declare those that it wishes to be used.

A terminal may indicate its capabilities to transmit. A terminal that receives transmission capabilities from the remote terminal may request a particular mode to be transmitted to it. A terminal indicates that it does not want its transmission mode to be controlled by the remote terminal by sending no transmission capabilities.

### 3.2.2.7 Round Trip Delay Determination Signaling Entity

It may be useful in some applications to have knowledge of the round trip delay between a transmit terminal and a receiving terminal. A mechanism is provided to measure this round trip delay. This mechanism may also be useful as a means to detect whether the remote terminal is still functioning.

### 3.2.2.8 Maintenance Loops Signaling Entity

Procedures are specified to establish maintenance loops. It is possible to specify the loop of a single logical channel either as a digital loop or decoded loop, and the loop of the whole multiplex.

### 3.2.2.9 Commands and Indications

Commands and indications are provided for various purposes: video/audio active/inactive signals to inform the user, and a fast update request for source switching in multipoint applications are some examples. Neither commands nor indications elicit response messages from the remote terminal. Commands force an action at the remote terminal while indications merely provide information and do not force any action.

A command is defined to allow the bit rate of logical channels and the whole multiplex to be controlled from the remote terminal. This has a number of purposes: interworking with terminals using multiplexes in which only a finite number of bit rates are available, multipoint applications where the rates from different sources should be matched, and flow control in congested networks.

### 3.2.2.10 Abstract Syntax Notation One (ASN.1) and X.691

In H.245, the message syntax is defined using a notation called Abstract Syntax Notation One (ASN.1). The ASN.1 defines a number of simple data types and specifies a notation for referencing these types and for specifying values of these types.

The ASN.1 notations can be applied whenever it is necessary to define the abstract syntax of information without constraining how the information is encoded for transmission in any way. It is particularly, but not exclusively, applicable to application layer protocols. The following is an example of a message definition by ASN.1 and the example message is the OpenLogicalChannel message.

ASN.1 deals only with the syntax and semantics of message specifications. The binary encoding of data structures is covered in other recommendations, notably X.690 (basic encoding rules or BER) and X.691 [18] (packed encoding rules or PER). BER allows data to be deciphered by systems that have general knowledge of ASN.1 but do not know the details of the specification used to form the data. In other words, the data types are encoded along with the data values. PER is much more efficient since only data values are encoded and the coding is designed with very little redundancy. This method can be used when both the transmitter and the receiver expect data to adhere to a known structure.

H.245 is implemented using the packed encoding rules. Since both sides of a call know that messages will conform to the H.245 specification it is not necessary to encode that specification into the messages. For decoding simplicity, the aligned variant of PER is used. This forces fields that require eight or more bits to be aligned on octet boundaries and to consume an integral number of octets. Alignment is done by padding the data with zeros before large fields.

H.223 is the ITU-T recommendation for data multiplexing protocol. As shown in the framework figure of 3G-324M, it is the real sending and receiving interface for all the video/audio/data applications during the call session. The recommendation defines the structures and routines for data transmission multiplexing. Therefore, it is the major focus of our investigation about how to improve the multimedia data transmission efficiency of 3G mobile networks.

### 3.2.3 H.223 Multiplexing and Demultiplexing

3G-324M defines three levels of H.223 transport in order to provide different degrees of error resilience.

Level 0, or baseline H.223, provides support for synchronization and bit stuffing. Level 0 allows 16 different multiplexing patterns to assemble media, control, and data packets. Multiplexing patterns are negotiated between the endpoints. The error resilience capabilities of Level 0 are limited. Bit errors can break the HDLC protocol, can interfere with bit stuffing, and can be the cause of flag emulations in the payload.

Level 1, defined in H.223 Annex A, provides a synchronization mechanism that considerably improves performance over error-prone channels. HDLC is replaced by a more robust framing and framing flag of larger length.

Level 2, defined by H.223 Annex B, is a further enhancement of Level 1, providing support for forward error correction (FEC) and including a larger header, which describes the contents of the multiplexed PDU.

Level 3, defined by H. Annex C and D, is a further enhancement of Level 2, providing the most robust set of mobile extensions to the basic specification. This annex specifies both a new adaptation layer and a multiplexing layer and provides several enhancements including error detection and correction, sequence numbering, automatic repeat request, and several retransmission schemes.

In H.223 every level comprises an adaptation, a multiplexing, and a demultiplexing layer. AL1, AL2, and AL3 are the three defined adaptation layers. AL1 is designed for data transfer and is typically used to transport user data and H.245 control messages. It relies on upper layers for error control and handling. This is shown in Figure 3.2.

AL2 provides an 8-bit cyclic redundancy check (CRC) and optional sequence numbering to allow loss detection. AL2 has the capability to handle AL service data units (SDUs) of variable length and is the preferred adaptation layer for the transport of audio data. AL3 is designed primarily for video and includes a 16-bit CRC and optional sequence numbering.

### 3.2.4 Summary

3GPP adopted H.324M with some mandatory requirements to form 3G-324M protocol stack for low bit rate multimedia communication in 3G circuit switched network. In this protocol stack, the H.245 recommendation is used for call control and signaling and the H.223 recommendation is used for multiplexing. These protocols work together closely to provide reliable service over an error prone wireless network.

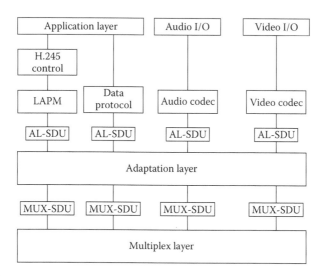

**FIGURE 3.2**
General structure of the H.223 module.

We have presented an efficient implementation of control and data multiplexing protocols H.245 and H.223, targeted for multipoint video user in both 3G and Internet environments. With a serialization approach to the multiplex descriptors, support for multiple media streams control function, our H.245 and H.223 implementations make the set-up and performing of video transmission much more convenient and efficient. With tests on the prototype system, the performance of our protocol implementation is satisfactory and stable.

Currently, we are investigating the transcoding and compatibility between 3G-324M terminals through heterogeneous environment. We'll try to make our protocol stack and video communication system compatible across 3G-324M, Wi-Fi, and WSN [29]. Thus, the video communication system can be performed over both 3G mobile and other networks, with the support of gateways.

## 3.3 Communication between Sensor and UMTS Network

Recent technological advances show us that a huge amount of low-power inexpensive video sensor devices could easily communicate with other networks like UMTS network and worldwide interoperability for microwave access (WiMAX) networks, merging them together in a hybrid network to bring the sensor video streaming to the terminal anywhere. With the rapid development of third generation telephony (Video Telephony), it is possible that different network domains that is, circuit switched networks (CSN) and packet switched networks (PSN), could interconnect with each other. To achieve pervasive video communication, the UMTS network foreseen to be the enabling technology for multimedia services with up to 64 kbps for a circuit switched network and 2 Mbps for a PSN, makes it feasible for visual communication over the wireless link. Compared with the UMTS infrastructure networks, the sensor networks build up a self-organizing ad hoc network to forward data packets to the sink nodes using multi-hop connections. Researchers at Berkeley developed embedded wireless sensor networking devices called motes, which were made publicly available commercially, along with TinyOS, an associated embedded operating system that facilitates the use of these devices. To date, wireless multimedia study has focused on robust video transmission over general PS wireless channels. Figure 3.1 shows an easily programmable, fully functional, relatively inexpensive platform for experimentation, and real deployment has played a significant role in the ongoing wireless sensor networks revolution. There is, however, a scarcity of work performed on delivering data from sensor wireless PSN or system to other CSN. The key issue for data delivery over a 3G network, which consists of several layers, is that it is not obvious how to achieve end-to-end optimality for data delivery although a single-layer performance can reach optimum. M.Y.Aal Salem did some work [31] about the interoperability framework for Sensor and UMTS networks;

however this work is based on the UMTS PSN, the video could not send to the mobile user directly. It is well known that the UMTS circuit switched video system is a QOS guaranteed system. Targeted to solve such a problem; we designed and implemented a prototype system, providing a novel architecture to communicate between the WSN and UMTS to move video from a sensor to a 3G handset user at low cost and power. Our previous work [1] mainly focused on the video conferencing for 3G wireless network and 3G-324M protocol implementation [2–6]. Figure 3.3 shows a sensor node in WSN.

### 3.3.1 Sensor Network Frameworks

Several approaches are suggested to allow communication between the gateway and the user. Three of these methods in current usage have significant disadvantages with respect to the main constraints related to sensor networks: power, consumption, and cost. However, the advantages of this approach, presented in this chapter for the first time, are to be available to all organizations with lower cost and lower power consumption.

The interoperability framework for sensor and UMTS network (IFSUN) approach, which will be explained later in this chapter, moves data from the gateway to the user at a lower cost and power by using the Universal Mobile Telecommunication System (OJMTS), which is standard for the Third

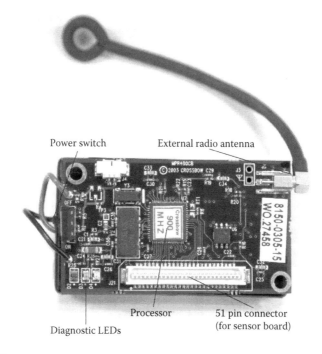

Power switch      External radio antenna

Processor      51 pin connector (for sensor board)

Diagnostic LEDs

**FIGURE 3.3**
Sensor hardware components and layout.

Generation Mobile System (3G). The integration of the mobile network infrastructure and sensor networks will reduce the cost of building a new infrastructure and enable the large-scale deployment of sensor networks.

### 3.3.2 UMTS Network Frame

The UMTS architecture consists of core network (CN), UTRAN (UMTS Terrestrial Radio Access Network), and user equipment (UE). The main function of the CN is to provide switching and routing of the packet traffic. It is further divided into CS and PS domains. The circuit-switched elements are comprised of the Mobile Services Switching Center (MSC), Visitor Location Register (VLR) and Gateway MSC. The PS elements are the Serving GPRS Support Node (SGSN) and Gateway GPRS Support Node (GGSN). The SGSN is responsible for mobility management and IP packet session management. It routes user packet traffic from the radio network to the appropriate GGSN and provides access to external packet data networks, such as the Internet and intranets. The UMTS core network has multiple SGSNs that are connected to multiple radio network controllers. This framework focuses on the packet-switched domain because it provides the Multimedia Message Service (MMS) and the Wireless Application Protocol (WAP). UTRAN consists of multiple base stations (Nodes B) and Radio Network Controllers (RNC) that provide the WCDMA air interface access method for user equipment. This communication network can carry many traffic types, from real-time CS to IP-based PS traffic. The RNC authorizes control functionalities for one or more Node Bs while the IUB is a logical interface between them. The RNC and its corresponding Node Bs are called the Radio Network Subsystem (RNS). There can be more than one RNS present in the UTRAN. The term UE refers to any device that has the capability to communicate with a UMTS network. This corresponds to our proposed architecture (IFSUN) for a wireless sensor network gateway, as depicted in Figure 3.4.

In a wireless environment, where bandwidth usage is significant, short address length and simplicity of user entry on limited keypads are the distinguishing features between various systems. The international mobile subscriber identity (IMSI) is used by any system that employs the GSM standard. The IMSI uses up to 15 digits, a 3-digit mobile country code (MCC), a 2-digit mobile network code (MNC), and a mobile subscriber identity number (MSIN) of up to 10 digits. The IMSI has been recognized as a better identifier than any other system.

### 3.3.3 Interoperability Sensor and UMTS Network

Here we propose an approach that will allow the different network standards to communicate with each other. The scope of this work is to develop a framework that will allow the integration of sensor networks into the fabric of other wireless networks. The framework has been divided into two parts: the UMTS portion and the gateway. The use of a UMTS network enables direct

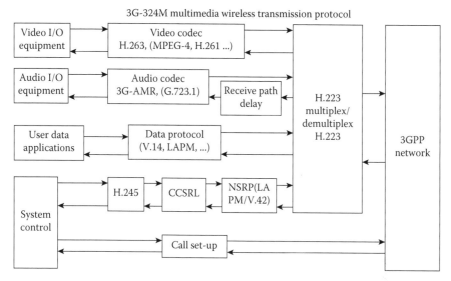

**FIGURE 3.4**
3G-324M structure.

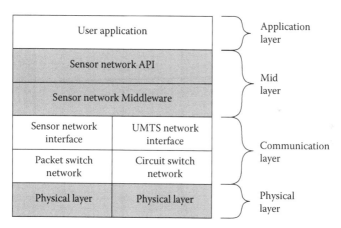

**FIGURE 3.5**
Sensor node and gateway architecture.

access of the sensor network, where the user can request data anywhere and anytime. The gateway is an expensive device and has a short lifetime.

### 3.3.4 Network Framework

The proposed architecture of a wireless sensor network gateway is shown in Figure 3.5. It is a flexible architecture that supports a range of communication technologies for sensors and user applications with no or minimum modification.

### 3.3.4.1 Network Interfaces

This is the communication layer. Two different communication technologies will be supported by the gateway: the wide range interface and the short-range communication interface. Examples of the short-range interface are Bluetooth, ZigBee, IEEE 802.16, 802.1 1, proprietary RF, and GPRS. The wide range is the UMTS interface.

### 3.3.4.2 Sensor Network Middleware

Sensor network middleware represents the central component of the gate-way architecture. This is the layer that divides communication between the sensors and the users, encapsulates the internal organization of the sensor network, and provides API function to the users.

### 3.3.4.3 User Application

The user application layer employs the API functions that are provided by the sensor network middleware. Depending on the application scenario, short-range wireless communication links or the wide-area network are used for communication between users and gateways.

### 3.3.5 The Solutions

This section considers the possible solutions incorporated in our framework for the different components of the proposed framework. Our framework solution is divided into a message solution and Internet working solution techniques, as explained in the following sections. The aim of these solutions is to receive data on demand at any time and anywhere throughout the country, or to share these data with other organizations worldwide by using Internet or e-mail technologies.

### 3.3.5.1 Message Solutions

This solution allows our framework to implement the SMS infrastructure in a sensor network using the WAP, which was chosen for its features. This solution allows the gateway to send the data that is gathered as SMS, which can easily be received by any mobile phone in any location. The Wireless Session Protocol (WSP) is used to transport the messages from the gateway to the MMSC and from the MMSC to the gateway. It is responsible for the general requirements, architecture, and functionality. The payload represents the SMS application layer PDU. See Table 3.1 for an example of PDU encoding and decoding.

The process for sending an SMS from an SMS client to another SMS client (data receiver) is explained below.

**TABLE 3.1**

Hexadecimal PDU Message

**Hexadecimal PDU Message**

0791 135613 13 13F3 1 1000A92402~56~910000AA27547419D40CE341D4721BOEEA8
1643850D84D0651D1655033ED065 1CB6D38A8078AD500

**7 Bit PDU Message (readable) is**

SMSC# 90100000

Sender: 63769074

TP-PID: 00 TP-DCS:OO

TP-DCS-popis: Uncompressed Text class:O

Alphabet: Default

Max Temp = 28 and Min Temp = 15

Length: 39

*Source:* Yufei Du, Fung Po Tso, Weijia Jia, "Architecture Design of Video Transmission between UMTS and WSN", International Conference on Ubi-media Computing, Lanzhou, P.R.C., 2008. With permission.

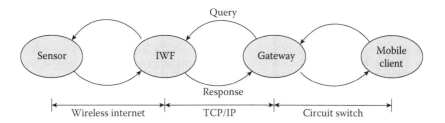

**FIGURE 3.6**

Interoperability sensor and UMTS network model. (From Du, Y., "UBI-Media" Conference of IEEE, 2008. With permission.)

Each gateway registers itself as a mobile unit in the Home Location Register (HLR) on the UMTS network. The gateway address is based on the Mobile Station ISDN Number (MSISDN) that is operated by the device. In many paging systems, users are assigned PINS that are used to authorize a caller to deposit a message. This addressing problem can be solved by adding this number to the gateway memory, or changing the gateway design to allow the USIM card to cooperate with it. After this, the gateway is ready for the next step. This is shown in Figure 3.6.

- Gateway sends SMS to the SMS server
- SMS server sends notification to recipient client
- Client fetches SMS from the SMS server
- SMS server sends delivery report to client.

### 3.3.5.2 Internet Working Solutions

The UMTS network is connected over the Internet. The Internet working solution of our framework: connecting the sensor network to the Internet via the UMTS network. The idea adopted in this solution is to obtain the data from any local network anywhere. As mentioned earlier, in the SMS addressing model, the SMS address format is compatible with Internet e-mail addresses, and the WAP gateway is responsible for converting the PDU or SMS message to HTTP format.

### 3.3.5.3 Power Consumption

Power consumption is highly critical in a sensor network. Location updating enables the HLR to keep track of the subscriber's current location. The sensor network is registered in the UMTS network and the location is then fixed. Here, the power consumption in the message and Internet working solutions can be minimized by preventing the location updating signals. Another solution is to use solar energy to supply power to the gateway.

### 3.3.6 Design of the Gateway of UMTS and Sensor Network

The wire and wireless sensor network has a role of gathering the sensor status data. The gateway does not only offer sensor status information to remote users but also receives sensor control messages from the external network.

Figure 3.7 shows the structure of the UMTS and Sensor Network Communication system proposed in this section. The system consists of the IWF, control server, and the web-based client, which is connected to gateway

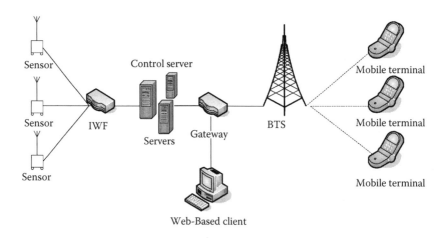

**FIGURE 3.7**
UMTS and sensor network communication system structure. (From Du, Y., "UBI-Media" Conference of IEEE, 2008. With permission.)

through the PSN. The gateway is connected to a circuit-switched network via UMTS mobile communication network. The IWF has a role of collecting data from the sensor network and converting the network protocol to connect the sensor network to the external network.

### 3.3.6.1 Control Server

The control server is the core unit to offer the web-based remote diagnosis and management service. The structure of the control server as shown in Figure 3.8 consists of a web server manager, a core engine, a protocol converter, and date base.

The web server provides a web client with web-based interface. The server-gateway communication manager is in charge of secure connection and access control between the control server and the gateway. The security manager

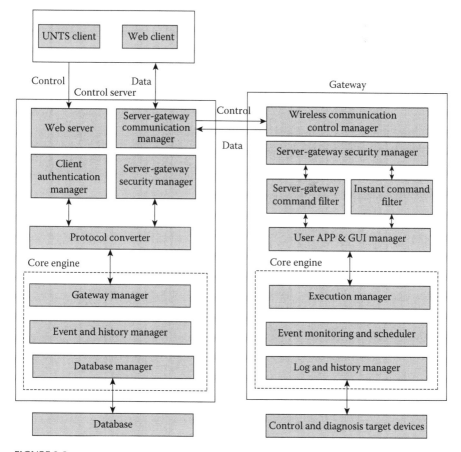

**FIGURE 3.8**
Software architecture for control server and gateway. (From Du, Y., "UBI-Media" Conference of IEEE, 2008. With permission.)

consists of a client authentication manager for web clients and server gateway security manager for the users. The protocol converter to exchange information among the different types of networks has a role of converting the HTTP protocol packet format on the web client side into point-to-point protocol (PPP) format on the vehicle gateway side and vice versa. The core engine consists of a gateway manager, an event and history manager, and a database manager. The gateway manager is in charge of managing the connection request, the gateway configuration and the information transmission and reception. Finally, the event and history manager has the role of an event analysis, a control command transmission and execution, a reception data analysis, and a log file generation.

### 3.3.6.2 Gateway in UMTS and Sensor Network System

The main role of gateway is to translate and execute control commands received from the UMTS network and web client. It cooperates with the sensor network to help the remote client diagnosis and management performed. It also provides the users with useful information that is sent from the control server. As shown in Figure 3.8, the gateway consists of a wireless communication manager, a server-gateway security manager, a command filter, core engine, and other application programs for users.

### 3.3.6.3 Sensor Node

In general, the sensor nodes are hardly collected and recharged for recycle after they are deployed in the sensor field. Therefore, the power consumption is one of the very significant constraints in the sensor nodes. However, the sensor nodes deployed for communication with UMTS network have a more important issue than the power consumption. The node structure for the UMTS-based sensor network is shown in Figure 3.9. It consists of a UMTS protocol processing module, a message processor, a sensor, and actuator management module.

### 3.3.7 Summary

UMTS users could conveniently and quickly access the sensor network. In this section, we have proposed the architecture for video transmission from WSN to UMTS networks that allow an ad hoc PSN to communicate with the CSN. However few researchers consider integration of the mobile network infrastructure a sensor network. This section presented our new approach, using the characteristics of sensor networks and mobile network infrastructure to deliver sensor network signals. Furthermore, we build up the platform and implement the video gateway. The advantages are that communicating between these two systems dynamically and intelligently can reduce the cost and increase the lifetime of sensor networks. It was also shown that this approach is suitable for all organizations and for gathering

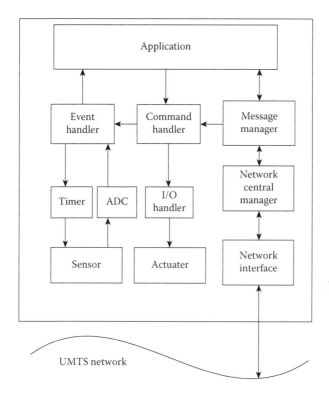

**FIGURE 3.9**
Sensor node architecture for sensor network. (From Du, Y., "UBI-Media" Conference of IEEE, 2008. With permission.)

data on demand. The feasibility and viability of the proposed method has been proven through initial experimental work.

On the perspective of wireless telecommunication, we aim at developing a system that can provide seamless multimedia and data communication provision among all types of IP and PSTN applications. For this communication environment, a robust and efficient multimedia gateway [30] is also required for universal compatibility and transcoding.

## 3.4 Optimization of Multimedia Gateway for Pervasive Communication

We present the design and implementation details of an efficient and robust multimedia gateway that enables ubiquitous multimedia wireless communication. Primarily, we studied current technological trends and realized that in order to support various network technologies and to handle high call traffic conditions, a robust and efficient gateway is required for the universal adaptability and transcoding. To achieve the performance improvement of the gateway, we focus on efficiently implementing the H.245, which is the main call handling and signaling protocol. In our method, sets of H.245 messages are compiled, used, and stored in a table so as to be reused for further incoming calls. After implementing this procedure, we have successfully tested our 3G-IP gateway that is robust enough for smoothly handling up to one million concurrent calls. It is also experimentally verified that our gateway provides the feature of invariant call setup time even in high traffic conditions. As ubiquitous communications among the heterogeneous networks are in demand today and our gateway will play a key role in this area.

### 3.4.1 Introduction

These days, mobile devices like iPods, personal digital assistant (PDAs), smartphones, tablet PCs, and high end mobile phones have become inexhaustibly essential in day-to-day life. It is because these equipments provide not only the features like phone calling, music playing, and photo and video capturing, but they also do have incredible features like seamless accessibility and connectivity that enables the modern life to be pervasively connected to any equipment around him or to any point in this world. These advanced communication devices possess the capability of connecting through various network types such as, infrared, Bluetooth, cellular networks, traditional landlines, and the Internet. On the mobile devices, the demand of multimedia applications such as video conferencing, video on demand (VOD), internet protocol television (Internet Protocol (IP) TV), mobile trading, online games, and geographical and vehicular information systems have become a general trend in today's networked lifestyle. Moreover, if we study the current technological trend, it is prominent that there is a high demand on the complete convergence and unification of the traditional telephone, mobile phone, and the Internet networks so as to provide any type of seamless services to subscribers on any device at any place or time.

This ubiquitous communication environment among the heterogeneous networks is in demand today. In this situation, a gateway is required for bridging these gaps between different network domains (e.g., PSTN and IP) [23–26]. As the ubiquitous communication is expected to grow rapidly and is researched extensively, we believe our gateway will play the key role in this sort of application.

Basically, communication networks can be classified into two categories, such as CSN and PSN. The telecommunication networks mainly use the circuit-switching technique due to its service reliability and provision of fixed bandwidth. On the other hand, the internet uses the packet-switching method in order to deal with the burst traffic pattern and working over a range of available bandwidth in the network.

Presently we observe that the mobile networks are in increased demand for using a packet switching network in order to meet their heavy demand of internet and multimedia applications. These bulky multimedia contents have eventually caused a largely growing traffic load over both mobile phone and internet networks. Moreover, as mentioned earlier, cell phones are now being used to transmit and receive not only voice but high quality real-time video and bulky data. From these observations we realized that in the next generation of networking, 3G and beyond wireless networks, and cross network multimedia service delivery will assume an increasingly important role. A multimedia gateway that facilitates establishing service sessions and delivery of multimedia services to users and terminals of differing capabilities is highly recommended. In this case, the supporting gateway must be giant, robust, and efficient enough for call handling. Otherwise the gateway faces not only the loss of QoS, but also a severe hazard of performance bottleneck.

There have already been some practical developments of gateways that successfully bridge the gaps among the traditional landline phone, 3G mobile, and internet networks in order to fulfill many advanced application aspects. The ITU-T provides the standards H.245, H.223, H.324M, and 3GPP's 3G-324M standards are the building blocks of an up-to-date multimedia gateway. There are some commercially available 3G gateways in the market. Radvision has developed a 3G gateway called Scopia. The Scopia 3G video gateway supports video telephony as well as video streaming between 3G-324M based mobile handsets or devices and IP-based video media servers. In the meantime, Tandberg and Dilithium have also developed similar gateways.

During this research, we have developed our own 3G-324M- and SIP-based 3G-IP gateway in the City University of Hong Kong. We aim at developing a gateway that is robust enough to meet the challenge of handling large numbers of calls concurrently and at the same time to ensure a good QoS to all the subscribers. In our research, we have given emphasis on efficient implementation of the H.245 protocol so as to enhance the efficiency and robustness of the gateway that can smoothly handle high traffic conditions with a satisfactory level of QoS.

### 3.4.2 Background

According to ITU-T standards, H.245 is a control channel protocol capable of conveying information for multimedia communication. In voice and video telephony as well as in VoIP, this subprotocol (of 3G-324M) is basically responsible for call initialization, setup, and for continuing the conversation. This protocol manages information exchange through a set of predefined messages. Some members of this message set are vital for the call initialization and setup, while some are responsible for continuing the conversation. So any attempt to improve the performance of the H.245 procedure is nothing but handling the message exchange process efficiently. As this

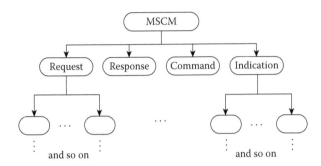

**FIGURE 3.10**
Hierarchical representation of the H.245 messages. (From Du, Y., "UBI-Media" Conference of IEEE, 2008. With permission.)

protocol is responsible for the call setup and conversation continuation, by efficiently handling of its message exchanges, will eventually lead to shorter call setup time and better conversation quality. Also in high traffic conditions, for handling large numbers of calls concurrently and to provide good conversational quality (the QoS), H.245 is the only vital protocol that should be taken care of properly.

According to the ITU-T X.691 recommendation, H.245 messages are initially in the form of ASN.1 and then they are converted into binary streams, when the call is initiated. After a requested terminal receives the bit stream, it reconstructs those messages back into meaningful ASN.1 text and then it sends proper responses to react to the requester terminal.

In the H.245 module, messages are defined in a tree-like structure as is depicted in Figure 3.10. This defines a general message type called, MultimediaSystemControlMessage (MSCM). MSCM further comprises four different types of special messages, namely, request, response, command, and indication. A request message corresponds to a specific action and requires an immediate response through a response message. A command message requires an action but no explicit response. An indication message contains information that does not require action or response.

A response message requires specific action and an immediate response. A command message requires an action but no explicit response. An indication message contains information that does not require action or response.

### 3.4.2.1 Characteristics of H.245 Messages in 3G System

The most common H.245 messages for a 3G video call are

1. TerminalCapabilitySet
2. TerminalCapabilitySetAck
3. MasterSlaveDetermination

4. MasterSlaveDeterminationAck

5. VendorIndentification

6. MultiplexEntrySend

7. MultiplexEntrySendAck

8. OpenLogicalChannel

9. OpenLogicalChannelAck

10. MiscellaneousCommand:VideoFastUpdatePicture

11. RoundTripDelayRequest

12. EndSessionCommand:disconnect

Here only messages 7 and 8 are dynamic because their content depends on messages sent by remote terminals. Normally, message 1 to message 9 are only exchanged during call setup phase and message 10 and 11 are used to maintain the call, and finally conversation is terminated using the EndSessionCommand:disconnect command.

To present a clear idea about the details of a message and its submessages, let's take TerminalCapabilitySet as an example. The message, TerminalCapabilitySet is used by sending the terminal to inform the receiving terminal about its multiplexer capabilities and also about its supported media codecs. An example of the hierarchical message structure of TermnalCapabilitySet is shown in Figure 3.11. Here, we may observe that most contents can be kept unchanged since they describe the capability of a multimedia mobile terminal (as they are fixed for that specific terminal). However, there is a field called "sequenceNumber," which is used to label instances of the TerminalCapabilitySet so that the corresponding response can be identified. If there appear multiple instances of TerminalCapabilitySet with the same content, then only the sequenceNumber field can easily be changed dynamically.

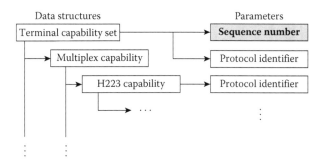

**FIGURE 3.11**
Message structure of TerminalCapabilitySet. (From Fung Po Tso, Yufei Du, Weijia Jia, "Design of an Efficient and Robust Multimedia Gateway for Pervasive Communication," *Wireless Telecommunications Symposium*, 2008. With permission.)

### 3.4.2.2 Suggestion of a Novel Procedure

Traditionally H.245 messages are encoded in a chunk-by-chunk style. Each chunk corresponds to a submessage under the main message. For each chunk of input, an encoding process must be called every time and the chunks are encoded serially. So in this serial and individual encoding process for each and every message eventually leads to a high cost of system time.

We note some important points here:

- Each main message and its submessages are the same for all the individual calls.
- From our experiment, we observe that some submessages under a main message are also repeated.
- Messages the like TerminalCapabilitySet request are no doubt the same for all calls. But still, the response is also very few in number; it is because, in reality, terminal capabilities are fixed and there are few numbers of different mobile terminals produced by different brands. So for most calls, these requests are the same and responses are just a little varied.

Hence, it seems that the earlier mentioned serial coding process is just an unnecessary killing of system's CPU time. In contrast, if there is a low-cost process that serves the same objective of producing the same encoded output then that process is definitely worth adopting. In this scenario, with the motivation of improving the call handling efficiency of the 3G gateway, we would like to suggest a more efficient method on an experimental basis. In this method, we consider the possibility of reusing a replicated encoded message set that is already compiled from an earlier call, unless the message data is changed. And if there is any change in the submessages then that part can easily be handled by dynamically compiling the part and updating the returned value in the whole encoded message string. We describe the method in detail as follows.

Considering the programming level of message encoding, it is not necessary to encode the whole H.245 message for each call, rather the produced binary stream can just be updated with minor changes in a few bits. Here the precompiled message streams may be saved in a lookup table and that can be accessed for each call instead of recompilation each time. Eventually it saves an appreciable amount of encoding compilation time.

A visual description of the above suggested reusable precompiled message-data set and its lookup table implementation method is compared to the traditional message encoding method as depicted in the above tables. Here we suppose that M1, M2, and M3 are H.245 messages and A, B, and C are their encoded binary streams, respectively. Table 3.2 shows the traditional approach for H.245 message encoding where each message is encoded

**TABLE 3.2**

Traditional Approach for H.245 Message Encoding

| Input | Process | Output |
|---|---|---|
| M1 | Encoding | A |
| M2 | Encoding | B |
| M2 | Encoding | B |
| M3 | Encoding | C |
| M1 | Encoding | A |
| M1 | Encoding | A |

*Source:* Tso, F. P., "WTS" Conference of IEEE, 2008. With permission.

**TABLE 3.3**

Suggested Table Lookup Approach for Efficient H.245 Message Encoding

| Input | Process | Output |
|---|---|---|
| M1 | Encoding + Reuse | A |
| M2 | Encoding + Reuse | B |
| M2 | Reuse | B |
| M3 | Encoding + Reuse | C |
| M1 | Reuse | A |
| M1 | Reuse | A |

*Source:* Tso, F. P., "WTS" Conference of IEEE, 2008. With permission.

individually. Table 3.3 illustrates our suggestion of an improved approach for H.245 message encoding in which the reuse of previous encoded bit streams is deployed along with a minor update by dynamic compilation process. Comparing Tables 3.2 and 3.3, we get exactly the same output but at least three extra encoding processes can be skipped. So this leads to a saving of at least 50% compilation time and thus provides extra free system resources and time for better quality conversation in the post-setup phase in this example.

### 3.4.3 Implementation Details of Lookup Table-Based Message Encoding

Basing on the above discussions and suggestions, here we present the technical details of the lookup table-based message processing approach. An algorithmic representation is shown in Figure 3.12 and its process flow is illustrated diagrammatically in Figure 3.13. Basically this algorithm consists of two major functions. In the first process, the system is initialized and then a set of messages that is, M(m) are (pre)compiled and followed by storing the return data (bit-streams) into a table. Here we defined M(m) as a set of frequently used messages. Each time, when the system comes

```
Algorithml: Table lookup message processing.
Input: H.245 ASN.1 messages m
Output: binary stream B(m)
  1. Initialize system
  2. Pre-compile a message set M(m) and store into a table T(m)
  3. For each ith message H.245 ASN.1 message m_i
  4. If m_i ∈ M(m) then
  5.    Retrieve bit stream B(m_i) from table T(m_i)
  6.       If replacement needed then
  7.          Replace changed filed(s) and return
  8.             Else return the bit stream
  9.    Else dynamically compile it and store it into table T(m)
 10.End for
 11.Repeat step 3 to 10
```

**FIGURE 3.12**
Table lookup message processing algorithm. (From Tso, F. P., "WTS" Conference of IEEE, 2008. With permission.)

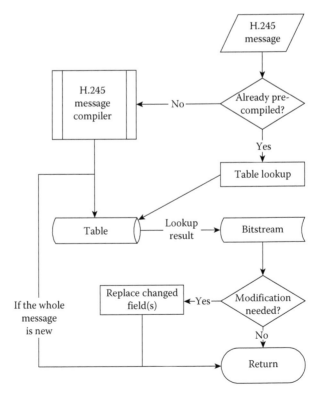

**FIGURE 3.13**
Table-lookup process flow. (From Tso, F. P., "WTS" Conference of IEEE, 2008. With permission.)

across a new message that is not in the set, then the message is added automatically to the database in the table for future use. In the second part of the algorithm, simple message matching is performed. The input of this algorithm is a single H.245 message, mi (mi is the *i*th message to be encoded), and the output is the corresponding bit stream B(mi). Looking from the programming point of view, during the system initialization, a H.245 message set is compiled into bit streams and then is stored in a table, T(m). For each H.245 message mi (to be encoded), it is looked for in the precompiled message database (i.e., table T(m)), if a match is found, then it is retrieved and returned. After that, the returned bit stream B(mi) is further checked if a replacement is needed, if no, B(mi) is returned directly, if yes, then the changed field(s) is dynamically compiled, replaced, and returned. However, if the message mi is not found, then it is just compiled and returned dynamically and is stored in the database for further use (for the next incoming calls).

One of the key issues of the lookup table approach is to find the efficient way to manage the table data storage and retrieval system. We propose the table should be managed in an index-based fashion. Each encoded message string is saved as an array in that table with an index number. So in the retrieval process, each entity is returned in terms of its reference number (integer). This approach is used because, it is a much less time consuming way to locate and return each table entry as compared to the traditional algorithms involving special key generation. Thus, an appreciable amount of time can be saved for data retrieval in the table and eventually leading to a big gain in the overall system performance. Figure 3.13 describes this procedure pictorially.

### 3.4.4 Performance Evaluation and Discussion

In this section, we evaluate the performance and efficiency of our method through experimental results. As we have already mentioned earlier, we have developed our IP-3G bidirectional call handling gateway. Our system leverages PC to PSTN/PC and vice versa video, voice, and text chatting-conferencing provision.

We have successfully tested our 3G-gateway for the compatibility test after embedding the currently proposed performance enhancement algorithm into it. We made phone calls to different brands and makes of 3G handsets and also to the Dilithium 3G Network Analyzer (model no. FXPAG-P42G). All attempts of 3G video calls were setup and conversations were held successfully. It proves that the proposed algorithm is fully compatible with the existing 3G protocol as well as with all commercially available 3G handsets or mobile devices too, as depicted in Figure 3.14.

Although the gateway is verified to be fully compatible and interoperable with SIP and 3G-324M protocol stack, we still have some difficulties in subscribing a T1 line from a local operator. Due to this constraint, we can

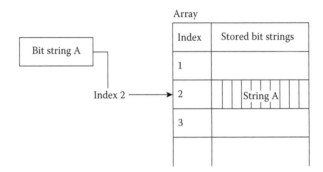

**FIGURE 3.14**
Array implementation for the lookup table creation and management. (From Tso, F. P., "WTS" Conference of IEEE, 2008. With permission.)

not conduct our experiments on a large scale in a real environment. For evaluating the performance of our gateway, we have carried out a systematic procedure for this experimentation. During this experiment, in order to track the call handling efficiency, we added two test modules namely "MessageFeeder" and "DataCollector" to our gateway. MessageFeeder is responsible for selecting a large number (say 10,000 at a time and then iteratively increasing in arithmetic progress) of H.245 message sets to be input to the 3G gateway. And the DataCollector is employed to record the time at the start of encoding the messages and the time corresponding to bit stream is returned at the end of the encoding process. It is practically obvious that these two factors can be compared so as to let us study the performance and efficiency of our 3G gate way in the view of handling high traffic incoming calls.

In our experiment, the number of messages we chose are ranging from 10,000 to 1,000,000. Here we take care of another important factor. In principle, the longer the message length the better will be the performance of our gateway, as the longer the time for compilation is saved through our message handling method. Some messages are of short length while some are longer in real time. In order to keep balance in the testing process, we have taken consideration of both these types of message structures and have chosen samples randomly, so that the collected data in the respective tables are nearly the average values, as depicted in Figure 3.15.

The proposed method is compared with the method that uses the traditional way that is, if there is a message then encode it and output it. The traditional procedure is time consuming as it follows this message-encode-output procedure repeatedly for each incoming message corresponding to each individual incoming call request. In contrast, we handle H.245 message processing through our lookup table-based method that can save the call setup and continuation time because of shortened message processing time and thus claims the efficiency of our procedure.

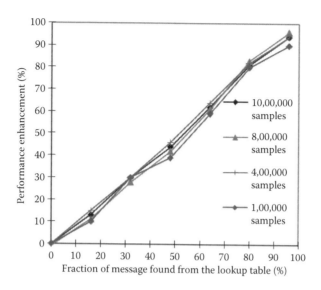

**FIGURE 3.15**
Experimental result of message processing in table lookup procedure. (From Tso, F. P., "WTS" Conference of IEEE, 2008.)

In our experiment, for the same set of messages, we calculated the time taken for message encoding in the traditional approach ($t_1$) and that in our proposed approach ($t_2$), respectively. The performance enhancement (in terms of reducing call setup time) can be calculated by the following formula:

$$\frac{t_1 - t_2}{t_1} \times 100\%.$$

If the result is positive, then it is obvious that we have achieved the enhancement of the performance.

The experimental result of performance enhancement in our lookup table method of H.245 message processing is illustrated in Figure 3.15. From this graph, two important observations can be noted. First, the performance remains almost unchanged when the number of calls is increasing. Second, the performance is linearly and proportionately improving with the fraction of messages found in the lookup table. That means the performance of the gateway increases when more calls encountered go through it.

The first observation just proves the very purpose of the currently introduced message processing approach that is, in this approach, the call handling performance of the 3G gateway remains invariant in spite of the number of concurrent calls increasing consistently to a large amount. So our procedure enables the gateway to successfully face the challenge of any amount of traffic load without degrading its performance.

Let us consider the second observation in the light of the discussion in Section 3.3. There are 12 common messages for a normal 3G video call where 10 of them are responsible for call setup and are almost the same in the case of all calls. So, these messages can be precompiled and can be stored in the lookup table and then those can be directly retrievable (or reusable) for the next incoming calls. Thus while processing the messages, the percentage of retrievable precompiled messages is 10/12 (83.3%). So, the performance improvement should also be predicted to be improved up to 83.3%. Our experimental results proves this prediction and it is visually obvious in Figure 3.15.

Thus, if we consider a physical interpretation, the second observation implies that the efficiency of call setup is improved appreciably. In other words, the call setup time is reduced greatly when the gateway handles a large number of calls.

Here another important advantage of our procedure may be predicted. An appreciable amount of improvement in the conversation quality (for the ongoing calls) can be achieved in our procedure. This is because our procedure also uses the same lookup table approach for handling the post connection messages too. So the system saves a lot of resources and time that may eventually enable a better quality in the ongoing conversation. Thus this suggested procedure claims better QoS. But, it is beyond the scope of our laboratory experiment to provide the data in support of this claim. However this claim can be proved when the gateway is deployed to handle high call traffic situation in real life applications, for example, by a commercial 3G service provider.

### 3.4.5 Summary

In this section we have presented our success of attaining appreciable improvement in robustness and efficiency of the multimedia gateway that is suitable for handling high call traffic in a telecommunication scenario. We adopt an efficient compile-store-reuse model for speeding up the process of encoding the ITU-T H.245 ASN.1 formatted messages in to a binary stream. In this method, H.245 sets of messages are compiled, used, and stored in a table. For encoding further incoming messages, the table is queried and the matching result is used instead of recompiling the messages again and again. Experimental results show that our algorithm greatly improves call handling efficiency of a 3G gateway. Thus, we have proved that our lookup table-based message encoding algorithm can provide stable performance and a satisfactory QoS assurance even under high traffic conditions. Also our method is tested to be fully compatible with existing 3G protocols.

## 3.5 Mobile Video Quality Measurements and Predictions

### 3.5.1 Introduction

This section addresses the video quality prediction from a source to a mobile terminal by using the information of video type and channel status. In particular, when a user requires a video clip from a source mobile device (Laptop, PDA. or Smart Phone), he has many methods to get it like using the interface of Wifi, 3G PSN, or 3G Circuit Switched Network, as depicted in Figure 3.16. The Wifi has more bandwidths and a lower cost but least coverage. The 3G Circuit Switched Network has lowest delay and largest coverage but cost much. In our work, an algorithm is given to predict the video quality. Then the user could choose the interface that has the best quality for video transmission.

### 3.5.2 Video Measurements

For video measurements, we use PEVQ as a video quality judgment standard. The PEVQ is an end-to-end measurement algorithm to provide mean opinion score (MOS) of the video quality of IPTV, streaming video, mobile TV, and video telephony. The scope for PEVQ is to measure degradations of the picture quality occurring through a network by analyzing the degraded video signal output from the network. This Quality-of-Experience (QoE) testing is based on modeling the behavior of the human visual tract and besides an overall quality MOS score (as a figure of merit) abnormalities in the video signal are quantified by a variety of performance indicator (KPIs), including peak signal-to-noise ratio (PSNR), distortion indicators, and lip-sync delay.

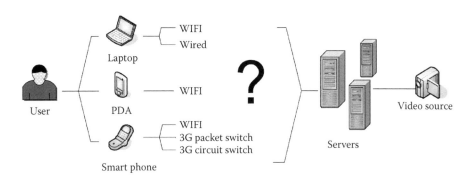

**FIGURE 3.16**
User requires the video stream from video source. (From Tso, F. P., "WTS" Conference of IEEE, 2008.)

Depending on the information that is made available to the algorithm, video quality test algorithms can be divided into three categories:

1. A Full Reference (FR) algorithm has access to and makes use of the original reference sequence for a comparison (i.e., a difference analysis). It can compare each pixel of the reference sequence to each corresponding pixel of the degraded sequence. FR measurements deliver the highest accuracy and repeatability but tend to be processing intensive.

2. A Reduced Reference (RR) algorithm uses a reduced side channel between the sender and the receiver that is not capable of transmitting the full reference signal. Instead, parameters are extracted at the sending side that helps predict the quality at the receiving side. RR measurements may offer reduced accuracy and represent a working compromise if bandwidth for the reference signal is limited.

3. A No Reference (NR) algorithm only uses the degraded signal for the quality estimation and has no information of the original reference sequence. NR algorithms are low accuracy estimates only, as the originating quality of the source reference is completely unknown. A common variant of NR algorithms does not even analyze the decoded video on a pixel level but only works on an analysis of the digital bit stream on an IP packet level. The measurement is consequently limited to a transport stream analysis.

PEVQ is a full-reference algorithm according to the FR above and analyzes the picture pixel-by-pixel after a temporal alignment (also referred to as "temporal registration") of corresponding frames of reference and test signals. PEVQ MOS results range from 1 (bad) to 5 (excellent).

In order to eliminate the effect of the encoder, we introduce the parameter of relative PEVQ rate, reference PEVQ, and destination PEVQ. The reference PEVQ is encoded video before the transmission compared with the original video. The destination PEVQ is output video compared with the original video. The destination PEVQ contains the transmission loss, so it must be lower than the reference PEVQ. The relative PEVQ rate is destination PEVQ over reference PEVQ. This is shown in Figure 3.17.

### 3.5.3 Video Quality Prediction

We did experiments with the ITU standard video. There are 18 videos and each video tests three times. The videos are sent from a Dilithium UMTS network analyzer (model no. FXPAG-P42G) to a 3G handset though circuit switched network.

Based on our experiment results, the explicit expression could not fit the data exactly. Then we tried implicit expression, it gives a good performance

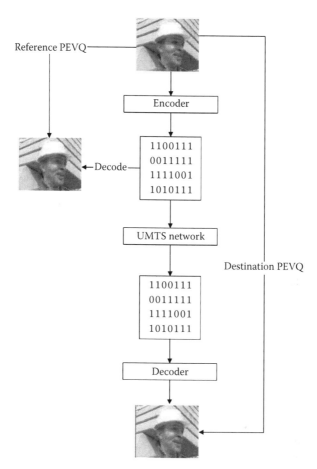

**FIGURE 3.17**
Video encode and decode process.

on result fitting. We proposed a trinary implicit polynomial to fit the data. As previously noted, the packet lost rate $x$, frame lost rate $y$, and relative PEVQ rate $z$ has a relation. We suppose using below function $\phi(x, y, z)$ to describe $x$, $y$, and $z$. $\beta_0$ to $\beta_9$ is coefficient.

$$\phi(x,y,z) = \beta_0 x^2 + \beta_1 y^2 + \beta_2 z^2 + \beta_3 xy + \beta_4 xz + \beta_5 yz + \beta_6 x + \beta_7 y + \beta_8 z + \beta_9 xyz + 1.$$

To effectively apply curve fitting to identify the optimal design with trinary implicit polynomial, a major obstacle exists—design optimization. The design optimization based upon curve fitting is a very complex and challenging problem due to its complex computation process with an implicit parametric model and many local minima. In addition, the design analysis or simulation process is usually a demanding and computation intensive process. An effective optimization algorithm, which can quickly lead to the global design optimum, is badly needed. Conventional numerical optimization algorithms are based on explicitly formulated design models. No explicit formulations are given and only implicit parametric models can be used [27]. Some intuitive design aspects, such as video quality measurements, can only be evaluated human interventions and cannot be modeled in mathematics and are incapable of handing an objective function with local optima. For searching global optimum, two approaches are widely used: least squares method (LSM) and simulated annealing (SA) [28]. These two methods do not require gradient information but need tens and thousands of iterations to converge. Due to the computation intensive nature of the curve fitting, the direct use of these two global optimization methods becomes less feasible.

### 3.5.4 Least Squares Method

The method of least squares (LSM) is used to solve overdetermined systems. Least squares is often applied in statistical contexts, particularly regression analysis.

A plot of $x$, $y$, $z$ against $-1$ (see below) shows that it cannot be modeled by a straight line, so a regression is performed by modeling the data by a parabola.

$$\beta_0 x^2 + \beta_1 y^2 + \beta_2 z^2 + \beta_3 xy + \beta_4 xz + \beta_5 yz + \beta_6 x + \beta_7 y + \beta_8 z + \beta_9 xyz = -1.$$

Where the dependent variable, $x$, is packet lost rate, the independent variable, $y$, is frame lost rate, and the independent variable, $z$ is PEVQ.

Place the coefficients, $x_i^2$, $y_i^2$, $z_i^2$, $x_i y_i$, $x_i z_i$, $y_i z_i$, $x_i$, $y_i$, $z_i$ and $x_i y_i z_i$ of the parameters for the $i$th row of the matrix $S$.

The values of the parameters are found by solving the normal equations

$$(S^T S)\hat{\beta} = -S^T [1\dots \ ]^T.$$

The matrix is well conditioned and positive definite, that is, it has full rank, the normal equations can be solved directly by using the Cholesky decomposition.

$$\hat{\beta} = -(S^T S)^{-1} S^T [1\dots \ 1]^T.$$

### 3.5.5 Simulated Annealing

SA is a generic probabilistic meta-algorithm for the global optimization problem, namely locating a good approximation to the global minimum of a given function in a large search space. It is often used when the search space is discrete. For certain problems, simulated annealing may be more effective than exhaustive enumeration—provided that the goal is merely to find an acceptable good solution in a fixed amount of time, rather than the best possible solution.

In the SA method, each point s of the search space is analogous to a state of some physical system, and the function $E(s)$ to be minimized is analogous to the internal energy of the system in that state. The goal is bring the system, from an arbitrary initial state, to a state with the minimum possible energy.

The neighbors of each state (the candidate moves) are specified by the user, usually in an application-specific way.

The probability of making the transition from the current state s to a candidate new state s' is specified by an acceptance probability function $P(e, e', T)$, that depends on the energies $e = E(s)$ and $e' = E(s')$ of the two states, and on a global time-varying parameter $T$ called the temperature. The pseudo code is shown in Figure 3.18.

### 3.5.6 Performance Evaluation

There are a total of 54 samples of video files. We use a different size of training sets to estimate the PEVQ expression. The training set is separated to 5, 10, 18, and 30.

```
1.   s := s0; e := E(s)                              //  Initial state,energy.

2.   sb := s; eb := e                                //  Initial "best" solution

3.   k := 0                                          //  Energy evaluation count.

4.   while k < kmax and e > emax                     //  While time remains and not good enough:

5.        sn := neighbor(s)                          //  Pick some neighbor.

6.        en := E(sn)                                //  Compute its energy.

7.        if en < eb then                            //  Is this a new best?

8.           sb := sn; eb := en                      //  Yes, save it.

9.        if P(e,en,temp(k/kmax)) > random()then //  Should we move to it?

10.          s := sn; e := en                        //  Yes, change state.

11.       k := k + 1                                 //  One more evaluation done

12.   return sb                                      //  Return the best solution found.
```

**FIGURE 3.18**
Simulated annealing pseudo code.

**TABLE 3.4**

Calculated Coefficient by Different Training Set with LSM Algorithm

| Training Set Size | 10 | 18 | 30 |
|---|---|---|---|
| $\beta_0$ | 0.1503 | 0.0528 | −1.38407 |
| $\beta_1$ | −0.0071 | 0.0778 | −0.90331 |
| $\beta_2$ | 0.0016 | 0.0005 | −0.12004 |
| $\beta_3$ | 2.4637 | 2.2441 | −0.43265 |
| $\beta_4$ | 0.4383 | 0.2818 | 0.244986 |
| $\beta_5$ | 0.6521 | 0.8361 | 0.418736 |
| $\beta_6$ | −1.4899 | −0.9349 | −1.22127 |
| $\beta_7$ | −1.8959 | −2.4403 | −0.10035 |
| $\beta_8$ | −0.3229 | −0.3236 | 0.05914 |
| $\beta_9$ | −0.8456 | −0.7946 | 0.628173 |

### 3.5.6.1 Least Squares Method

In order to adopt the LSM, the minimum training set size is 10. The obtained model was tested with data that is in the training set after the training process. The coefficient calculated from the training set is listed in Table 3.4 and shows the experiment with different results. The experimental result of LSM is shown in Figures 3.19–3.21 and Table 3.5.

### 3.5.6.2 Simulated Annealing

A portion of the training data and change in the error for the training process are show in Table 3.6. The training set size is larger and the mean squared error root is minimized for these values. This shows that the LS is a good algorithm to solve the problem.

As seen from the Figures 3.22–3.25, the estimation becomes more accurate when the training set is larger. There is not much difference with both algorithms when the training set is more than 18, which is the number of the sample video.

## 3.5.7 Summary

Based on the aforementioned analysis, we present here how these results can be further exploited toward developing a video quality prediction method base on a network status and original frame lost rate. Hence, consider an unknown video clip in a specific network environment and predict its PEVQ that better describes its quality after transmission and encode, decode. Thus the algorithm can successfully predict the video quality. Consequently, the service provider can predict analytically video service quality that satisfy specific quality levels at a pre-encoding state.

We will try to make our protocol stack and video communication system compatible across 3G-324M, Wi-Fi, and H.323. Thus, the video communication

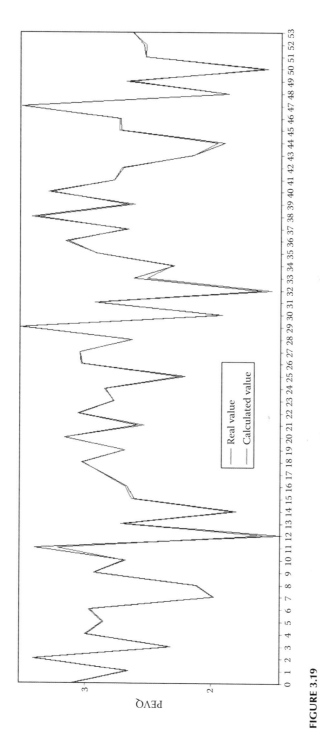

**FIGURE 3.19**
Real value and calculated vale comparison with SA algorithm (training set = 10).

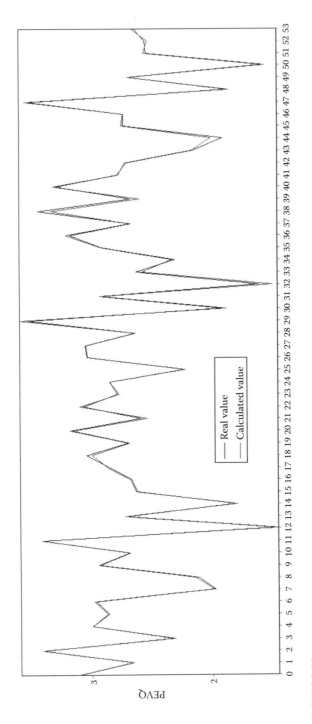

**FIGURE 3.20**
Real value and calculated vale comparison with LSM (training set = 18).

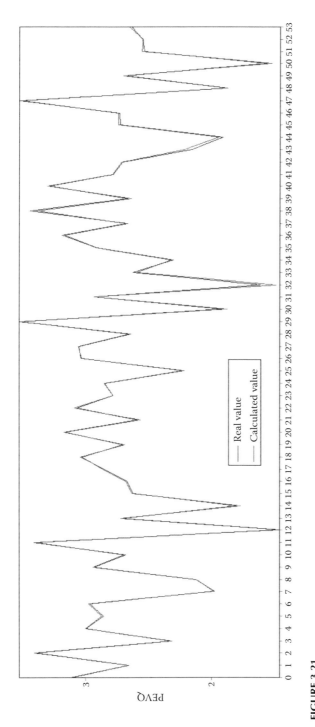

**FIGURE 3.21**
Real value and calculated vale comparison with LSM (training set = 30).

**TABLE 3.5**

Results of LS Algorithm with Different Training Sets

| Training Set Size | Loops | Calculation Times (H:M:S:MS) | Start Temperature | Root Mean Squared Error (RMSE) | Sum of Squared Error (SSE) |
|---|---|---|---|---|---|
| 5 | 505 | 00:00:01:281 | 4.52262 | 0.280969 | 3.868242 |
| 10 | 283 | 00:00:00:984 | 2.57655 | 0.168316 | 1.246539 |
| 18 | 269 | 00:00:00:813 | 2.23726 | 0.037804 | 0.051448 |
| 30 | 295 | 00:00:00:672 | 0.12073 | 0.001481 | 5.26E-05 |

**TABLE 3.6**

Calculated Coefficient by Different Training Set with SA Algorithm

| Training Set Size | 5 | 10 | 18 | 30 |
|---|---|---|---|---|
| $\beta_0$ | −1.38407 | 0.137138 | 0.001477 | −0.02762 |
| $\beta_1$ | −0.90331 | 0.059266 | 0.083278 | 0.01848 |
| $\beta_2$ | −0.12004 | 0.00745 | 0.041985 | 0.043541 |
| $\beta_3$ | −0.43265 | 1.662551 | 1.395316 | 2.196734 |
| $\beta_4$ | 0.244986 | 0.330581 | 0.213052 | 0.282333 |
| $\beta_5$ | 0.418736 | 0.410387 | 0.488881 | 0.715724 |
| $\beta_6$ | −1.22127 | −1.23667 | −0.68023 | −0.77176 |
| $\beta_7$ | −0.10035 | −1.50019 | −1.52025 | −1.9891 |
| $\beta_8$ | 0.05914 | −0.3144 | −0.44648 | −0.48048 |
| $\beta_9$ | 0.628173 | −0.47825 | −0.47317 | −0.79836 |

system can be performed over both 3G mobile and internet networks, with the support of gateways. In order to support more users and improve the call quality, the 3G gateway should be composed of the following unit: call setup server for connecting the 3G network to an IP network, transcoding server for the multimedia in different networks, register server for VoIP user, and AAA server for billing.

## 3.6 Conclusions

Our work mainly focused on the video communication for 3G wireless network, 3G-324M protocol implementation, and UMTS-based cross platform communication. However few researchers consider integration of the mobile network infrastructure and sensor ad hoc networks. This chapter presented our new approach, using the characteristics of sensor ad hoc networks and mobile network infrastructure to deliver sensor network signals. The

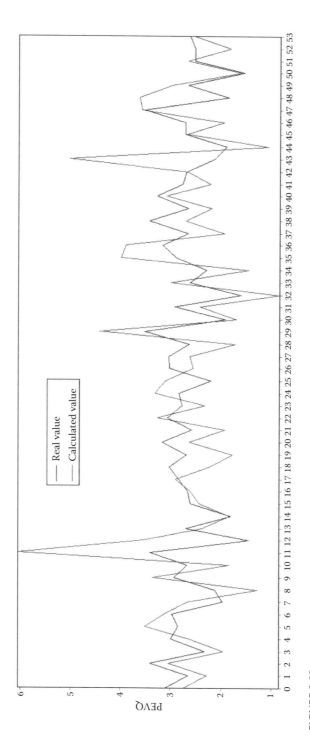

**FIGURE 3.22**
Real value and calculated vale comparison with SA algorithm (training set = 5).

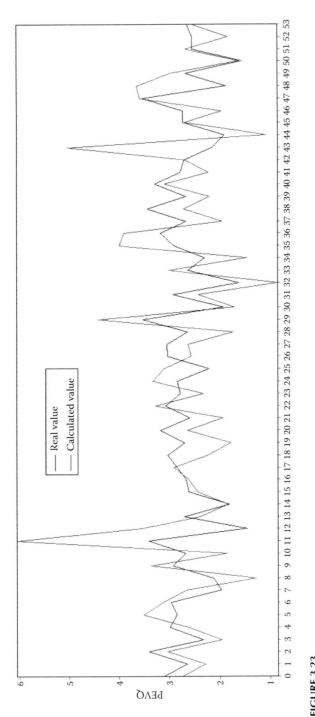

**FIGURE 3.23**
Real value and calculated vale comparison with SA algorithm (training set = 10).

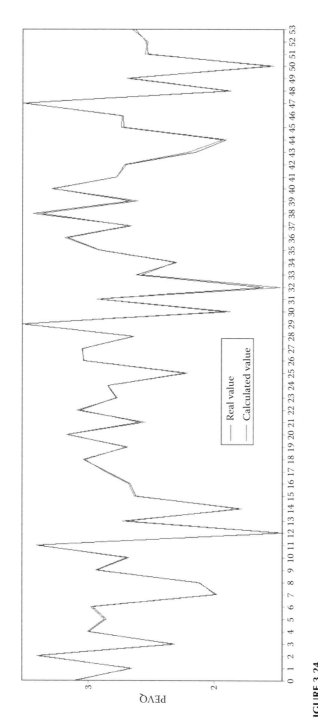

**FIGURE 3.24**
Real value and calculated vale comparison with SA algorithm (training set = 18).

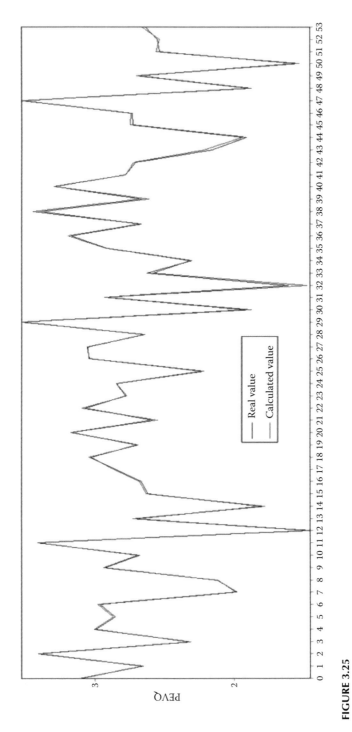

**FIGURE 3.25**

Real value and calculated vale comparison with SA algorithm (training set = 30).

advantages are that communicating between these two systems dynamically and intelligently can reduce the cost and increase the lifetime of sensor networks. It was also shown that this approach is suitable for all organizations and for gathering data on demand. The feasibility and viability of the proposed method has been proven through initial experimental work. As shown in this chapter, delivering sensor network signals is much more complex than the mere translation of message encoding and protocols. However, we are currently working on building a new framework to achieve the goals of video communication between mobile networks and sensor networks, suitable for a range of commercial and other different applications and with a short transmission range.

### 3.6.1 Circuit-Switched Mobile Video

3GPP adopted H.324M with some mandatory requirements to form 3G-324M protocol stack for low bit rate multimedia communication in 3G CSN. In this protocol stack, H.245 recommendation is used for call control and signaling and H.223 recommendation is used for multiplexing. These protocols work together closely to provide reliable service over an error prone wireless network.

Currently, we are investigating the transcoding and compatibility between 3G-324M terminals through the heterogeneous environment. We'll try to make our protocol stack and video communication system compatible across 3G-324M, Wi-Fi, and H.323. Thus, the video communication system can be performed over both 3G mobile and internet networks with the support of gateways.

### 3.6.2 UMTS Video Gateway Implementation

The 3G gateway enables the multimedia communication cross PS and CSN providing more exciting service to users. We have presented the technical challenges we faced during our development. First of all we looked at the importance of service and system reliability. Moreover, we examined the interoperability issues of the gateway. Furthermore, we discussed the scalability and interconnectivity and some suggestions are given. Finally we studied the efficiency problem of the gateway, with the possible improvement for H.245 and H.223, respectively. In the future, our main work will concentrate on improving system efficiency.

### 3.6.3 Optimization to Make the 3G Video Gateway Efficient and Robust

We present our success of attaining appreciable improvement in robustness and efficiency of the multimedia gateway that is suitable for handling high call traffic in a telecommunication scenario. We adopt an efficient compile-store-reuse model for speeding up the process of encoding the ITU-T H.245 ASN.1 formatted messages in a binary stream. In this method, H.245 sets of messages are compiled, used, and stored in a table. For encoding further incoming messages, the table is queried and the matching result is used instead of recompiling the messages again and again. Experimental results show that our algorithm greatly improves call handling efficiency of a 3G gateway. Thus, we have proved that our lookup table-based message encoding algorithm can provide stable performance and a satisfactory QoS assurance even under high traffic conditions. Also our method is tested to be fully compatible with existing 3G protocols.

### 3.6.4 Communication between Sensor and UMTS Network

UMTS users could conveniently and quickly access sensor network. We have proposed architecture for video transmission from WSN to UMTS networks, which allow an ad hoc PSN communicating with a CSN. However few researchers consider integration of the mobile network infrastructure a sensor network. Our new approach, use the characteristics of sensor networks and mobile network infrastructure to deliver sensor network signals. Furthermore, we build up the platform and implement the video gateway. The advantages are that communicating between these two systems dynamically and intelligently can reduce the cost and increase the lifetime of sensor networks. It was also shown that this approach is suitable for all organizations and for gathering data on demand. The feasibility and viability of the proposed method has been proven through initial experimental work.

### 3.6.5 Mobile Video Quality Measurements and Predictions

We present how these results can be further exploited toward developing a video quality prediction method base on network status and original frame lost rate. Hence, consider an unknown video clip in a specific network environment and predict its PEVQ that better describes its quality after transmission and encode, decode. Thus the algorithm can successfully predict the video quality. Consequently, the service provider can analytically predict video service quality that satisfy specific quality levels at a pre-encoding state.

We'll try to make our protocol stack and video communication system compatible across 3G-324M, Wi-Fi, and H.323. Thus, the video communication system can be performed over both 3G mobile and internet networks, with the support of gateways. In order to support more users and improve the call quality, the 3G gateway should be composed by the following unit: call setup server for connecting the 3G network to IP network, transcoding server for transcoding the multimedia in different network, register server for VoIP user, and AAA server for billing.

# References

1. W. Jia, F. P. Tso, L. Zhang, and Y. Du, "Videoconferencing for 3G Wireless Networks," *Handbook on Mobile Ad Hoc and Pervasive Communications,* American Scientific Publishers, Valencia, CA, 2006.
2. W. Jia, F. P. Tso, and L. Zhang, "Efficient 3G-324M Protocol Implementation for Low Bit Rate Multi-point Video Conferencing," *Journal of Networks,* Academy Publisher, Issue 5, 2006.
3. M.-C. Yuen, J. Shen, W. Jia, and B. Han, "Simplified Message Transformation for Optimization of Message Processing in 3G-324M Control Protocol," Proc. of 2005 International Conference on Computer Networks and Mobile Computing (ICCNMC 2005), 64–73. LNCS 3619, Zhangjiajie, China.
4. W. Jia, B. Han, H. Fu, J. Shen, and C. Yuen, "Efficient Implementation of 3G-324M Protocol Stack for Multimedia Communication," IEEE 11th International Conference on Parallel and Distributed Systems, July 20–22, 2005, (FIT), Fukuoka, Japan.
5. W, Jia, J. Shen, and H. Fu, "Object-Oriented Design and Implementations of 3G-324M Protocol Stack," Proc. of ICA3PP 2005, LNCS3719, pp. 435–41, 2005, Melbourne, Australia.
6. B. Han, H. Fu, J. Shen, P. O. Au, and W. Jia, "Design and Implementation of 3G-324M: An Event-Driven Approach," Proc. IEEE VTC'04, Fall, 2004, Los Angeles, CA.
7. ITU-T Recommendation H.223: "Multiplexing Protocol for Low Bit Rate Multimedia Communication," http://www.itu.int/ITU-T/
8. ITU-T Recommendation H.223: Annex A: "Multiplexing Protocol for Low Bit Rate Multimedia Mobile Communication Over Low Error-Prone Channels," http://www.itu.int/ITU-T/
9. ITU-T Recommendation H.223: Annex B: "Multiplexing Protocol for Low Bit Rate Multimedia Mobile Communication Over Moderate Error-Prone Channels," http://www.itu.int/ITU-T/
10. ITU-T Recommendation H.223: Annex C: "Multiplexing Protocol for Low Bit Rate Multimedia Mobile Communication Over Highly Error-Prone Channels," http://www.itu.int/ITU-T/
11. ITU-T Recommendation H.245: "Control Protocol for Multimedia Communication," http://www.itu.int/ITU-T/
12. ITU-T Recommendation G.723.1: "Dual Rate Speech Coder for Multimedia Communication Transmitting at 5,3 and 6,3 kbit/s," http://www.itu.int/ITU-T/
13. ITU-T Recommendation H.263: "Video Coding for Low Bitrate Communication," http://www.itu.int/ITU-T/
14. ITU-T Recommendation H.324: "Terminal for Low Bitrate Multimedia Communication," http://www.itu.int/ITU-T/
15. 3GPP TR 26.111 V5.0.0 (2002–06) Codec for Circuits Switched Multimedia Telephony Service; Modifications to H.324 "http://www.3gpp.org/
16. 3GPP TS 26.110: "Codec for Circuit Switched Multimedia Telephony Service; General Description, "http://www.3gpp.org/"
17. 3GPP TR 26.911: "Codec for Circuit Switched Multimedia Telephony Service; Terminal Implementors Guide (Release 4), "http://www.3gpp.org/

18. ITU-T Recommendation X.691: "Information Technology: ASN.1 Encoding Rules: Specification of Packed Encoding Rules (PER)," http://www.itu.int/ITU-T/
19. ISO/IEC 14496-2: "Information Technology: Coding of Audio-Visual Objects: Part 2: Visual."
20. 3GPP TS 26.071: "General Description," http://www.3gpp.org/
21. 3GPP TS 26.090: "Transcoding Functions," http://www.3gpp.org/
22. 3GPP TS 26.073: "Adaptive Multi-Rate (AMR); ANSI C Source Code," http://www.3gpp.org/
23. 3GPP TS 24.008: "Mobile Radio Interface Layer 3 Specification; Core Network Protocols-Stage 3," http://www.3gpp.org/
24. 3GPP TS 27.001 V4.11.0 (2003–2009). General on Terminal Adaptation Functions for Mobile Terminals http://www.3gpp.org/
25. 3GPP TS 29.007: General Requirements on Interworking Between the Public Land Mobile Network (PLMN) and the Integrated Services Digital Network (ISDN) or Public Switched Telephone Network (PSTN) http://www.3gpp.org/
26. 3GPP TS 23.108: Mobile Radio Interface Layer 3 Specification Core Network Protocols; Stage 2 (Structured Procedures) http://www.3gpp.org/
27. Grossmann, I. E., ed., *Global Optimization in Engineering Design,* Norwell, MA: Kluwer Academic Publishers, 1996.
28. Ingber, L. and Rosen, B., "Genetic Algorithms and Very Fast Simulated Re-annealing: A Comparison," *Math. Comput. Modelling* 16 (11): 87–100, 1992.
29. Y. Du, F. P. Tso, and W. Jia, "Architecture Design of Video Transmission Between UMTS and WSN," International Conference on Ubi-Media Computing, Lanzhon, P.R.C, 2008.
30. F. P. Tso, Y. Du, W. Jia, "Design of an Efficient and Robust Multimedia Gateway for Pervasive Communication," Wireless Telecommunications Symposium, Pomona, CA, 2008.
31. M. Y. Aal Salem, D. Everitt, and A. Y. Zomaya, "An Interoperability Framework for Sensor and UMTS Networks," ICWMC Guadeloupe, French Caribbean, 2007.

# 4

# A New Trust and Reputation-Based Chain Model in Peer-to-Peer Networks

Sinjae Lee and Wonjun Lee

## CONTENTS

Computer security has attracted significant attention in recent times. This is reflected in the fact that numerous mechanisms, both existing and new, such as encryption, sandboxing, digital rights management, reputation-based systems, and firewalls have been employed to enhance the security of peer-to-peer (P2P) networks. Among the abovementioned mechanisms, a reputation-based mechanism that is an active self-organized mechanism without a controlling entity, is particularly useful for automatically recording, analyzing, and even adjusting the metrics of reputation, trust, and credibility among peers; it enables a system to self-adjust according to the changes in credibility. This kind of a self-adjusting system is suitable for the anonymous, dynamic, and variable P2P environments. With the aid of the resource chain model (RCM), which is a novel reputation-based trust model, it is possible to effectively and efficiently identify a resource location while maintaining the system security. The objective of this study is to verify whether or not the RCM enhances the successful download rate when the number of transactions and the number of malicious nodes are large. The results of the study revealed that the RCM improved the performance metrics of a P2P network. Therefore, this approach provides an efficient method for finding the best resource chain in a P2P community.

## 4.1 Introduction

Existing peer-to-peer (P2P) applications can be classified into one of the following three categories: file sharing, distributed processing, and instant messaging. The focus of our study will be on the P2P applications for file sharing since it is the most common application for P2P networks [6]. Reputation-based systems have been researched extensively by many, and some of these researchers have constructed reliable theoretical models. Among these is a simple model that is almost exclusively based on the trust feedback and credibility metrics of the P2P community; this model is used to decide the next course of action [1]. Transaction histories are stored in trust vectors, and the number of significant bits in each trust vector is denoted by an integer assigned to it. After each action, the most significant bit is replaced by the latest result, and the history bits are moved to the right of the vector. We calculate trust and distrust ratings on the basis of the bits in the trust vector.

### 4.1.1 Reputation-Based Trust Model

In a more complex model [3], the designer included additional factors in order to enhance the robustness of the reputation-based system. In addition to the feedback $S(u,i)$ that denotes peer $u$'s feedback of peer $i$ and credibility $Cr(p(u,i))$ that denotes the credibility of $p(u,i)$, other transaction factors ($TF(u,i)$: transaction context factor) like the transaction number $I(u)$ and transaction scale were included for improved transaction control. Further, the community context factor $CF(u)$ is also included to easily adapt the model to different situations. The following is the general metric of this model [2,7]

$$T(u) = a * \frac{\sum_{i=1}^{I(u)} S(u,i) * Cr(p(u,i)) * TF(u,i)}{r!(n-r)!} + b * CF(u)$$

$$\text{new\_credibility} = \text{old\_credibility} * (1 - \text{Trans\_Factor}).$$

The first part of this general metric is used to collect all the transaction information and the second part is used to compensate for the effect of the community context factors on the final reputation result. This reasonably effective model successfully collects all the useful information of the P2P environment.

One major drawback of models like the ones mentioned above is that they tend to concentrate on the relationship between a node and its direct neighbors. Usually, the path to a desired resource, known as the resource path, is obtained from numerous peers and their neighbors; such a resource path affords the most useful information that could be employed during future searches. Thus, we try to make an efficient use of this resource chain; the entire resource chain is strengthened if the end of the chain provides reliable service and it is weakened in the event of an unsuccessful download. Our model is based on this reasoning.

---

## 4.2 The Approach

The major aspects of our approach are that our model is a best effort service and that it does not include the guaranteed service feature. In a P2P network, our model always tries to choose the most suitable node from among a set of possible peers for the purpose of downloading. In our approach, the reliability of a node is judged by considering credibility of the node. We present the following scenario to indicate the chain of events when a node enters a new P2P community.

Internet protocol (IP) and media access control (MAC) addresses provide a peer with a unique identity. Therefore, as a node enters a P2P network, it broadcasts or multicasts connection requests and waits for replies from the rest of the peers. After a sufficient number of replies are received, the node can choose neighbors from among the peers that have replied to the node's connection requests. This choice is made on the basis of IP addresses that is, neighbors across sufficiently different IP address ranges are chosen so that peers from different IP groups can be accessed. Since making this choice is just the beginning of the search process, it is possible that there are malicious users among the neighbors chosen. This is taken care of by the fact that a node's neighbor table can be updated to add new neighbors and delete malicious nodes on the basis of subsequent transactions. We will discuss this in the update credibility. Network neighbors can be obtained as explained above. Later, the desired resource can be found by sending out search queries to the node's neighbors and the neighbors' neighbors. These search paths, for example, the path from the node that sends out the request to the node at which the desired resource is found, can sometimes comprise several long chains. Therefore, the best chain from among such candidate chains is found on the basis of the existing knowledge about the peer network, and attempts are made to establish a connection between the node that sends out the request and the node at the end of the best chain in order to obtain the desired resource. In the course of establishing this connection, it is not a concern if the credibility of the node at the end of the best chain is lower than a certain threshold value, which would make it seem less trustworthy. It is sufficient to ensure that the most reliable peer among all the available peers is chosen. The degree of truthfulness for certain P2P nodes is calculated using their credibility. After a destination is selected for downloading, the desired resource can then be downloaded from that node.

When the requesting node's attempt to download a resource from the destination node is successful, the credibility between the starting node and the destination node is increased. On the other hand, if the transaction is not successful the credibility between nodes is decreased. Such updating is the most critical phase in our model because it is efficient and self-adjusting.

## 4.3 The Resource Chain Model (RCM)

### 4.3.1 Assumptions

In order to simplify the explanation of our model [4,5], we would like to make the following assumptions. The number of neighbors is eight, and the length of the search is eight. This means that the longest requesting length is eight. A record of a node's neighbors is maintained in a list called the neighbor list. The nodes are ordered on the basis of their credibility; the higher a node's credibility, the higher the position it holds. In addition to these assumptions, we assume that we have a formula to evaluate the relationship between the nodes. For $0 < i < 9$, $Ni = $ Neighbor (N0, 1). This means that the distance between nodes $Ni$ and N0 is 1.

### 4.3.2 Working the RCM

Consider a node N0 that seeks to know where it can download File X in the peer network. Since N0 is only aware of a limited number of nodes by virtue of their being its neighbors, it transmits the request for file X to all of its neighbors. We assume that those nodes that already know the node that hosts File X will send a reply to N0 without sending out any more forwarding requests. Here, we assume that N1, N2, N3, and N4 know the location of the node that hosts File X. Further, we assume that we have new nodes that are previously unknown to node N0. They will forward N0's request for file X to their neighbors. Here, we assume that N5, N6, N7, and N8 do not know the location of the node that hosts File X. Therefore, they forward the request to their neighbors. N0 will wait for their replies for a stipulated period of time. While N0 waits, the other nodes forward their findings, and some of these node runs would probably inform N0 of the location of file X. At the end of the stipulated waiting period, each node that has sent out file requests to its neighboring nodes would have received several replies. Therefore, the temp table for each node has to be changed.

## 4.4 Simulation

### 4.4.1 Assumptions

After the requesting node obtains responses from many assigned neighbors, they are added to the neighbor list of the requesting node and each neighbor is given a credibility number of 0.5, which would be the neighbor's initial credibility. The neighbor list can be updated when the requesting node sends out resource-related queries to the neighbors and when it later begins transactions. One of the problems that could be faced is that for a given requesting node, it is difficult to establish the number of reliable neighbors. Another problem is regarding the search depth. When the requesting node does not know the node where the resource is located, it would send out search queries to its neighbors. The requesting node's neighbors will have a depth of 1. Further, if the neighbors do not know the location of the resource and forward the node's requests to their own neighbors, these indirect neighbors will have a depth of 2. Therefore, the request can be forwarded to any depth ranging from 3 to the length of the resource chain L.

It is desirable that there are only a reasonable number of neighbors at each stage of the search so that each node can refer to its limited set of neighbors and compile the required information without a considerable amount of computing effort. Further, it is desirable that the length of the resource chain is not very large as to affect the search efficiency. At the same time, it is also desired that such a resource chain can span almost all the nodes in the P2P community.

The scale of the P2P community T_Scale can be obtained in the following manner by considering the number of neighbors for each node $N$ and the length of a resource chain $L$.

$$1 + N + N^2 + N^3 + N^4 + \ldots + N^{L-1} = \frac{N^{L-1}}{N-1} = \text{T\_Scale}.$$

In the case where $N$ is sufficiently large, the above equation can be approximated to the following equation

$$N^{L-2} = \text{T\_Scale}.$$

The value of T_Scale is predicted using suitable methods such as statistical methods. We can then achieve a balance between $N$ and $L$ by using the relationship between the two numbers, and the scale of our community. Considering the relationship between the different numbers and our computational abilities, we assume that $L = 5$, $N = 8$, and T_Scale = 500.

### 4.4.2 Simulation Data Collection

We used a simulation tool developed in-house to collect five different groups of data. In the test conducted on each group, the total number of nodes was the same but the number of malicious nodes in the community was varied. For each test, we let the system perform 1000 transactions, and for each transaction, we collected all the possible download paths. The number of effective resource chains—the resource chains that were not affected by the presence of the malicious nodes in the system—was calculated by considering these download paths. We determined the percentage of effective resource paths out of the total number of possible download paths; the percentage of effective resource paths denotes the security level of the current system. As the number of transactions increased, a corresponding increase in the number of malicious acts was observed and some of the malicious nodes were even identified. These malicious nodes that were identified pose a lesser threat to the system security compared to those that are not identified since they had been identified and had even been blocked by other nodes when the system employed our dynamic reputation control mechanism. See Table 4.1 for the collected data after the test.

### 4.4.3 Simulation Data Analysis

An analysis of the trend lines obtained using the abovementioned simulation revealed that our model was sufficiently reliable in improving the security of a P2P community. We will elaborate on our model's advantages in the following three sections.

#### 4.4.3.1 Efficiency

In order to compare the data collected from the test, we used third-order polynomial trend lines to demonstrate the variations in the system security level. From Figure 4.1, it can be observed that the level of system security increases with the number of transactions, and the curve finally appears to converge at the peak system security, which corresponds to a 100% successful transaction rate in the P2P community. This conclusively suggests that the RCM was able to enhance the security of the P2P network automatically without a central controlling entity.

**TABLE 4.1**

Collected Data

| Group | Total Number of Nodes | Malicious Nodes | Transactions | Figure |
|---|---|---|---|---|
| 1 | 500 | 30 | 1000 | Figure 4.1 |
| 2 | 500 | 40 | 1000 | Figure 4.2 |
| 3 | 500 | 50 | 1000 | Figure 4.3 |
| 4 | 500 | 75 | 1000 | Figure 4.4 |
| 5 | 500 | 100 | 1000 | Figure 4.5 |

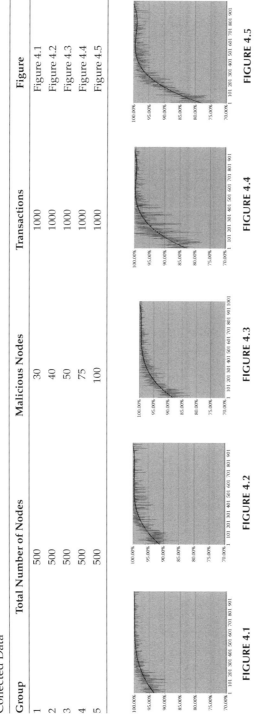

FIGURE 4.1

FIGURE 4.2

FIGURE 4.3

FIGURE 4.4

FIGURE 4.5

### 4.4.3.2 Quick Convergence

From an analysis of the trend lines observed during the simulation of all the different data groups, the following conclusion can be drawn: after the number of transactions reached a value of ~500, the percentage of successful transactions was close to the peak of 100%. This means that the system security level is converging at the peak level only after ~500 transactions in a P2P community with a node scale of 500. This suggests that after each node sends out a request and completes one transaction, the system can self-adjust itself to achieve a relatively enhanced system security. Therefore, it can be proposed that the rapid convergence of our model's system security level on account of its self-adjusting ability is a reliable indicator of its efficacy.

### 4.4.3.3 High Stability

The best feature of our model, as inferred from the data analysis, was that the system security levels were converging at the peak after ~500 transactions, irrespective of the number of malicious nodes that existed in the system. Our data groups were carefully selected so as to observe how the efficiency of our system's self-adjusting ability varied with an increase in the number of malicious nodes. Such an observation can be made from Figure 4.2. The RCM that we have introduced attempts to update the chain credibility instead of the node-to-node trust updating that is usually performed by existing methods and the use of our model affords enormous improvement in the system stability.

### 4.4.4 Comparison with Complain-Only Model

We also compared our model with the complain-only model, a model that concentrates on the credibility relationship between two nodes as against all the resource chains considered in the RCM. We chose the same P2P community with a scale of 500. Further, we select the number of malicious nodes existing in the P2P community at the beginning to be 50. We perform a total of 500 transactions since our model's performance reaches the maximum possible security level after ~500 transactions, as revealed by the analysis discussed above. In addition, after obtaining all the data, we assign a trend line to each data group to indicate the time variations in the system security levels. A comparison of the two models is shown in Figure 4.6.

From Figure 4.6, it can be observed that there was a considerable difference in the performance of both the models after 500 transactions. After 500 transactions, it can be observed that the system security level in the RCM, as indicated by its trend line, is sufficiently reliable and also that it was continuously improving. In contrast, the complain-only model appeared to need more transactions to provide a reliable system security level.

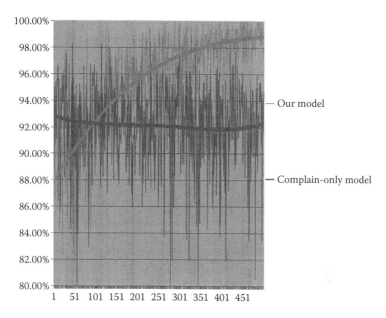

**FIGURE 4.6**
Comparison of performance of the RCM and complain-only model.

### 4.4.5 Evaluation

The test results have clearly revealed that our dynamic, reputation-based trust model RCM, enhanced the system security level after a certain number of transactions were performed in the P2P community. An analysis of each trend line, which describes the trends in the system security level, revealed that the security of the P2P community improved with time. We were also able to observe that, on account of its rapid self-adjusting property as inferred from the trend line, our system performed well even when there were many malicious nodes existing in the system. After having compared the performance of our model with that of some existing models, we can assert that our model affords significant advantages over the existing models and that it is very effective in automatically enhancing system security by using the concept of self-adjusting chain credibilities.

## 4.5 Conclusion

It could be concluded, from the above discussion, that our model successfully addresses the P2P reputation problem. Our model can locate a malicious resource chain after each transaction and weaken it to prevent it from affecting subsequent transactions. Communicating nodes also promote each other's transaction reputations if the transaction between them is successful; this serves as an incentive and encourages everyone in the P2P network to be honest and benefit from the whole P2P community. This fundamental concept of our model would serve the P2P file-sharing system well. In the course of our study the following conclusions were drawn.

- Our model can locate the resource chain with malicious nodes after each transaction and weaken the whole chain to prevent it from affecting subsequent transactions.
- Communicating nodes also promote each other's transaction reputation if the transaction between them is successful.

Therefore, our model can be applied to a P2P file-sharing system to enhance the security and ensure automatic self-adjustment. On account of our model's automatic information collection capability and reputation updating capability in different system environments, it can be applied to P2P communities to enhance their trustworthiness.

## References

1. K. Aberer and Z. Despotovic. 2001. Managing trust in a peer-2-peer information system. Paper presented at the Proceedings of the Ninth International Conference on Information and Knowledge Management, Atlanta, Georgia.
2. V. Bhat. 2004, December. Reputation management in peer-to-peer systems. Technical report from the Department of Computer Sciences, University of Texas at Austin.
3. E. Damiani, D. Vimercati, and S. Paraboschi. 2002, November 18–22. A reputation-based approach for choosing reliable resources in peer-to-peer networks. Paper presented at the Proceedings of the 9th ACM Conference on Computer and Communications Security, CCS2002, Washington, DC.
4. S. Lee, S. Zhou, and Y. Kim. 2007, July 30–August 1. P2P trust model: The resource chain model. Paper presented at the Proceedings of the 8th ACIS International Conference on Software Engineering, Artificial Intelligence, Networking and Parallel/Distributed Computing, SNPD2007, Qingdao, China.
5. S. Lee, K. Park, and Y. Kim. 2007, August 9–12. The new P2P reputation model based on trust management system. Paper presented at the US-Korea Conference on Science, Technology, and Entrepreneurship, UKC2007, Reston, Virginia.
6. Y. Wang and J. Vassileva. 2003, September. Trust and reputation model in peer-to-peer networks. Paper presented at the Proceedings of IEEE Conference on P2P Computing, Linkoeping, Sweden. University of Saskatchewan, Canada.
7. L. Xiong and L. Liu. 2003, June. A reputation-based trust model for peer-to-peer ecommerce communities. Paper presented at the IEEE Conference on Electronic Commerce CEC2003, Newport Beach, CA.

# 5

## Bounding Data Aggregation for Ubiquitous Applications within Wireless Sensor Networks

**Sami J. Habib**

### CONTENTS

We have formulated the data collection process within wireless sensor networks as a scheduling problem to alleviate workload on the data collection centers. A sensor device has three tasks: sensing data, processing data, and transmitting the collected data. We modeled the sensor's three tasks as a task-flow graph (TFG), and then we combined all TFGs for all sensors within the Wireless Sensor Network (WSN) as a super task-flow graph (STFG). We utilized two scheduling algorithms, as soon as possible (ASAP) and as late as possible (ALAP), to order all tasks within the STFG subject to the data collection centers' limitations. The experimental results provide excellent boundaries on how early and how late to schedule all sensors' tasks with the variation of the number of data collection centers.

### 5.1 Introduction

Many ubiquitous applications for a wireless sensor network (WSN) have emerged, such as battlefield surveillance, disaster relief, border control, and environmental and habitat monitoring. These ubiquitous applications are executed by a large set of sensors, which are distributed over a wide physical area. The benefits, which are offered by these ubiquitous applications and their WSNs, are easily integrated and embedded within the surroundings and require minimal human interventions. The WSN supports continuous data collection from a distributed network of stationary and mobile sensor devices deployed in the surroundings. The collected data can be processed either locally (within the individual sensors) or globally (in other central nodes, such as base stations), and fused into an interactive virtual environmental monitoring model that can be accessed remotely by humans.

We strongly believe that the deployment of WSN offers several advantages, and solves some of the spatial and temporal data collection. The WSN can cover a larger physical area with minimum human effort than it currently handles by the manual coverage process (spatial drawback). Also, WSN offers an automatic periodical data collection, which corresponds to all possible events occurring in the monitoring area (temporal drawback). Due to the nonstop data collection process by WSN, it presents a challenging problem to the collection centers (or base stations [BS]). On a regular basis, sensors transmit their collected data to a BS; however, the BS has certain process and storage capacities. Eventually the BS will reach its limit with respect to the processing capacity if there are too many sensors in the WSN transmitting data. At the same time, the BS will reach its limit with respect to the storage capacity if the ubiquitous application of WSN captures multimedia data, such as a collection of video clips, audio, or still-images. The multimedia data requires a lot of memory especially for high-quality resolutions.

In this chapter, we present a model for data aggregation based on mapping all tasks within the wireless sensor network into a super task-flow graph (STFG), and a scheduling methodology that regulates the three tasks within each sensor (sense, process, and transmit) in order to balance the workload at the base stations. The sense task estimates a physical phenomenon, such as a temperature, light intensity, humidity, and so forth. The process task manages all activities within a sensor. The transmit task sends the collected data to the base station. We utilized two algorithms: as soon as possible (ASAP) and as late as possible (ALAP) to determine boundaries on how early and late to schedule all sensor tasks, keeping in mind the threshold on the BS with respect to the process and storage capacities.

## 5.2 Structure of a Sensor and Its Related Work

The main components of a sensor consist of a sensing unit, processing unit, communication unit, and power unit as shown in Figure 5.1.

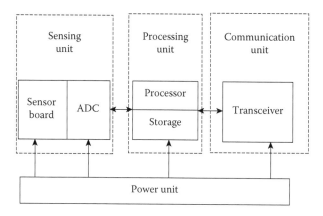

**FIGURE 5.1**
An overview of a sensor device.

The main functionality of the sensing unit is to closely estimate parameters, such as measuring the current temperature or detecting a specific gas in the monitored area. An analog signal is generated by the sensing unit, which is corresponding to the observed parameters. Then, the analog waveform is digitized by an analog-to-digital converter (ADC), which then is delivered to the processing unit for further analysis [4]. The sensing unit is considered a bottleneck because the sensing technologies are following at a slower pace than those of the semiconductors [14].

The processing unit plays a major role in managing the tasks within the device and collaborating with other sensors to achieve the predefined tasks. There are currently several families of the processing unit, including a microcontroller, microprocessor, and field programmable gate array (FPGA). A nonvolatile memory and interfaces, such as analog-to-digital converters (ADCs), can be integrated onto a single integrated circuit [4,9]. The processing unit needs storage for saving all its executed tasks and packaging all collected data for later transmission.

The communication unit is comprised of a transceiver, which acts as the connection between a sensor device and the rest of the network. The transceiver can be implemented in three communication schemes, including either an optical, infrared, or radio frequency (RF). The optical communication consumes less energy than RF and it provides a high security. However, the optical communication requires a line-of-sight and it is sensitive to atmospheric conditions [12]. The infrared, like optical communication, needs no antenna, but it is limited in its broadcasting capacity. The RF is the most easy to use, but it requires an antenna (an extra hardware component needs to be added into the sensor device).

The power unit is a major weakness of the sensor networks. Any energy preservation schemes can help to extend a sensor's lifetime. Batteries used in sensors can be categorized into two groups: rechargeable and nonrechargeable. Often in harsh environments, it is impossible to recharge or change a battery. Current sensors are developed to be able to renew their energy from solar or vibration energy [4,14]. The alkaline battery has a wide voltage range and large physical size while the lithium battery provides a constant voltage supply but with very low nominal discharge currents. The nickel metal hydride battery can be recharged but with a significant decrease in energy density [4].

The data fusion has occupied the interests of many researchers and there are many published papers. Here we present a sample of these papers. However, none of these have attempted to formulate the data fusion as a scheduling problem and try to find its boundaries.

Joint directors of laboratories' (JDL) model [1] is a conceptual data fusion model, which is defined as three levels of fusion: object, situation, and threat refinement. Also, the model provides a common framework and terminology for the data fusion community to unify concepts and terminology. The object refinement level defines the objects identified in the sensing area. Situation refinement defines the relationships between objects and events. The threat refinement projects future threats, vulnerabilities, and opportunities for operations.

The work in paper [8] describes current state of data fusion schemes for wireless sensor networks such as: Kalman filtering [2], blind beamforming [6], transferable belief model [13], filter-based techniques [3], and linear mean square estimator [10]. The methods in [2,6,10,13] are all based on the JDL model for data fusion; however, they are only designed to work at level one, which refers to the object refinement.

The work in paper [16] focused on identifying a route for a mobile agent through a subset of sensor nodes, which indicate the presence of a target by utilizing the signal strengths of the sensor nodes. They formulate the mobile agent routing problem in the distributed sensor network as a combinatorial optimization problem involving the cost of communication and path loss. Their objective is to maximize the sum of a signal energy received at the visited nodes while minimizing the power needed for communication and path losses. One of the most important aspects of mobile agents that was not addressed by this work is the security.

In paper [15], the researchers developed a computation technique based on a particle swarm optimization (PSO) method for obtaining the optimal power scheduling scheme for data fusion in a wireless sensor network when the distributed sensors collected and observations are correlated.

Trying to make sense of the observed data is the focus of the paper [5], where the authors proposed a three-layer observer network architecture, consisting of the application (top), services, and sensors (bottom). The goal of the observer network is to automate the reasoning of accepting or rejecting data.

A routing algorithm for gathering correlated data in the sensor network is proposed by Luo [11], which is based on an adaptive fusion Steiner tree. A routing tree among sensors, which have data to be collected, is constructed keeping in mind the fusion cost, transmission cost, and data structure.

## 5.3 Modeling Data Aggregation within WSN

The tasks, which are performed by a sensor, consist mainly of three operations: sense, process, and transmit. We model these three tasks within a sensor as a task-flow graph (TFG) as shown in Figure 5.2. Each task ($S_i$, $P_i$, and $T_i$) has an execution time, which indicates the duration of time to complete the task and these values are deterministic for now. However, the starting and finishing times for each task are unknown, and our goal is to schedule the starting and finishing times with an acceptable load to be handled by the base station. Thus, scheduling all the tasks within the WSN should coordinate among sensors and we have proposed to combine all TFGs into a STFG as illustrated in Figure 5.3. The graph, STFG (V,E), consists of vertices (tasks) and a set of direct edges representing the flow of tasks. The STFG is an acyclic graph.

To start the process of scheduling, we need to insert two dummy nodes representing the starting and finishing of all tasks. These two dummy nodes (SN and FN) require zero execution time.

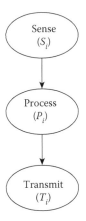

**FIGURE 5.2**
A task-flow graph (TFG) for a sensor *i*.

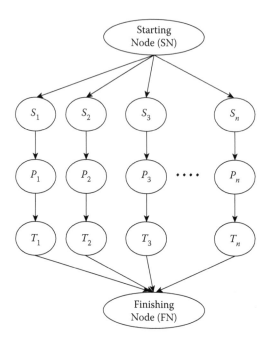

**FIGURE 5.3**
An example of a super task-flow graph (STFG) consisting of *n* sensors.

## 5.4 An Overview of Scheduling Algorithms

There are a number of algorithms for scheduling tasks, subject to a number of different goals, such as the reduction in overall execution of the tasks, or the utilization of minimal resources during the execution of the tasks. Here we present two algorithms: ASAP and as ALAP to schedule all the tasks within the WSN. These two algorithms are mainly employed to determine the earliest and latest boundaries within those tasks to be scheduled [7]. The objective is to assign either the earliest state or latest state to a node taking into consideration its predecessor or successor nodes.

In Figure 5.4, we illustrate the ASAP scheduling algorithm according to Gajski et al. [7], where it assigns an ASAP label $F_i$ to each node $v_i$ of STFG, thereby scheduling the starting execution of $v_i$ with the earliest possible state (time = 0). In lines 1–4, all nodes are initialized, and the function $\text{Pred}_{vi}$ determines all nodes, which are immediate predecessors of the node $v_i$. The vector $F$ contains the earliest starting execution times for all nodes. In lines 5–9, all the residual nodes, which have predecessor nodes, will be assigned starting times for execution. In line 7, the function *all-nodes-scheduled* returns true if all the nodes in set $\text{Pred}_{vi}$ are scheduled. In line 8, the function MAX determines the node with the longest execution value (finishing execution time—starting execution time) from the set of predecessor nodes for $v_i$.

```
ASAP_Algorithm(G(V,E))

        1. for vᵢ ∈ V do
        2. if Pred_vi = Øthen
        3. Fᵢ = 0;
        4. V = V - {vᵢ};
        5. while V ≠ Ødo
        6. for vᵢ ∈ V do
        7. if all nodes scheduled(Pred_vi, F)then
        8. Fᵢ = MAX(Pred_vi, F);
        9. V = V - {vᵢ};
        10. return (F)
```

**FIGURE 5.4**
An overview of the ASAP algorithm.

```
ALAP_Algorithm(G(V,E),T)

        1. for vᵢ ∈ V do
        2. if Succ_vi = Øthen
        3. Lᵢ = T;
        4. V = V - {vᵢ};
        5. while V ≠ Ødo
        6. for vᵢ ∈ V do
        7. if all nodes scheduled(Succ_vi, L)then
        8. Lᵢ = MIN(Succ_vi, L);
        9. V = V - {vᵢ};
        10. return (L)
```

**FIGURE 5.5**
An overview of the ALAP algorithm.

In Figure 5.5, we illustrate the ALAP scheduling algorithm according to Gajski et al. [7], where it assigns an ALAP label $L_i$ to each node $v_i$ of STFG, thereby scheduling the starting execution of $v_i$ into the latest possible state. In lines 1–4, all nodes are initialized and the function $Succ_{vi}$ determines all nodes, which are immediate successors of the node $v_j$. The vector $L$ contains the latest starting execution times for all nodes. In lines 5–9, all the residual nodes, which have successor nodes, will be assigned starting times for execution. In line 7, the function all-nodes-scheduled returns true if all the nodes in set $Succ_{vi}$ are scheduled. In line 8, the function MIN determines the node with the longest execution value (finishing execution time—starting execution time) from the set of predecessor nodes for $v_j$.

## 5.5 Computational Results

We have chosen a WSN consisting of 10 sensors to test our proposed scheduling model, and ASAP and ALAP algorithms. Table 5.1 illustrates the execution times for 10 sensors that comprise the WSN.

We ran two experiments over the WSN, where in both experiments we assumed that one base station was available to receive all the transmissions from the 10 sensors. In the first experiment, the base station is restricted to receive one transmission at a time, where in the second experiment; we relaxed to five transmissions at a time. In the second experiment, we also assumed that the five transmissions arrived at different frequency channels to avoid collisions.

In Figures 5.6 and 5.7 we illustrated the ASAP and ALAP outcomes from our program for experiment 1. All scheduled tasks by the two schedules were executed in 12 states (the number of rows from top to bottom), where 10 and 1 tasks are executed in state 1 at ASAP and ALAP respectively. By looking at the ALAP schedule in Figure 5.7, we noticed that fewer tasks are executed in parallel than in the ASAP schedule.

In Figures 5.8 and 5.9 we illustrated the ASAP and ALAP outcomes from our program for experiment 2, where the BS could receive five transmissions at a time. All scheduled tasks by the two schedules are executed in four states, where 10 and 5 tasks were executed in State 1 at ASAP and ALAP respectively. By looking at both ASAP and ALAP schedules in Figure 5.8 and Figure 5.9 we noticed that the number of tasks were executed in parallel about the same but at different States. At State 2, the 10 tasks were scheduled by both algorithms; at state 5, the 5 tasks were scheduled by both algorithms.

Figures 5.10 through 5.13 depict the timing diagrams for the first experiment and second experiment respectively. The *x*-axis represents the time

**TABLE 5.1**

Tasks' Execution Times

| Sensor | Sense Time | Process Time | Transmit Time |
|--------|-----------|--------------|---------------|
| 1 | 2 | 2 | 5 |
| 2 | 6 | 4 | 2 |
| 3 | 3 | 3 | 4 |
| 4 | 3 | 2 | 1 |
| 5 | 5 | 5 | 5 |
| 6 | 10 | 10 | 10 |
| 7 | 1 | 2 | 3 |
| 8 | 2 | 1 | 5 |
| 9 | 6 | 4 | 2 |
| 10 | 3 | 3 | 4 |

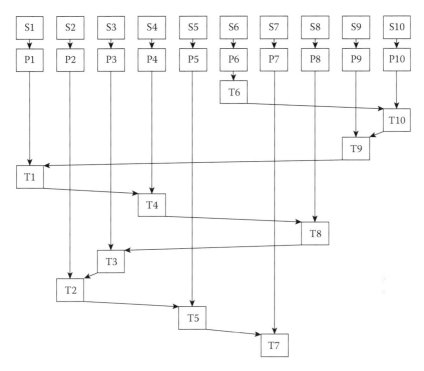

**FIGURE 5.6**
All tasks are scheduled according to ASAP with one transmission accepted by BS at a time.

and *y*-axis represents the sensors' tasks. The three gray-scaled colors (black, light gray and dark gray) represent the sense, process, and transmit tasks respectively.

## 5.6 Conclusion and Future Work

In this chapter, we introduced a novel data-aggregation model for wireless sensor networks. The new model combined all tasks within the wireless sensor network into a STFG, and then we utilized two scheduling algorithms (ASAP and ALAP) to determine the best workload on the base stations.

We are following two tracks in future work. The first track is a theoretical one, which is to find out if our model of data aggregation can be scaled up for a large wireless sensor network. The second track is a practical one, which comprises of experimenting with many base stations with different capacities.

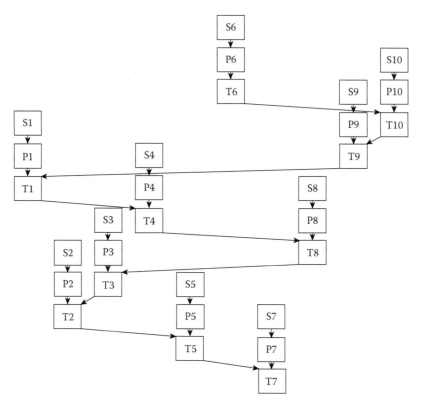

**FIGURE 5.7**
All tasks are scheduled according to ALAP with one transmission accepted by BS at a time.

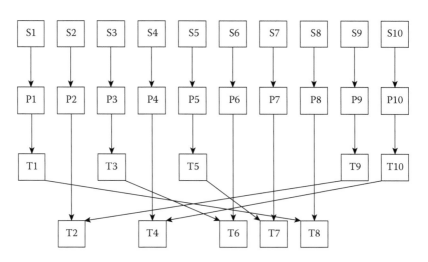

**FIGURE 5.8**
All tasks are scheduled according to ASAP with 5 transmissions accepted by BS at a time.

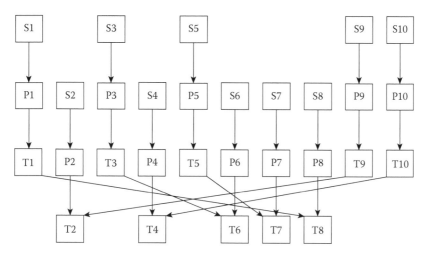

**FIGURE 5.9**
All tasks are scheduled according to ALAP with 5 transmissions accepted by BS at a time.

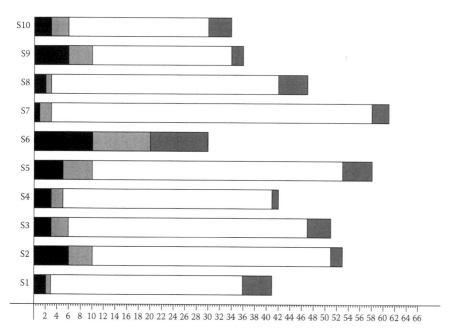

**FIGURE 5.10**
A timing diagram for ASAP schedule for the first experiment.

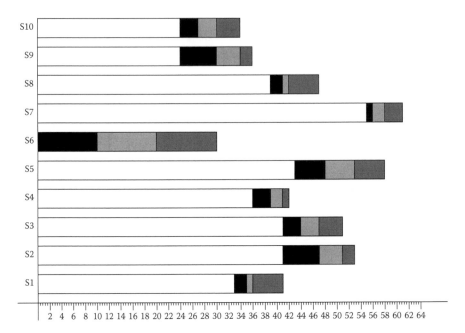

**FIGURE 5.11**
A timing diagram for ALAP schedule for the first experiment.

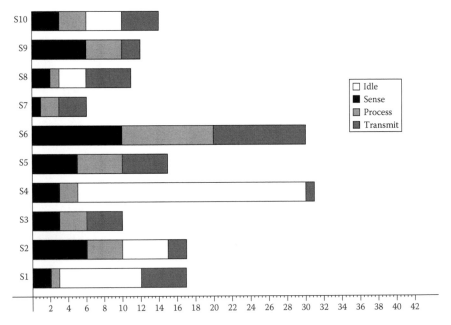

**FIGURE 5.12**
A timing diagram for ASAP schedule for the second experiment.

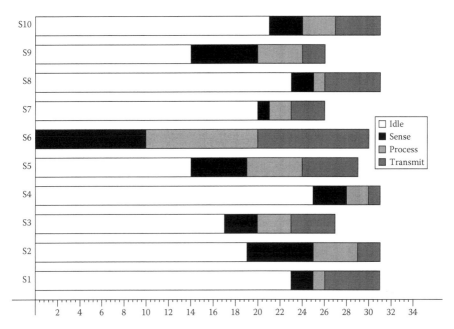

**FIGURE 5.13**
A timing diagram for ALAP schedule for the second experiment.

# References

1. R. Aguilar-Ponce, J. McNeely, A. Baker, A. Kumar, and M. Bayoumi. 2007. Multisensor data fusion schemes for wireless sensor networks. In the Proceedings of the IEEE International Conference on Application-Specific Systems, Architectures and Processors, 136–41.

2. A. Andrews and M. Grewal. 2001. *Kalman Filtering: Theory and Practice Using Matlab,* 2nd ed. New York: John Wiley & Sons.

3. F Argenti and L. Alparone. 2000. Filterbanks design for multisensor data fusion. IEEE *Signal Processing Letters* 7 (5): 100–103.

4. F. Bennett, D. Clarke, J. Evans, A. Hopper, A. Jones, and D. Leask. 1997. Piconet: Embedded mobile networking. *IEEE Personal Communications* 4 (5): 8–15.

5. L. Bodrozic, D. Stipanicev, and D. Krstinic. 2007, October 8–11. Data fusion in observer networks. In the Proceedings of the 4th IEEE International Conference on Mobile Ad-hoc and Sensor Systems, Pisa, Italy.

6. J. C. Chen and K. Yao. 2002, March. Source localization and beamforming. *IEEE Signal Processing Magazine,* 30–39.

7. D. Gajski, N. Dutt, A. Wu, and S. Lin. 1991. *High-Level Synthesis.* Boston, MA: Kluwer Academic Publishers.

8. D. Hall and J. Llinas. 1997. An introduction to multisensor data fusion. *Proceedings of the IEEE* 85 (1): 6–23.

 9. P. Horowitz. 1996, November. New Technological Approaches to Humanitarian Demining. Report JSR-96-115, JASON, The MITRE Corporation, McLean, Virginia.
10. X. Li, Y. Zhu, and C. Han. 2000, July. Unified optimal linear estimation fusion part I: Unified models and fusion rules. In the Proceedings of the International Conference on Information Fusion, 1–8.
11. H. Luo, J. Luo, Y. Liu, and S. Das. 2006, October. Adaptive data fusion for energy efficient routing in wireless sensor networks. *IEEE Transactions on Computers* 55 (10): 1286–99.
12. U. Okorafor and D. Kundur. 2005. Efficient routing protocols for a free space optical sensor network. In the Proceedings of the 2nd IEEE International Conference on Mobile Ad Hoc and Sensor Systems, Washington DC.
13. P. Smets 2000, July. Data fusion in the transferable belief model. In the Proceedings of the 3rd International Conference on Information Fusion, vol. 1, PS21–PS33.
14. F. Stajano and A. Jones. 1998. The thinnest of clients: Controlling it all via cellphone. *Mobile Computing and Communications Review* 2 (4): 1–8.
15. T. Wimalajeewa and S. Jayaweera. 2007. PSO for constrained optimization: Optimal power scheduling for correlated data fusion in wireless sensor networks. In the Proceedings of the 18th IEEE International Symposium on Personal, Indoor and Mobile Radio Communications, Athens, Greece.
16. Q. Wu, N. Rao, J. Barhen, S. Iyengar, V. Vaishnavi, H. Qi, and K. Chakrabarty. 2004, June. On computing mobile agent routes for data fusion in distributed sensor networks. *IEEE Transactions on Knowledge and Data Engineering* 16 (6): 740–753.

# Part II

# Ubi-Media Middleware

# 6

## The Research of a Multiplayer Mobile Augmented Reality (MiMAR) System and Its Applications

Jiung-Yao Huang, Ming-Chih Tung, and Chung-Hsien Tsai

### CONTENTS

This chapter presents a research that enables the user to join a networked virtual environment with a wearable computer. Different from the notebook computer, the wearable computer is the computer system that allows the user to operate it while moving. Consequently, the legacy keyboard and mouse are not feasible input devices for the wearable computer to interact with the networked virtual environment. To navigate the virtual world other positioning devices (such as the GPS receiver is often used as the outdoor positioning device) are required for the wearable computer user. However, the external positioning device will cause temporal and spatial synchronization issues between the wearable computing system and the conventional networked virtual reality (VR) system. This paper investigates these issues and presents solutions to them. The tips to implement such a system are also given. The study shows linking up the wearable computing system and network virtual environment as the augmented reality environment. The system supporting this environment is called a multiplayer mobile augmented reality (MiMAR) system. Finally, an ongoing 3-D virtual campus project is given at the end to demonstrate the result of this research.

## 6.1 Introduction

Pervasive computing (PvC) has already become the computing trend of the next century and, consequently, the research of a wearable computer has attracted significant attention. The famous propaganda slogan of the Nokia Company–Connecting People (science and technology come from human nature all the time) proclaims the interaction among the scientific, techno-logical products and people. Due to the immature development of the scien-tific and technological products in the past, there were passive operational relations between the computer and people. With the progressive improve-ment of relevant technology in the semiconductor, the trend of a computer becomes smaller, cheaper, and more quickly, even more intelligent [1]. Steve Mann [2] considered mobile computing "it belongs to the personal space, controlled by the users, and operated interdynamic continuation at the same time (constancy)." In other words, it is always on and always accessible. The computer that supports this characteristic is called a wearable computer.

This chapter presents research to couple the networked virtual environment and mobile augmented reality into a multiplayer mobile augmented reality (MiMAR), so that all the distributed users of the same physical environment can increase interaction with each other. That is, a user of the MiMAR sys-tem can become aware of and interact with other MiMAR users in the same physical environment even though they are visually obstructed from each other. With the MiMAR system, the wearable computers are continuously exchanging messages on the background that allow you to unobstructedly collaborate with a long distance partner as if they were next to you.

## 6.2 Related Works

The major technologies to achieving a ubiquitous network virtual environ-ment are the network virtual reality, PvC, and mobile augment reality. The framework of these technologies are the sensor network, the seamless com-munication, and the wearable computing. This section will briefly outline the previous research focused on these issues.

Recent research on seamless networks [3,4] proposed a "fast hand off" and a "smooth hand off" model to accommodate mobility across heterogeneous networks that use standard protocol (i.e., IP). These researches attempt to seamlessly and automatically hand off the user information from one access point to another. They also minimize handoff-related delays and packet loss, so that the network session can be persistent among the different access points. An important application of the seamless communication technology is the sensor network environment. Ardizzone et al. [5] successfully deploys large amount of powerful wireless sensor networks and uses the standard IEEE 802.11 (Wi-Fi) radio channel to monitor transient events in physical environments. Ryan et al. [6] use high level control to cooperate several small unmanned aerial vehicles (UAVs) with sensors to form a mobile sensor network for remote sensing. Their research goal is to execute tasks such as obtaining sensory measurement, border patrol, search and rescue, surveillance, communications relaying, and mapping of hostile territory over a large area. Akyildiz and Kasimoglu [7] proposed a wireless sensor and actor networks (WSANs) to achieve seamless communication among distributed static sensor networks. The WSAN is a group of sensors and actors linked by wireless media to perform distributed sensing and actuation tasks. In such a network, sensors gather information about the physical world, while actors make decisions and then perform appropriate actions in the environment, which allows a user to effectively sense and act at a distance.

The wearable computing is the last ingredient of a ubiquitous network virtual environment. As Starner pointed out [8], a wearable computer should possess the following features: portable while operational, hands-free use, sensors, proactive, and always on, always running. Among these features, "portable while operational" is the essential distinction between the wearable computer and the legacy computer, such as a desktop computer and the laptop. The wearable computer can be used while the wearer is on the move. The hands-free requirement focuses on how to avoid holding any input device while in operation. However, some wearable computing applications may use special input devices that are tied to a user's hands, such as chording-keyboards, dials, and joysticks. For example, the Australian Institute of Marine Science (AIMS) designed a so-called WetPC [9] underwater computer to explore how the divers perform their underwater activities. To facilitate the divers to manipulate the WetPC, a five-button KordPad was specially designed for their use. The wearable computer may be equipped with sensors to perceive surrounding information for the user. For example, DeVaul et al. [10] designed the MIThril 2003 wearable computing research platform equipped with a wide range of body sensors, such as pulse oximetry, respiration, blood pressure, EEG, blood sugar, humidity, and $CO_2$ sensors, for example, to monitor the wearer's physiological status for biomedical or social network research applications. In 1998, IBM-Japan announced a wristwatch computer, called a Linux Watch [11], which looks like an ordinary watch but uses Linux X11 graphics, and offers Bluetooth wireless connectivity. The Linux watch also embodies accelerometer and vibrator sensors to detect the wearer's motion. Bauer et al. presented a collaborative wearable system called NETMAN [12,13], with remote sensing capability for network maintenance. Their work has proven that remote sensing capability can significantly improve communication and cooperation of highly mobile computer technicians.

In this chapter, a user who joins the virtual environment through his desktop computer is called the desktop player. Meanwhile, the mobile player is the user who uses a wearable computer or a notebook to wirelessly login to the virtual environment and to interact with others while he is moving in the physical world. In the following sections, the feasibility of allowing the wearable computer user to join the network virtual environment is discussed based on the background technologies of the MiMAR system. The conceptual architecture of the MiMAR system along with research issues are then followed in the next section. Finally, the implementation of a prototyping MiMAR system is given in the last section.

## 6.3 Background Technologies

Virtual reality (VR) is the technology that allows a user to interact with a simulated synthetic world generated by computer graphics. Cooperative virtual reality (CVE) is the extension of VR technology by using the network to connect multiple VR systems into an unified virtual environment. This network and computer-generated synthetic environment is aimed to enable geographically distributed users to be immersed in the same virtual space and to interact with each other in real time. The study of CVE is a popular research topic since 1983. However, only a few application areas gained successful breakthrough achievement, such as Massively Multiplayer Online Game (MMOG) [14] and military simulation [15]. One of the key reasons is that players of current NVE systems are required to sit in front of a computer to participate in the activities and such limitation significantly reduces applications of NVE. Along with the development of computing technologies presently, people are currently demanding to interact on computers anytime and anywhere. This demand brings up the evolution of PvC [16] or ubiquitous computing [17].

The original goal of PvC is to embed network computing nodes throughout the physical environment to provide sensible computing service to the user. In other words, PvC attempts to augment the man-machine computation to provide seamless computing anywhere and anytime. The PvC encompasses a wide range of research topics including distributed computing, mobile computing, sensor networks, human-computer interaction, and artificial intelligence. Furthermore, the research of mobile computing combined with VR technology is called mobile augmented reality (MAR) [18].

Mobile augmented reality is the integration of augmented reality and mobile computing systems into one. The significance of MAR is to augment the capability of mobile computing by adding friendly interface from augmented reality technology. Or, we may say that MAR frees the legacy augmented reality from the desktop with the help of mobile computing. Mobile augmented reality allows 3-D graphics to be overlapped with real images while the user is navigating inside of the real world. Among the researches of mobile augmented reality, MARS [18] and ARQuake [19] are two well-known systems. Those two systems allow users to interact in an environment with a "pervasive" graphical user interface. That is, the user can see a 3-D object on top of, say, any wall within his FOV, without actually installing a screen on that wall. This capability promotes the augmented reality to a more friendly and daily application. However, one drawback of mobile augmented reality is that it still isolates the user from each other. The user can only interact with the environment alone and unknown to the existence of other players. That is, it lacks the interoperability among users of mobile augmented reality system. With such capability, users of mobile augmented reality systems can perform social interaction such as playing a team game or performing a work collaboration [20].

## 6.4  The Concept of Multiplayer Mobile Augmented Reality (MiMAR)

MiMAR is designed to enable the player of the mobile augmented reality system to join the conventional network virtual environment. To achieve this goal, the MiMAR architecture is composed of two types of servers. One is the multiuser server, called the game server, and the other is the mobile support server (MSS). The MiMAR is the study of adding multiplayer capability to MAR. The basic idea of MiMAR is integrating NVE with MAR to enable the distant users of MAR systems to interoperate with each other.

### 6.4.1  The Server Architecture

Similar to the design goal of CVE [21], the MiMAR environment aims to provide an illusion to the distributed users that they all observe the same things and interact with each other within the virtual space. This goal implies that each player has to share his status and event with others within the virtual space and for them to realize his existence by a graphic rendering engine. This dynamic state sharing is achieved by message exchange among players of the virtual world. There are three general approaches to implement this message exchange mechanism among hosts of a network virtual environment. They are centralized repository, frequent state regeneration, and dead reckoning [21]. Due to the instability of the wireless signal, the frequent state regeneration approach can easily flood the wireless bandwidth on the mobile host. Further, the instability of the wireless signal may also cause the dead reckoning method to produce jittering remote avatars on the mobile host. Therefore, by considering the computing resource of the mobile player, MiMAR adopts the centralized repository approach to design its server system. In addition, the implementation of the centralized repository can be broadly classified into three techniques including the file repository, the repository in server memory, and the virtual repository [21]. Due to the latency of the instability of the wireless signal, the transmission overhead induced by the server on the mobile host has to be minimized. Hence, the method of repository in server memory is implemented to achieve prompt message exchange between the desktop players and the mobile players.

The use case diagram of UML illustrates the MiMRA architecture (Figure 6.1). The MiMAR architecture can be roughly divided between the server side and the client side. The server side contains the login server and the game server; whereas the client side includes the login agent, a 3-D virtual space scene file, and the application agent. The user has to register a user name with a password to the login server before being allowed to join the 3-D virtual space. After successful registration, the login server will validate the user information whenever a player wants to enter the 3-D virtual space. The game server is responsible for exchanging dynamic shared states and events between the desktop and the mobile players. On the client side, the login agent provides an interface for the user to register and login to the 3-D virtual space. For each registered player of the 3-D virtual space, the login server will return a validation key to the login agent. This key will then activate the application agent to enter the 3-D virtual space. The application agent then takes over all the interactions until the player leaves the 3-D virtual space.

To reduce the transmission delay between the server host and the client host, the message pushing methodology is realized on the game server. There are two known technologies (i.e., push and pull [22–23]) to implement a message passing mechanism between the server and the client. The pull approach demands the client to specifically request a message from the server; while the push method allows the server periodically and automatically to send a

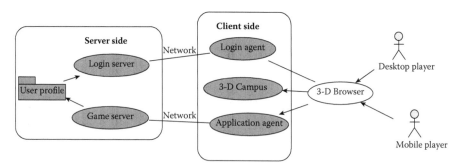

**FIGURE 6.1**
The use case diagram of the MiMAR system.

message to the client. Another benefit of the push approach is its low over-head on the client host. Further, this pushing method allows the server to control and filter message flows to each client host.

Finally, in order to reduce the bandwidth consumption, the game server is implemented with a packet aggregation method. The packet aggregation functions to accumulate several packets into one before messages are trans-mitted and is able to reduce the number of headers on each transmission [24]. According to the study from Singhal and Zyda [21], this savings can achieve 50% of the required bandwidth depending upon different types of network virtual environments. There are three aggregation policies generally consid-ered: timeout-based transmission policy, quorum-based transmission policy, and hybrid transmission policy [21]. Considering the wireless bandwidth and latency, the timeout-based transmission policy was realized on the game server. In summary, to reduce the transmission time and bandwidth con-sumption at the same time, the game server periodically collects messages from clients and then pushes aggregated packets to each client.

### 6.4.2 The Mobile Support Server

The most challenging issue of designing the MiMAR system was to enable the mobile player to interact with other players in the same virtual world. Due to the signal instability and the limited bandwidth of the wireless net-work, messages from the server may easily jam the wireless network band-width if the mobile player is directly connected to the game server. Further, if the mobile player is directly receiving a message from the game server, the limited computation power of the mobile device will be unable to proceed with interactive messages as the number of simultaneous players increases. To solve these problems, a MSS is designed as a data mediator between the game server and the mobile player as shown in Figure 6.2. The MSS prin-cipally adjusts the message flow between the game server and the mobile player.

In addition to the prior research issues of the network virtual environment [21], there are still other issues needed to be solved before the mobile player can interact with a shared virtual space. These issues can be broadly classi-fied into three categories. One is the instability and bandwidth problem of the mobile network. The second issue is the limited computation resources of the mobile device. The last one is the data correlation between the geo-graphical coordinate system and the Cartesian coordinate system [25]. These three problems further lead to more subproblems as depicted in Figure 6.3.

- Mobile networking: The ultimate goal of MiMAR is to allow both desktop players and mobile players to interact with each other in the shared space. Since the desktop device is often connected by a fixed location network, there is not much to be discussed due to signal stability of such a network. On the other hand, the mobile player can

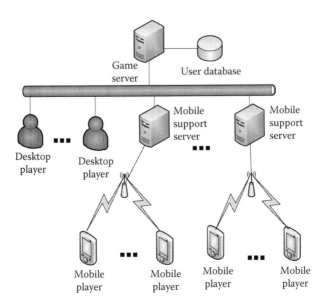

**FIGURE 6.2**
MSS to support mobile players.

only join through the wireless network and interact while he is moving around. The available choices of wireless network include GPRS, 3G, WiMAX, WiFi, and so forth. Due to the fragility of the wireless signal while the user is moving, the issue is how to keep a logical link between the mobile player and the server alive while the signal strength is still within an acceptable threshold. However, this chapter does not concentrate on the network technologies of any specific wireless network, but only the logical linking problem is explored. The MSS periodically detects the connecting status to re-login the mobile player or dead reckoning missing data.

- Data filtration: The mobile player uses a mobile device, such as a notebook or a wearable computer, to navigate the 3-D virtual space. Due to the limitation of power consumption, the mobile computer has inferior computing resources and insufficient network bandwidth. For this reason, the issue focuses on how to filter messages to reduce the computing load of the mobile device. To achieve this purpose, part of rendering related computing work is shifted to the MSS, such as the visibility of objects, the level of detail of object appearance, and the realism of object animation.

- Data mediation: Since the mobile player uses the location sensing device (such as GPS) to navigate the virtual world and animate his remote avatar, this location information has to be converted into the

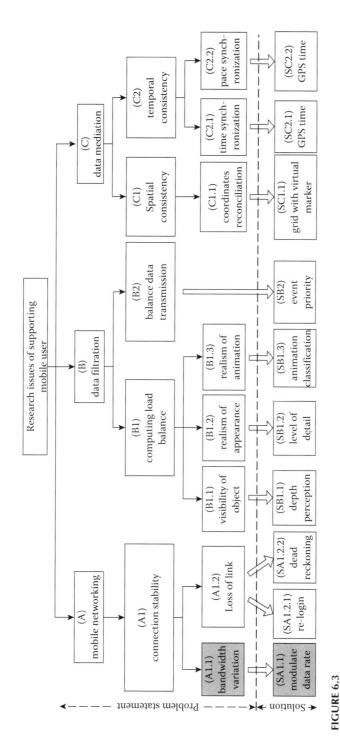

**FIGURE 6.3**
The MSS research issues.

position data recognizable by the virtual world. However, the data from the GPS is expressed in a geographical coordinate system, whereas the virtual world uses Cartesian coordinate system, this conversion may cause spatial and temporal inconsistency between the mobile player and the desktop player. Additionally, because the mobile player is moving in the physical world, the difference of these two coordinate systems may further corrupt the causal order relationship of events within the shared virtual world. As a result, the last concern is how to harmonize coordinates and events between these two coordinate systems. The MSS uses the GPS time to synchronize messages between the mobile player and itself. To the advanced state, the grid approach to translate data between the two coordinate systems as discussed previously is implemented in MSS.

The following section will present a typical application of MiMAR system—a network augmented reality campus in National Taipei University (NTPU), Taiwan.

## 6.5 Three-Dimensional Virtual Campus

Other than the interaction issues discussed in the previous session, the implementation approach will also affect the performance of the MiMAR environment. The most straightforward implementation allows the mobile device to communicate with the multiuser server directly and to implement a proposed solution on the multiuser server. However, this type of implementation will overload the server and, then, decrease the performance of the MiMAR environment. To avoid this problem, as illustrated in Figure 6.4, a mediate server called a MSS was designed to realize the proposed solutions. The MSS performs as a data mediator between the multiuser server and the mobile player. For the multiuser server, MSS is a desktop player that controls multiple avatars inside the virtual world. On the other hand, the mobile player treats the MSS as a special purpose server that shares some of its computation load.

The MSS is implemented on the Windows XP platform. To further verify the validity of the proposed solutions, the MSS is realized in the existing networked virtual environment called NTPU 3-D virtual campus [26]. The 3-D virtual campus is a network virtual environment of National Taipei University (NTPU), Taiwan. By adding MSS, both the mobile players and desktop players are allowed to join this 3-D campus environment. The position of the mobile player is derived by the GPS receiver. The mobile device (i.e., a wearable computer or a notebook) will translate and transmit its received GPS data to the server. The server then will forward the received data to other players for

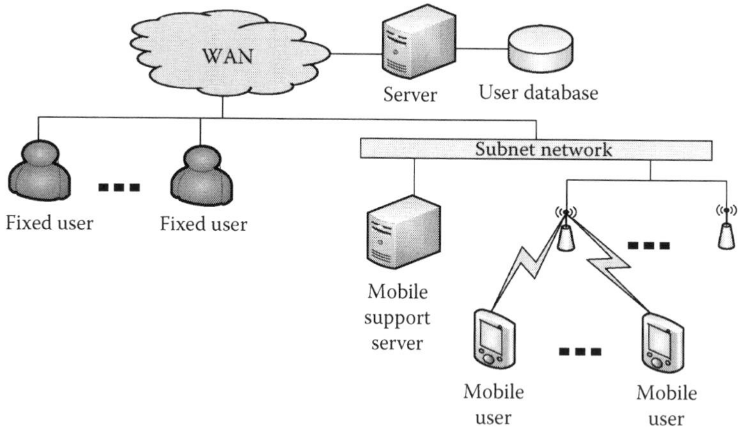

**FIGURE 6.4**
The architecture of the M³R environment.

them to remote render his avatar. Consequently, the NTPU 3-D campus project allows more vivid interactive experiences when the user is navigating this
overlapped virtual and physical world. The movement of the mobile player
changing with the VR picture is shown in Figure 6.5.

Voice over IP capability is additionally embedded in the 3-D virtual campus to empower live chatting among players. When a user wants to chat with
another player on the user list, he can directly click that player's name. The
system will then launch the Skype [27] software to connect to that specific
player. For example, if the user "sennin32" wants to voice chat with "annhei-
long," for example, he can click the receiver's name "sennin32" on the right
subwindow or on top of the avatar. As shown in Figure 6.6, a calling notification will pop-up on the receiver's browser. The receiver can decide whether
to accept or deny this call by clicking buttons on the pop-up window.

The NTPU 3-D virtual campus demonstrates a new application of MAR
research. This exposition shows a promising direction of integrating PvC and
CVE with MAR. The proposed system structure combines the advantage of
WLAN, wired network, and the polling mechanism to achieve a pervasive,
proactive, mobility, and wearable virtual campus environment.

## 6.6 Conclusion

This chapter studies the techniques to merge mobile computing into the
network virtual environment. The integrated environment is referred
to as MiMAR and it enables the user to wear a mobile device to interact
with the conventional desktop player in the shared virtual space. Further,

**FIGURE 6.5**
Snapshot of the mobile device.

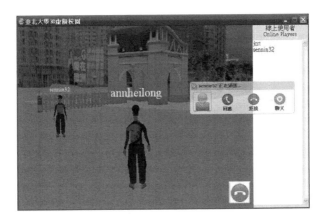

**FIGURE 6.6**
Live chatting through Skype software.j

an architecture to support the MiMAR environment is presented in this chapter. To fully sustain the mobile player, MiMAR includes a MSS as the data mediator between the mobile player and CVE. The MSS is designed to reduce the computational load on the mobile device while the mobile player is interacting with the mixed reality environment. Three resource-related issues of MSS are pointed out: mobile networking, data filtration, and data mediation.

Although this chapter demonstrated the possibility of designing a network mixed reality environment for both the desktop player and the mobile player, more advanced research issues are exposed for further studies. The ultimate goal of the mobile device is an optical see through mobile augmented reality system running on a wearable computer. The rotation of the mobile player that changes the corresponding 3-D display of the virtual space is another important issue. In a MiMAR system, the digital compass is used to detect the rotation of the mobile player. The digital compass has a well-known inaccurate and unstable problem. Conclusively, other auxiliary direction-sensing devices require more progressive exploration. Currently, the research for using the optical flow technology to detect the mobile player's rotation is under investigation. Finally, the performance of the MSS is another important topic that deserves further exploration. Experiments to explore this issue are underway currently.

# References

1. B. J. Rhodes. 1997, October. The wearable remembrance agent: A system for augmented memory. Paper presented at the First International Symposium on Wearable Computers, Cambridge, MA, 123–28.
2. S. Mann. 1998, November. Humanistic intelligence: WearComp as a new framework for intelligent signal processing. *Proceedings of the IEEE* 86 (11): 2123–51.
3. A. Mohammad and A. Chen. 2001, November 25–29. Seamless mobility requirements and mobility architectures. Global Telecommunications Conference, GLOBECOM 2001. *IEEE* 3, 1950–56.
4. N. Badache and D. Tandjaoui. 2002, September 9–11. A seamless handoff protocol for hierarchical Mobile IPv4. 4th International Workshop on Mobile and Wireless Communications Network, 651–55.
5. E. Ardizzone, M. La Cascia, G. Lo Re, and M. Ortolani. 2005. An integrated architecture for surveillance and monitoring in an archaeological site. Proc. ACM Workshop on Video Surveillance and Sensor Networks, 79–86.
6. A. Ryan, M. Zennaro, A. Howell, R. Sengupta, and J. K. Hedrick. 2004, December 14–17. An overview of emerging results in cooperative UAV control. In Proc. IEEE Conf. on Decision and Control, Atlantis, Paradise Island, Bahamas, 6027.
7. I. F. Akyildiz and I. H. Kasimoglu. 2004, October. Wireless sensor and actor networks: Research challenges. *Ad Hoc Networks Journal* 2 (4): 351–67.
8. T. Starner. 2001, July–August. The challenges of wearable computing: Part 1 and 2. *Micro, IEEE* 21 (4): 44–67.
9. WetPC. 1997–2005. Available at http://wetpc.com.au/html/technology/wearable.htm
10. R. DeVaul, M. Sung, J. Gips, and A. Pentland. 2003. MIThril 2003: Applications and architecture. The 7th IEEE International Symposium on Wearable Computers (ISWC'03), 4–11.
11. C. Narayanaswami, N. Kamijoh, M. T. Raghunath, T. Inoue, T. M. Cipolla, J. L. Sanford, et al. 2002, January. IBM's Linux watch: The challenge of miniaturization. *IEEE Computer* 35 (1): 33–41.

12. M. Bauer, T. Heiber, G. Kortuem, and Z. Segall. 1998, October 19–20. A collaborative wearable system with remote sensing. Proceedings Second International Symposium on Wearable Computers (ISWC'98), Pittsburgh, PA, 10–17.

13. R. W. Brodersen and B. C. Richards, 1995, April. InfoPad: The Design of a Portable Multimedia Terminal. Proceedings 2nd International Workshop on Mobile Multi-Media Communications, Bristol University, Bristol, England.

14. M. R. Minw, J. Shochet, and R. Hughston. 2003, October. Building a massively multiplayer game for the million: Disney's Toontown Online. *Computer in Entertainment* 1 (1): 15–35.

15. IEEE Std. 1516.1. 2000. IEEE Standard for Modeling and Simulation (M&S), High Level Architecture (HLA)–Federate Interface Specification, i–467.

16. D. Saha and A. Mukherjee. 2003, March. Pervasive computing: A paradigm for the 21st Century. *IEEE Computer,* 25–31.

17. M. Weiser. 1993, July. Some computer science problems in ubiquitous computing. *Communications of the ACM,* 75–84.

18. T. Höllerer, S. Feiner, T. Terauchi, G. Rashid, and D. Hallaway. 1999, December. Exploring MARS: Developing indoor and outdoor user interfaces to a mobile augmented reality system. *Computers and Graphics* 23 (6):779–85.

19. W. Piekarski and B. Thomas. 2002. ARQuake: The outdoor augmented reality gaming system. *Communications of the ACM* 45 (1): 36–38.

20. G. Kortuem, M. Bauer, T. Heiber, and Z. Segall. 1999. Netman: The design of a collaborative wearable computer system. *ACM/Baltzer Journal on Mobile Networks and Applications (MONET),* 4(1): 49–58.

21. S. Singhal and M. Zyda. 1999. *Networked Virtual Environments Design and Implementation.* Boston, MA: Addison Wesley.

22. N. Nakamura, K. Nemoto, and K. Shinohara. 1994, October. Distributed virtual reality system for cooperative work. *NEC Research and Development* 35 (4,): 403–9.

23. V. Anupam, C. Bajaj, and D. Schikore. 1994. Distributed and collaborative visualization. *IEEE Computer,* 27(7): 37–43.

24. S. K Singhal. 1996. Effective remote modeling in large-scale distributed simulation and visualization environments. PhD thesis, Stanford University, CA.ftp://reports.stanford.edu/pub/cstr/reports/cs/tr/96/1574/CS-TR-96-1574.pdf

25. J. Y. Huang, M. C. Tung, H. C. Keh, J. Wu, K. H. Lee and C. H. Tsai. 2009, July. Interaction wearable computer with networked Virtual Environment, Human-Computer Interaction, Part III, HCII 2009, LNCS 5612:741–751.

26. J.Y. Huang, M. C. Tung, H. C. Keh, J. J. Wu, K. H. Lee and C. H. Tsai. 2009. A 3D Campus on the Internet – A Networked Mixed Reality Environment. Transactions on Edutainment II, LNCS 5660, Springer: 282–298.

27. Skype. 2009. Available at http://www.skype.com

# 7

## Finger Gesture Interaction on Large Tabletop for Sharing Digital Documents among Multiple Users

Yue Shi, Chun Yu, and Yuanchun Shi

### CONTENTS

Sharing digital documents among multiple users on a large tabletop is one of the typical application scenarios for tabletop systems. In this chapter, we elaborate on the guidelines of designing a user interface for this application scenario. A set of finger gestures are then defined within the interface that includes single finger operation for selecting, moving, and delivering a digital document among multiple users; double finger operation for resizing and rotating a digital document; trajectory operation for creating and deleting a user work area; and creating, deleting, and copying a digital document. A prototype tabletop application called JuTable is developed to explore these gestures. A discussion on learnability, usability, and the naturalness of our interface design for this application scenario is last.

## 7.1 Introduction

Tabletop systems augment information display and direct manipulation onto conventional table functions. The tabletop surface can create a virtual environment for users based on computer vision and sensing techniques; tabletop systems can provide users with a more natural way (such as touching and writing with a pen) to interact with the virtual environment. With the development of technology, the tabletop system will have a wide range of uses and be beneficial.

One of the typical application scenarios of the tabletop is sharing documents among multiple users. Compared to sharing photographs on relatively small tabletops, sharing document applications requires larger tabletop surfaces that makes some techniques no longer available. Though some research [9–11] has been done on sharing photographs and other application scenarios, the solution cannot be used to solve the problem of sharing digital documents among multiple users on the large tabletop. Though many interactive techniques on a tabletop have been proposed [4–8], no one has systematically designed an interface for this application.

Our goal is to give a solution to this application scenario. We explore guidelines for the design of an interface, list what kinds of finger gestures are required; and develop a prototype application, JuTable, to evaluate our ideas. In our design, finger trajectory operation techniques are used for complicated finger gestures as we find that they provide a more natural way for users to interact with the tabletop systems. Specifically, we design a radar map widget to facilitate delivering documents.

Section 7.2 is an overview of the related work. In Section 7.3, after analyzing the scenario of sharing documents on a large tabletop, we propose the user interface design guidelines and then define a set of gestures that the interface would support under the guidelines. Section 7.4 describes the implementation of the finger gestures and explores them in the prototype application, JuTable. In Section 7.5, we collect the users' feedback and have a discussion on the learnability, usability, and the naturalness of our interface design. Finally in Section 7.6, we summarize our work and give some suggestions on future research.

## 7.2 Related Works

There are many tabletop systems in the research world such as DiamondTouch [1], SmartSkin [2], and Microsoft Surface [3]. Based on these tabletop systems, many studies on designing freehand gestures have been carried out, including cooperative gestures [4], multifinger and whole hand gestural interaction techniques [5], interaction techniques for transition in the status of an electronic document [6], bimanual continuous gestures [7], and rotation and translation techniques [8].

As for tabletop applications, solutions have been proposed for some application scenarios, for example, story sharing around the table [9]; tabletop sharing of digital photographs for the elderly [10]; and browsing, sorting, and sharing digital images [11].

Though a lot of work has been done, no one has yet given a complete systematic interface design for the tabletop application of sharing digital documents among multiple users and solving the problems raised by this scenario.

## 7.3 User Interface Design

### 7.3.1 Design Considerations

For designing a user interface, first we should know what the user will do when sharing documents. A group of people want to share documents for a discussion. First, everyone will get his or her own working area on the table. Second, they should be able to operate their own digital documents as needed. Third, during the discussion, users may want to share their documents with others. From the analysis above we can understand the user interface design guidelines for sharing digital documents as follows:

- **Support user working area**
  When a user joins the discussion, he can have his own working area on the table. When he leaves, he can delete his working area.

- **Support single user operation on documents**
  A user can read, annotate, and manipulate the digital documents (such as moving, resizing, and rotating).

- **Support documents sharing operation among multiusers**
  A user can deliver his own document to others when he wants to share his document.

### 7.3.2 Gestures Needed

According to the guidelines listed above, a set of interaction gestures are proposed:

- Create/Delete user working area, Select, Create, Copy, Annotation
- Delete, Move, Resize, and Rotate digital document
- Deliver document to other working areas

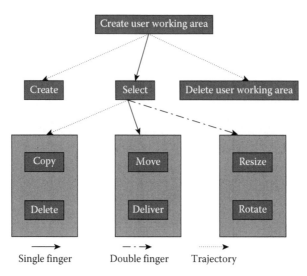

**FIGURE 7.1**
Finger gesture set.

Among these gestures, annotating a document is better implemented with a pen. Other 10 gestures can be seen as finger gestures. As for a user, the finger is the most natural and convenient way to manipulate documents. These 10 gestures also have a dependent relationship with each other. Some gestures must be done before others. Figure 7.1 shows the set of gestures and their temporal relationship represented by the arrows. For example, creating a user working area is the basic gesture of them. After that, three gestures: creating a document, selecting a document, and deleting a user work area can be done. The other six gestures: copying, deleting, moving, delivering, resizing, and rotating a document, can only be done after selecting the document.

## 7.4 A Prototype Application

In this section we will describe a prototype application, JuTable. The application is based on the tabletop system uTable developed by our lab.

### 7.4.1 uTable

The uTable is a multipoint, direct-pen-input tabletop system that can be expanded to be touch sensitive. Figure 7.2 shows what the uTable looks like. The size of a single table is 100 cm × 75 cm × 80 cm. Multiple uTables can be seamlessly combined into a larger one. This is considered a four-in-one table. There is a projector, an infrared camera, and infrared light bulbs in the

**FIGURE 7.2**
uTable tabletop system.

uTable. Infrared light bulbs are set to extend the laser pen input to be finger touch sensitive as infrared light can be reflected from the surface when a finger touches on the tabletop. The photo taken by the infrared camera is disposed to determine the location of the fingers touching on the surface.

### 7.4.2 JuTable

JuTable is the tabletop prototype application for sharing documents by multiple users around a table.

We tried to reduce the learning affordance to the user and achieve the cooperative work in a natural way. To make the virtual environment similar to the real world, some GUI elements such as buttons and tool bars are ignored.

The JuTable supports finger gestures in three ways: single finger operation, double finger operation, and finger trajectory operation. As shown in Figure 7.1, each gesture is pointed to one of the three kinds of arrows according to its operation type. The solid line represents the single finger operation, the dotted line represents the finger trajectory operation, and the dashed line represents the double finger operation.

Figure 7.3 shows the software interface of JuTable. Each block is a user working area. Users operate the digital documents in their own working area. The tiny grids in the working area is a radar map widget for users to deliver documents to others.

#### 7.4.2.1 Single Finger Operation

Single finger operation requires the user to put a single finger on the surface, move the finger, and lift it up. This operation can be used to accomplish three gestures: Select, Move, and Deliver.

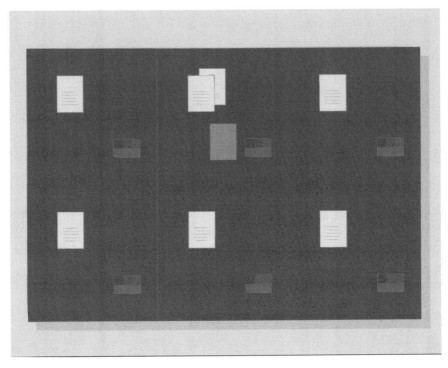

**FIGURE 7.3**
The software interface of JuTable.

- **Select**
  The user selects a document by touching it with a single finger. Then the color of the document will get darker. Figure 7.4a shows the Select gesture.

- **Move**
  To move a digital document, a user should first touch the document, then drag the document on the surface, and last lift up the finger and complete the gesture. Figure 7.4b shows the Move gesture.

- **Deliver**
  On a large tabletop, it is very hard to touch the user seated far away, so users cannot deliver a document hand to hand. Throwing the document on the table will face the problem of inaccuracy. To deliver a digital document to the other user, we use the working area mapping method. In each user's working area, there is a radar map widget mapping all the working areas on the table. See the tiny grids in Figure 7.5b. When one user, A, wants to deliver a document to another user, B, he can move the document and stop at the grid in the map representing the working

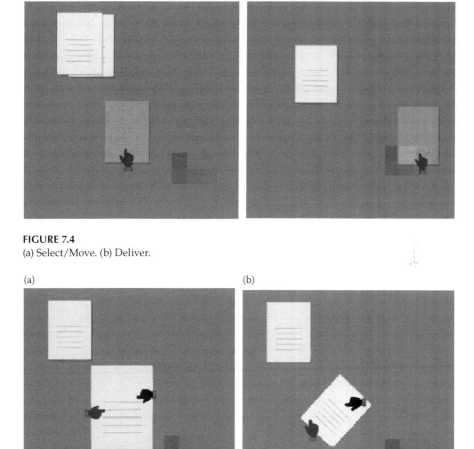

**FIGURE 7.4**
(a) Select/Move. (b) Deliver.

**FIGURE 7.5**
(a) Resize. (b) Rotate.

area of B, and then the document will be delivered to user B's working area. Using this method, a user can deliver documents accurately to any user he wants without moving out of his own working area.

### 7.4.2.2 Double Finger Operation

Double finger operation means putting two fingers on the surface, moving, and lifting up. Meanwhile, the distance and angle between the two fingers will be changed. The change of the distance and angle will affect the document's size and angle. This operation can be used to accomplish Resize and Rotate gestures.

- **Resize**
  To resize a document, a user should touch it using two fingers and change the distance between them; correspondingly, the size of the document will be changed. Figure 7.5a shows the Resize gesture.

- **Rotate**
  To rotate a document, a user should touch on it using two fingers, hold on one and move the other, and then the document will be rotated. Figure 7.5b shows the Rotate gesture.

### 7.4.2.3 Finger Trajectory Operation

The five remaining gestures are more complicated than the others, which cannot be realized in the real physical world. Here we use the trajectory operation to accomplish these five gestures. A user can use his finger to draw some lines on the surface, and then some special trajectory patterns will be recognized by the software and be comprehended as finger gestures. Generally speaking, users will feel more comfortable and natural to interact with the tabletop by drawing lines than using the hand shape or pushing virtual buttons. So we choose to use the finger trajectory operation to accomplish the five remaining gestures.

There are three special trajectories defined by the software: One Line, Two Parallel Lines, and Two Cross Lines all of these meanings are obvious. Figure 7.6 shows the three symbols.

We leverage the automata principle to recognize the trajectory patterns. The auto-machine has four states: None, One Line, Delete, and Create. Figure 7.7 shows the auto-machine. When a user draws a line on the surface of the table, the slope and the correlation coefficient of the line will be calculated. Then in the auto-machine an event will be created depending on the begin point, end point, slope, and correlation coefficient of the line. There are four kinds of events including Draw a line, Not a line, Parallel, and Cross.

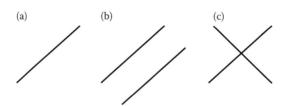

**FIGURE 7.6**
(a) One Line. (b) Two Parallel Lines. (c) Two Cross Lines.

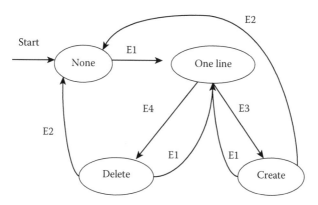

**FIGURE 7.7**
Auto-machine for recognizing trajectories: (E1) Draw a line. (E2) Not a Line. (E3) Parallel. (E4) Cross.

- **Create User Working Area**
  This gesture is the basic finger gesture of the set. When a line is drawn on the blank area of the surface, a user working area will be created. At the same time, the state of the auto-machine of the table will become One Line. Figure 7.8d shows the Create user working area gesture.

- **Delete User Working Area**
  When no document is selected, drawing two cross lines in the user working area will delete it. The state of the auto-machine of the working area will become Delete. Figure 7.8e shows the Delete user working area gesture.

- **Create**
  When no document is selected, drawing two parallel lines in the user working area will create a new digital document. The state of the auto-machine of the working area will become the Create. Figure 7.8a shows the Create gesture.

- **Copy**
  When a document is selected, drawing two parallel cross lines in the user working area will copy the document selected. The state of the auto-machine of the working area will become Create. Figure 7.8b shows the Copy gesture.

- **Delete**
  When a document is selected, drawing two crossed lines in the user working area will delete the document. The state of the auto-machine of the working area will become Delete. Figure 7.8c shows the Delete gesture.

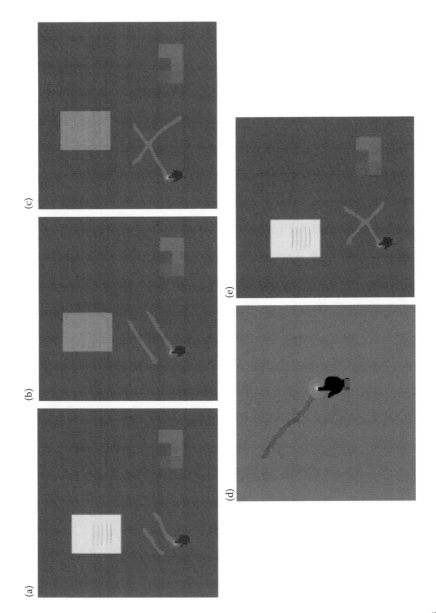

**FIGURE 7.8**
(a) Create. (b) Copy. (c) Delete. (d) Create user working area. (e) Delete user working area.

## 7.5 Discussion

According to the user feedback for this prototype application, we give some comparative discussion to similar work from three aspects: learnability, usability, and naturalness.

- **Learnability**
  To get used to this interface, users only need to remember three symbols to complete the five gestures accomplished by finger trajectory operation. Other gestures are just the same as the gestures people do when they share real documents with others. What is more, the three symbols are only constructed by lines that are easy to remember. Users can learn how to use this interface easily and quickly.

- **Usability**
  According to the three guidelines proposed, the set of finger gestures can satisfy the users' interactive needs. In addition, the working area mapping method designed for delivering a digital document makes this gesture easily and accurately used, especially when the table is very large and not easy for users to hand documents to each other. Throwing the document would be considered inaccurate or imprecise. Users will benefit from the convenience of the interface.

- **Naturalness**
  All of the gestures can be done in the user's own working area, which makes sharing the digital document experience comfortable. As we have discussed above, the gestures accomplished by single and double finger operations are just as what people do in their daily lives with real paper documents. In addition, the symbols used by finger trajectory operations are very simple and in accordance with the users' habits. For example, we use crossed lines to represent the meaning of deleting. Users will feel comfortable and natural using the interface.

## 7.6 Conclusion

In this chapter, we study the typical application scenario of the tabletop system: sharing digital documents on large tabletops among multiple users, give a user interface design using finger gestures, and explore the design in the prototype application, JuTable. Novel aspects of our work are: we propose a systematic design of the user interface for the application scenario we studied. We then use a finger trajectory operation that provides users with a more natural interactive paradigm. We design a radar map widget to deliver the digital document on large tabletops.

We hope our work will provide a reference for designing a user interface for tabletops. In the future, we will design more finger gestures for the convenience of interaction and work out a high-level, user interface design framework for the tabletop system into which our work in this chapter will then be integrated.

## References

1. P. Dietz and D. Leigh. 2001, November 11–14. DiamondTouch: A multi-user touch technology. In Proceedings of the 14th Annual ACM Symposium on User Interface Software and Technology, UIST'01, 219–26, Orlando, FL.
2. J. Rekimoto. 2002, April 20–25. SmartSkin: An infrastructure for freehand manipulation on interactive surfaces. In Proceedings of the ACM CHI 2002 Conference on Human Factors in Computing Systems Conference, CHI2002, 113–20, Minneapolis, MN.
3. Microsoft Surface http://www.microsoft.com/surface/index.html
4. M. R. Morris, A. Huang, A. Paepcke, and T. Winograd. 2006, April 22–27. Cooperative gestures: Multi-user gestural interactions for co-located groupware. Conference on Human Factors in Computing Systems, CHI2006, 1201–10, Quebec, Canada.
5. M. Wu and R. Balakrishnan. 2003. Multi-Finger and whole hand gestural interaction techniques for multi-user tabletop displays. Annual ACM Symposium on User Interface Software and Technology, UIST'03, 193–202, Vancouver, BC, Canada.
6. M. Ringel, K. Ryall, C. Shen, C. Forlines, and F. Vernier. 2004, April 24–29. Release, relocate, reorient, resize: Fluid techniques for document sharing on multi-user interactive tables. Proceedings of ACM CHI 2004 Conference on Human Factors in Computing Systems, CHI2004, 1441–44, Vienna, Austria.
7. M. Wu, C. Shen, K. Ryall, C. Forlines, and R. Balakrishnan. 2006. Gesture Registration, relaxation, and reuse for multi-point direct-touch surfaces. Proceedings of the 1st IEEE International Workshop on Horizontal Interactive Human-Computer Systems, TABLETOP'06, 185–92, Washington, DC.
8. M. S. Hancock, F. D. Vernier, D. Wigdor, S. Carpendale, and C. Shen. 2006. Rotation and translation mechanisms for tabletop interaction. Proceedings of the 1st IEEE International Workshop on Horizontal Interactive Human-Computer Systems, TABLETOP'06, 79–88, Washington, DC.
9. C. Shen, N. Lesh, and F. Vernier. 2003. Personal digital historian: Story sharing around the table. *ACM Interactions* (March–April): 15–22.
10. T. Apted, J. Kay, and A. Quigley. 2006, April 22–27. Tabletop sharing of digital photographs for the elderly. Conference on Human Factors in Computing Systems, CHI2006, 781–90, Quebec, Canada.
11. O. Hilliges, D. Baur, and A. Butz. 2007, October 10–12. Photohelix: Browsing, sorting and sharing digital photo collections. Proceedings of the 2nd IEEE International Workshop on Horizontal Interactive Human-Computer Systems, TABLETOP07, 87–94, Newport, RI.

# 8

## Integrating Spatial Audio, Visual, and Gestural Interfaces for Supporting Situated Collaborative Work

Takeshi Nakaie, Takayuki Koyama, and Masahito Hirakawa

## CONTENTS

Ubiquitous computing, or pervasive computing, is an increasingly popular technique for developing new human-computer interfaces and computer applications. Computers watch what happens in a physical space where people are and serve them with meaningful information. Here, most of the systems rely on a visual media for user navigation, while there are some other interesting interaction media. In this paper we present a collaborative multimodal system toward spatial presence and perceptual realism of sounds. A two-dimensional sound display called Sound Table is its central component, in which 16 speakers are mounted so that all of the users have the sound experience in common. A stick-shaped device is provided as a means for interaction with the system. The system recognizes manipulation of the device in a 3-D space over Sound Table and then gives users feedback through both visual and auditory channels.

## 8.1 Introduction

There is a need for computing systems that naturally support users in their daily life. One of the most promising approaches toward this is ubiquitous computing, which is considered an integration of mobile computing and pervasive computing [1]. In mobile computing, much effort has been devoted to the design and implementation of computing architectures that enable anywhere and anytime computing. Pervasive computing takes another approach aimed at enhancing interaction between the user and the computer, or between users through computer(s), by taking the user's situation into account. Our interest goes toward this pervasive computing aspect.

Most of the ubiquitous (or pervasive) computing systems express information visually. It is obvious that vision plays an important role in interaction but is not the only channel. We focus on audition, which is another important channel through which the user perceives his or her environment.

Several auditory pervasive computing systems have been proposed. We classify those systems into four sections according to their size: gadgets, boxes, tables, and rooms. The first three are analogous to tabs, pads, and boards, respectively, proposed by Weiser [2], while the last one is particular to auditory pervasive computing.

Gadgets are the smallest machines of embodied virtuality [3–6]. The best example would be museum guide systems. The movement of a visitor triggers a guiding message or a sound related to that artifact(s) nearest the visitor. It is mentioned that such auditory information is presented in most cases to each individual visitor through a headphone or earphone they would wear.

Boxes are small tangible computers [7]. They are for the personal use as gadgets are, but differ from gadgets in that their spatial manipulation, such as rotation in a physical space, is allowed.

Tables are three foot in size multimodal computers [8]. Interaction and collaboration with others in a shared physical space are encouraged. They may support a video displaying facility as well, so that users can share their ideas with each other through both visual and auditory channels.

Rooms are indoor, ambient computing machines [7,9,10]. For users, this room-sized interaction space is too large to entirely monitor information. Audio cues to the users stay in the peripheral background to avoid too much attention. In other words, they form an ambient soundscape that supports opportunistic interactions and awareness.

In this paper, we propose a table-type, shared sound display, called Sound Table, in which 16 speakers are mounted in a $4 \times 4$ layout. Sounds are presented at any position on the table by controlling the loudness of the speakers.

It should be noted that all of the users surrounding the table will experience the same sounds whatever their position. This helps the users work collaboratively toward a common goal. A stick-type input device is provided. The system tracks the motion of the device (i.e., user's gesture) in a 3-D space over Sound Table, and then projects computer-generated graphical images onto a surface of Sound Table. Users can interact lively with the system through both visual and auditory channels. Application of the system to collaborative computer music creation is also presented.

## 8.2 Related Work

Context-aware auditory techniques have been successfully applied to museum guides [3,6,7], tourist and pedestrian navigation [5,11,12], and games [13,14] thanks to great advances in mobile and sensing technologies such as the PDA, portable music player, and GPS systems. Location is considered a key context in these systems.

The gpsTunes [5] is a gadget-type handheld computer that guides a user to a desired target by varying the volume and balance of the music played. The user knows the distance to the target by the volume of the music and attempts to move toward the sound source keeping the music in front.

Jones and Jones [11] propose a similar navigation-by-music approach to gpsTunes. Another interesting function is a "look around" cue that pans the music quickly several times to the left and right headphone. This is a prompt for the user to glance in both directions and notice an object of interest such as an interesting building, café, or visitor attraction.

The ec(h)o [7] is an auditory museum guide system. The movement of a visitor triggers an ambient soundscape that is made of sounds related to artifacts near the visitor. The visitor enjoys navigating the exhibit through ambient sounds that are dynamically created. The system also provides a box-type, tangible user interface, called ec(h)o cube. This device works within a specified zone in close proximity (around a meter or just over three feet) to an artifacts display. The visitor can make a selection by holding the cube in front of him or her and rotating it to the left or right, resulting in the generation of an audio message associated with the action.

Meanwhile, sound feedback is quite important in collaborative work environments where users work in parallel toward a shared goal. A simple system configuration (i.e., a pair of stereo speakers), however, causes difficulty in identifying to whom each of the sounds belongs.

Hancock et al. [8] present experiments on the use of nonspeech audio at an interactive multiuser tabletop display under two different setups: (1)

localized sound where each user has his or her own speaker, and (2) coded sound where users share one speaker but the waveform of the sounds are varied so that a different sound is played for each user. This could be one practical solution to business-oriented applications, but is not sufficient for sound-centric applications (e.g., computer music).

An experiment by Ogi and colleagues using multichannel speakers like ours was researched [15]. It is specially designed for a CAVE-like 3-D immersive system and is rather complex and expensive. Even more important, sound processing must be carried out beforehand for every possible position of sound sources and the audience. Certain sound data is selected and then presented to the user at the time of interaction with the system. Meanwhile, in our trial, sound processing is executed on the fly.

Transition Soundings [16] and Orbophone [17] are specialized interfaces using multiple speakers for interactive music making. A large number of speakers are mounted in a wall-shaped board in Transition Soundings, while Orbophone houses multiple speakers in a dodecahedral enclosure. Both systems are deployed for sound art and their goal is different from ours.

## 8.3 System Overview

### 8.3.1 Organization of the System

The system we propose in this paper is designed to provide users with effective visual and auditory feedback in performing collaborative work. Figure 8.1 shows a physical setup of the system. The system is organized by Sound Table as its central equipment, a pair of cameras, a video projector, and a PC (not shown) [18].

Sound Table is a table in which 16 speakers are equipped in a $4 \times 4$ matrix, as shown in Figure 8.2. It is of 90 cm width and depth, and 73 cm of height. The distance between two adjacent speakers is 24 cm. Two 8-channel audio interfaces (M-AUDIO FireWire 410) are equipped to the PC (Apple Mac Pro), and connected to Sound Table through a 16-channel amplifier. The top of Sound Table is covered by a white cloth so that computer-generated graphical images can be projected onto it. The video projector that is mounted over Sound Table is provided for this purpose.

The user expresses his or her commands by manipulating a stick device over Sound Table. Figure 8.3 shows the device whose base unit is Nintendo Wii Remote [19]. The Wii Remote has unique and powerful capabilities as a gesture controller, and many research trials of using it have been investigated (e.g., for music composition [20]). We customized it by newly attaching an infrared LED at its head. The position of the LED in a 3-D space

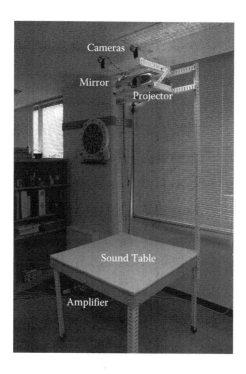

**FIGURE 8.1**
System setup.

**FIGURE 8.2**
Sound Table. (From Nakaie, T., Koyama, T., and Hirakawa, M. *Development of Collaborative System with a Shared Sound Display*, 14–19, 2008. With permission.)

is estimated by analyzing a pair of stereo images taken from two cameras over Sound Table. Here the LED is covered by a semitransparent cap to have the light diffused, as shown in the figure. In consequence, even though the device is turned below to the table, its position can be sensed by the cameras.

(a)                                              (b)

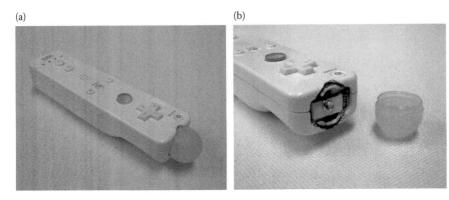

**FIGURE 8.3**
A stick-type input device. (From Nakaie, T., Koyama, T., and Hirakawa, M. *Development of a Collaborative System with a Shared Sound Display*, 14–19, 2008. With permission.)

## 8.3.2 Processing Flow

Figure 8.4 shows a processing flow of the system. We adopted MAX/MSP with Jitter as a tool for implementing software modules.

The system receives inputs from a pair of cameras and an accelerometer provided in the stick-type input device (i.e., Wii Remote). 3-D position and motion tracking data are analyzed to recognize a user's gesture. We define five primitive gestures as will be explained in the section below. Here, the task of identifying gestures is separated from that of their semantic interpretation in a target application so that development of the application can be made easier. The result of interpreting a gesture is expressed in a message and sent to audio and visual output processing parts. Here we adopt Open Sound Control (OSC) protocol [21] for message communication among the components.

In receiving the message, the system executes polyphony control to sound multiple notes at one time. Sound is then generated and comes out through 16 speakers. Meanwhile, graphical objects are generated in OpenGL and projected on the surface of Sound Table.

## 8.3.3 Primitive Gestures

We define five gestures that are simple, natural, and powerful so as to be used in a variety of applications. They are tap, sting and release, attack, flick, and tilt as explained below (see Figure 8.5).

Tap is a gesture of touching the head of the stick device down onto the tabletop. It specifies the 2-D position on the table and corresponds to a mouse click operation. Internally, when a tap gesture is recognized, a message with its coordinate values is generated.

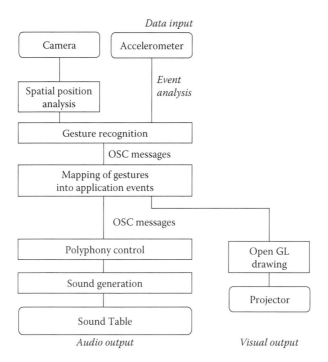

**FIGURE 8.4**
System processing flow.

Sting and release is achieved by bringing the stick device with keeping its head on the table. It corresponds to a mouse drag operation. Messages with 2-D coordinate values of the focus point are generated at certain intervals during the drag gesture.

Attack is a gesture of swinging the device downward quickly as is seen in the drum play. 2-D coordinate values of the attack point, height of the device at the lowered position, and velocity of the attack motion are considered as message parameters.

Flick is a gesture of handling a virtual object on the table with a light quick blow. This gesture is expressed internally by a message with parameters of device position and motion velocity.

Tilt is a gesture of tilting the stick device. This gesture should be carried out while pushing a button on the device. Posture data of the stick device in 3-D and its height from the tabletop are sent with a message. Here, though it is possible to have acceleration along three axes by the use of an accelerometer provided in Wii Remote in reality, we don't take the yaw angle into account since it is not considered essential this time. Furthermore, this can help reduce a gesture recognition error.

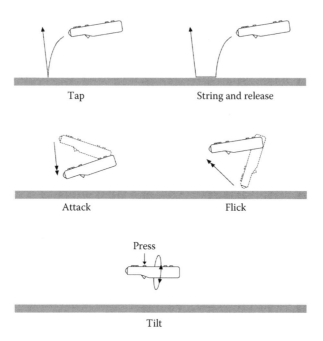

**FIGURE 8.5**
Five primitive gestures.

In the next section, we will explain a mechanism of input processing (i.e., motion capturing) and output processing (i.e., sound control) in detail. An application of the gestures into music performance will be demonstrated in Section 8.5.

## 8.4  Implementation

### 8.4.1  Capturing of Gestures

Spatial and motion properties of the stick device in a 3-D space are extracted by the system. Here, there are several possible object tracking techniques. Capacitive sensing with a mesh-shaped antenna, or SmartSkin, is a smart approach [22]. However, the sensing area is limited to the surface of the sensing table. Use of a supersonic sensor is another possibility. It can catch the height information, but doesn't work properly in a multiplayer collaboration environment. Magnetic tracking technology shows good performance in preciseness. However, we were afraid of the influence of magnetism caused by the speakers. Also it is rather expensive.

Finally we decided to apply a vision-based technique. The spatial (3-D) position of a stick device is obtained by looking at the infrared LED marker attached at its head. The height of the device over the table is estimated by applying a triangulation method to the stereo images taken from two cameras to which IR filters (infrared pass and visible light blocking filters) are attached.

Meanwhile, the motion of the stick device is captured by an accelerometer provided in Wii Remote. This means that it is needed for the system to have correspondence between motion (sensor-based) data and position (vision-based) data. However, images that are taken from cameras are in grayscale and no additional features are available. There is no simple way to identify which stick device it is. The system can easily make a misinterpretation when multiple users participate in a session and multiple devices are overlapped vertically over Sound Table.

In order to solve this problem, the system keeps comparing 3-D motion features that are extracted from camera input images with those from the accelerometer, and decides one-to-one correspondences between LED marker images and stick devices. Even if a correspondence is wrong at some point, it can be corrected properly.

## 8.4.2 Positioning of Sounds

### 8.4.2.1 Algorithm

Sound Table is designed to be used by multiple users. Multiple different sounds may appear at the same time, where each of the sounds is associated with an individual user's action. It is essential for the system to help users identify to whom the sound objects belong.

While 2-channel stereo and 5.1-channel surround sound frameworks for creation of a virtual sound space are available and common in these days, the best "spot" for listening is fixed. If the listener is out of the spot, a reality of the sound space cannot be maintained any more.

Sound Table adopts 16 speakers to overcome this drawback (see Figure 8.2). Sounds can be positioned exactly where they should appear on the table. The feel of sound presence is maintained wherever the user stands.

Here the distance between two adjacent speakers is 24 cm. Needless to say, if a sound comes out just at the position where a speaker is placed, it is not satisfactory. The system controls the loudness of each speaker so they can feel the sound at any position irrelevant to the speaker positions. We extendedly apply the well-known, sine-cosine pan law provided for regular stereo sound management to this spatial (speaker array) sound management.

The conventional sine-cosine pan law is explained as follows: For a pair of two adjacent speakers, the loudness of one speaker is determined by multiplying the input by the sine of a control value, and that of the other speaker

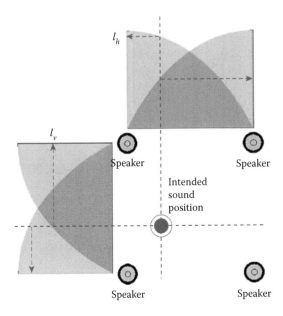

**FIGURE 8.6**
Sine-cosine pan law in a 2-D sound field. (From Nakaie, T., Koyama, T., and Hirakawa, M. *Development of a Collaborative Multimodal System with a Shared Sound Display*, 14–19, 2008. With permission.)

is by multiplying the input by the cosine value. Energy (i.e., loudness of a sound) is maintained as constant regardless of the position.

This law is applied to both horizontal and vertical directions of a speaker matrix (see Figure 8.6). Suppose, for example, the compensated results in horizontal and vertical directions for an upper left speaker are $l_h$ and $l_v$, respectively. Loudness for the speaker is calculated by $l_h \times l_v$.

Every sound object can be moved continuously from one place to another by controlling the loudness of each of the speakers properly. The reason why we adopt such a multi-speaker table configuration instead of using a speaker prepared in Wii Remote is a demand for keeping sound objects at their own positions where they are instantiated and generated. Eventually sounds are distributed throughout the table, which is natural and effective in our daily activities.

### 8.4.2.2 Evaluation

The performance of sound positioning control explained above is evaluated by analyzing how accurate the sound positioning is that is, errors in distance between the simulated sound position and the perceived sound position. Five university students (four male and one female) participated in the experiments. Subjects were requested to stand by the Sound Table in an upright position, and then heard test sounds and answered in their positions. The number of test sounds ranged from 1 to 5, and 125 trials were carried out in total. Here, eight-second long instrumental music segments (piano, bass, hihat cymbal, snare drum, and synthesizer) were prepared for the experiments.

We examine the performance in terms of mobility of sound object(s). Figure 8.7 shows the error distribution where the simulated sound position is stable and does not change during the trial. Horizontal and vertical axes represent error rates in horizontal and depth directions, respectively, where their values are normalized by the distance between two adjacent speakers. That is, the value 1.0 means that its error is equal to the distance between speakers (i.e., 24 cm). In summary, the average error in horizontal direction is 0.27 (6.5 cm) and that in depth direction is 0.42 (10.0 cm) [23].

Figure 8.8 shows the error distribution in the case when the position of simulated sounds moves. The average error in horizontal direction is 0.52 (12.5 cm) and that in depth direction is 0.72 (17.3 cm).

As seen in the figures, in both cases, the performance in depth direction is worse than that in horizontal direction. This is natural since we have ears at both sides of our head. And in the case when sounds move, subjects found difficulties in identifying their positions. In fact, the error distribution is rather sparse.

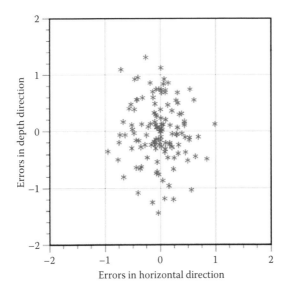

**FIGURE 8.7**

Distribution of errors (no sound position movement). (From Nakaie, T., Koyama, T., and Hirakawa, M. *Development of Collaborative Multimodal System with a Shared Sound Display*, 14–19, 2008. With permission.)

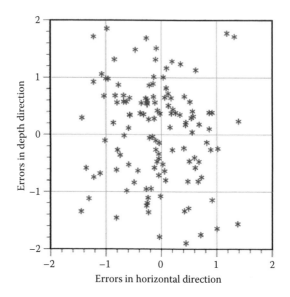

**FIGURE 8.8**
Distribution of errors (with sound position movement). (From Nakaie, T., Koyama, T., and Hirakawa, M. *Development of Collaborative Multimodal System with a Shared Sound Display,* 14–19, 2008. With permission.)

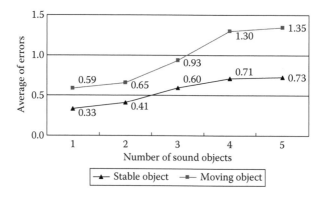

**FIGURE 8.9**
Influences of the number of sound objects. (From Nakaie, T., Koyama, T., and Hirakawa, M. *Development of Collaborative Multimodal System with a Shared Sound Display,* 14–19, 2008. With permission.)

Next we analyze the influence of the number of sounds that are played at one time. The result is shown in Figure 8.9, where the horizontal and vertical axes denote the number of simultaneous sounds and the average of errors, respectively. The average error becomes gradually higher as the number of simultaneous sounds increases and those two error patterns look similar.

Experiments showed that the proposed scheme works with reasonable performance, though further studies are needed to investigate, for example, the influence of the test sound and moving speed of sound objects.

## 8.5 Application to Computer Music Creation

Computer music is a particularly good example of applications at which the sound feedback plays an important role. In this section we will demonstrate how the proposed system is applied to collaborative computer music creation.

Here, there have been futuristic table-based computer music creation tools that adopt tangible and gestural interfaces. Sound Rose [24] and the system talked about by Davidson and Han [25] are musical table systems with touch sensitive interfaces. The user creates music by touching the table with finger(s). Block Jam [26], Audiopad [27], reacTable [28], and AudioCubes [29] are tangible tabletop instruments for computer music, where manipulation of plastic blocks on a table corresponds to music creation.

The system we propose in this paper uses a table as a means for interaction as is similar to the above systems. However, a music sequence is specified by 3-D motion of a stick-type device over the table, rather than a touch or an arrangement of music blocks on the table. Furthermore, our system is capable of outputting sounds at different positions on the table. This helps each of the players to sense others' performance, which is essential for collaborative music creation.

We have implemented several interface expressions to explore fundamental capabilities of the system. Here, as explained above, operation commands for music creation are implemented by defining semantic descriptions for the abovementioned gestures. For example, we assigned the following five commands as to the trial demonstrated in Figure 8.10.

Drop (implemented by tilt) is an operation to newly create a sound object. If the user tilts the stick device with keeping a button pressed, a sound object is generated and spilled over the table. Each of the sound objects is visualized with a certain color and shape as is seen in the figure. Their size and loudness depend on the position of the stick device. The higher the spatial position, the larger the circle size and the louder the generated sound.

Play and stop (implemented by attack) is an operation to switch between play and stop of a sound. When the device is swung down over an object that is sounding, the sound is terminated. At the same time, its color becomes black so you can see the change. If the gesture is applied again, the object restarts the sound.

Change (implemented by tap) is an operation to change sound sources. A different color is assigned to each of the sound sources. At each time when a tap gesture is placed on a sound object, its color and sound source change.

(a) (b)

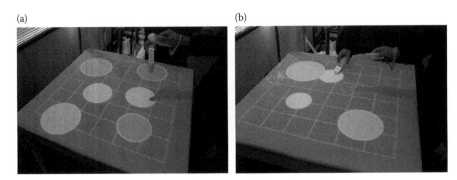

**FIGURE 8.10**
Creating music with the system.

**FIGURE 8.11**
Plane-based music creation interface. (From Nakaie, T., Koyama, T., and Hirakawa, M. *Development of Collaborative Multimodal System with a Shared Sound Display*, 14–19, 2008. With permission.)

Move (implemented by sting and release) is an operation to move a sound object to any position on the table. Needless to say, the associated sound moves along the drag gesture.

Remove (implemented by flick) is an operation to delete a sound object. When the user flicks a certain sound object, it flows out of the table with graphical effects (i.e., the object is broken in segments, see the picture on the right of Figure 8.10).

Players jointly work to create music by manipulating a number of sound objects on Sound Table. Here the music created is not the same as we enjoy hearing in everyday life. It is considered a mashup created out of pieces of two or more songs, or a soundscape composed of sound segments including songs of birds, chirping of crickets, blasts of wind, and the sound of water drops.

Figure 8.11 shows another implementation of the music creation interface (an auditory Voronoi diagram) in which a plane-based expression is adopted instead of the point-based one explained above. Players can enjoy reforming the scenery of sounds by applying the sting and release and change operations. The sting and release operation is used to move the position of a sound (a Voronoi cell), while the change operation serves to change its musical and graphical properties.

## 8.6 Conclusion

In this chapter, we explored spatial presence of sounds for supporting situated collaborative work. The shared tabular sound display called Sound Table allows multiple users to have a feeling that sounds are here and there with the help of visual feedback on the table. A stick-type input device is provided for manipulation of dynamic (i.e., sound and visual) objects.

Application of the proposed multimodal system to collaborative computer music creation was also presented. Users enjoy creating music collaboratively through auditory, visual, and gestural interaction.

This work represents an important first step bringing multimodal, multiuser interaction to spatial sound display. There remain several open questions, however, which need to be addressed. First, more detailed experiments in sound positioning control are needed. Second, exploration of the utility of the system still remains. Third, we will investigate how exactly the information is shared and how social protocols between people may evolve with the usage of the system. Fourth, it is interesting to explore principles of good sound visualization design.

## References

1. K. Lyytinen, and Y. Yoo 2002, December. Issues and challenges in ubiquitous computing. *Communications of ACM* 45 (12): 62–65.
2. M. Weiser. 1991, September. The computer for the twenty-first century. *Scientific American*, 94–104.
3. L. Terrenghi, and A. Zimmermann. 2004. Tailored audio augmented environments for museums. Proc., ACM IUI, 334–36.
4. F. Mueller, and M. Karau. 2002. Transparent hearing. Proc., ACM CHI2002, 730–31.
5. S. Strachan, P. Eslambolchilar, and R. Murray-Smith. 2005. gpsTunes controlling navigation via audio feedback. Proc., ACM MobileHCI'05, 275–78.
6. E. Bruns, B. Brombach, T. Zeidler, and O. Bimber. 2007, April–June. Enabling mobile phones to support large-scale museum guidance. *IEEE MultiMedia* 14 (2): 16–25.

7. R. Wakkary, and M. Hatala. 2006. ec(h)o: Situated play in a tangible and audio museum guide. Proc., ACM Conf. on Designing Interactive Systems, 281–90.

8. M. S. Hancock, C. Shen, C. Forlines, and K. Ryall. 2005. Exploring non-speech auditory feedback at an interactive multi-user tabletop. Proc., Graphics Interface, 41–50.

9. R. Jung, and T. Schwartz. 2007. Peripheral notification with customized embedded audio cues. Proc., Int'l Conf. on Auditory Display, 221–28.

10. A. Kainulainen, M. Turunen, and J. Hakulinen. 2006. An architecture for presenting auditory awareness information in pervasive computing environments. Proc., Int'l Conf. on Auditory Display, 121–28.

11. M. Jones, and S. Jones. 2006, July–August. The music is the message. *ACM Interactions* 13 (4): 24–27.

12. A. Kainulainen, M. Turunen, J. Hakulinen, and A. Melto. 2007. Soundmarks in spoken route guidance. Proc., Int'l Conf. on Auditory Display, 107–11.

13. K. Cater, R. Hull, T. Melamed, and R. Hutchings. 2007. An investigation into the use of spatialised sound in locative games. Proc., ACM CHI Extended Abstracts, 2315–20.

14. I. Ekman, L. Ermi, J. Lahti, J. Nummela, P. Lankoski, and F. Mäyrä. 2005. Designing sound for a pervasive mobile game. Proc., ACM Int'l Conf. on Advances in Computer Entertainment Technology, 110–16.

15. T. Ogi, T. Kayahara, M. Kato, H. Asayama, and M. Hirose. 2003. Immersive sound field simulation in multi-screen projection displays. Proc., Eurographics Workshop on Virtual Environments, 135–42.

16. D. Birchfield, K. Phillips, A. Kidane, and D. Lorig. 2006. Interactive public sound art: A case study. Proc., Int'l Conf. on New Interfaces for Musical Expression, 43–48.

17. D. Lock, and G. Schiemer. 2006. Orbophone: A new interface for radiating sound and image. Proc., Int'l Conf. on New Interfaces for Musical Expression, 89–92.

18. T. Nakaie, T. Koyama, and M. Hirakawa. 2008. A Table-Based Lively Interface for Collaborative Music Performance, Proc. Int'l Conf. on Distributed Multimedia Systems, 184–89.

19. http://www.nintendo.com/wii

20. S. Toyoda. 2006. Sensillum: An improvisational approach to composition. Proc., Int'l Conf. on New Interfaces for Musical Expression, 254–55.

21. M. Wright, and A. Freed. 1997. Open sound control: A new protocol for communicating with sound synthesizers. Proc., Int'l Computer Music Conf., 101–4.

22. J. Rekimoto. 2002. SmartSkin: An infrastructure for freehand manipulation on interactive surfaces. Proc., ACM CHI2002, 113–20.

23. T. Nakaie, T. Koyama, and M. Hirakawa. 2008. *Development of Collaborative Multimodal System with a Shared Sound Display*. Paper presented at UBI-Media Computing and Workshop Proceedings, sponsored by IEEE, 14–19.

24. A. Crevoisier, C. Bornand, S. Matsumura, and C. Arakawa. 2006. Sound rose: Creating music and images with a touch table. Proc., Int'l Conf. on New Interfaces for Musical Expression, 212–15.

25. P. Davidson and J. Y. Han. 2006. Synthesis and control on large scale multi-touch sensing displays. Proc., Int'l Conf. on New Interfaces for Musical Expression, 216–19.

26. H. Newton-Dunn, H. Nakano, and J. Gibson. 2002. Block jam. Proc., ACM SIGGRAPH2002 Conf., 67.

27. J. Patten, B. Recht, and H. Ishii. 2006. Interaction techniques for musical performance with tabletop tangible interfaces. Proc., ACM Conf. on Advances in Computer Entertainment Technology.

28. S. Jorda, G. Geiger, M. Alonso, and M. Kaltenbrunner. 2007. The reacTable: Exploring the synergy between live music performance and tabletop tangible interfaces. Proc., ACM Conf. on Expressive Character of Interaction, 139–46.

29. B. Schiettecatte, and J. Vanderdonckt. 2008. AudioCubes: A distributed cube tangible interface based on interaction range for sound design. Proc., Int'l Conf. on Tangible and Embedded Interaction, 3–10.

# 9

## Pervasive Carpet Encoding for Active Knowledge Semantic Surfaces

Kamen Kanev and Nikolay Mirenkov

## CONTENTS

This chapter is devoted to computer-assisted, human interactions with the surrounding world that are based on ubiquitous surface encoding of physical entities and backgrounds. It introduces the concept of *newsputers* as digitally enhanced printouts, allowing direct, real-time access to computer-based multimedia information through printed interfaces. Newsputers are applicable to conventional printed materials such as newspapers, books, and magazines as well as to surfaces of various objects and environment components enhancing their functionalities and allowing transparent access to complementary information and services. The newsputer concept is extended to active knowledge semantic surfaces for ubiquitous direct interaction environments with self-explanatory components and potential self-awareness functionality.

## 9.1 Introduction

Through the centuries human society has evolved a long way and presently, along with product manufacturing, services, and information are playing more and more important roles. Indeed, in the postindustrial society, fundamental knowledge is highly valued and considered an asset, often surpassing practical know-how and technical skills. Since the 1970s, information processing technologies have brought fundamental changes in the way societies work. Creation, dissemination, and employment of information are now vital for all economic, political, and cultural activities. An information society [1], with information as a primary commodity, has emerged and access, creation, and dissemination of information have been affirmed as fundamental human rights.

Since the invention of typography, printed materials have been and still remain one of the most common means for information dissemination. Well-defined publishing procedures and copyright laws govern production, distribution, and access to books, magazines, and newspapers. Worldwide library networks archive printed documents and manage and make accessible the heritage of accumulated human knowledge. Obviously, printed materials can only capture and preserve static information, leaving out live music, voices, and video. For the latter, however, the new digital multimedia has opened virtually unlimited possibilities. Presently, many libraries are gradually shifting from paper-based archival to digitization of documents and management of digital records, e-books, and online information services [2]. Nevertheless, we believe that paper is here to stay for quite a while. Due to the aggressive policies of current publishers, the existing vast paper-based information sources are actually increasing at very high rates. Many cultural issues and traditions are also favoring paper and in fact there are still many people who prefer to use printed materials rather than computer screens.

In our attempt to bring together the best of both worlds, we propose a new pervasive infrastructure for integration and mixed use of printed and digital information. Our approach is based on enhancement of new and exiting publishing content in printed or digital form with visible or invisible (for the human eyes) codes. We envisage a ubiquitous environment of code recognition enabled devices that perform a diversity of actions based on the codes and thus bring a new enhanced functionality of the printed materials.

In Sections 9.2 and 9.3, traditional and electronic newspapers, cross-media linking, and the newsputer concept are considered. In Section 4, we describe a software environment for printout code attachment, management and linking to corresponding functions or pieces of information. In Sections 9.5 and 9.6, some details of the environment implementation and related experiments are provided. Section 9.7 is dedicated to active knowledge and semantic surfaces, and finally in Section 9.8 our vision for future works is outlined.

## 9.2 Traditional Versus Electronic Newspapers

We have now entered the digital millennium and traditional printing and distribution of books, magazines, and newspapers have to adjust. On one hand, environmental conscious movements are exerting pressure to protect the natural resources employed in paper production. On the other hand the ubiquitous multimodal access to digital information sources drives up the demand for interactive publications with more dynamic content. Evidently printed materials in their traditional form could not satisfy such a demand and that is why publishers are presently offering complementary electronic versions of their materials. NewsStand Inc. [3], for example, provides subscriptions and access to the electronic versions of several dozens of newspapers from different countries and in different languages. All these e-newspapers, however, are said to appear exactly as their printed counterparts and do not thus offer any special advantages, apart from the instant delivery and paper savings.

A true e-newspaper could indeed support more advanced functionality that is not available in traditional printed newspapers. Some online e-newspapers, for example, link many of the photos, included in the printed version to video clips that are playable in the e-version. Published rapidly changing information, such as weather forecasts, currency exchange rates, stock prices, and so on is often linked in the e-version to some external online information sources that is periodically updated. In the e-newspapers additional assistive support like integrated access to online dictionaries with word descriptions, pronunciations, and translations in different languages could also be supplied. Despite all extras that e-newspapers might be in a position to offer, their presentation is still confined to displays with limited resolution and size. Conventional displays require constant power supply and the cost per unit area is far too high than the one for printed materials. Ideally, when adopting a new e-newspaper model, we should first try to preserve all the attractive features that traditional paper documents offer, and second introduce new advanced functionalities that are only available for digital documents.

## 9.3 Cross-Media Linking and the Newsputer Concept

One possible approach could be to keep using the paper for what it is best suited for, namely to show information content that is mostly static and would rarely need reprinting. All dynamic content, on the other hand, could be provided on a separate, much smaller display. The main challenge of such an approach is how to bring together the printed and the digital content in a single easy to operate system. The progress in camera-based document image analysis [4] stimulates the development of OCR-related approaches

[5,6] that allow text-based digital content linking but do not support image and position-based links. Other widely exploited methods for linking digital content to printed documents are based on barcodes [7,8]. Indeed, in Japan many cell phone models are equipped with cameras and QR-code [9] recognition software that allows instant online access to bar-coded URL addresses incorporated in printed documents as shown in Figure 9.1. QR-codes as all other barcodes, however, require fairly large dedicated space and thus cause significant visual disturbance.

More advanced approaches based on digital watermarking [10] and carpet encoding [11,12] that seamlessly blended with the printed document content are also available. For example, if the CLUSPI [12,13] carpet encoding method is employed, the visual appearance of a digitally enhanced newspaper changes so little that readers would hardly feel any difference. The publishers would, therefore be free to adopt and gradually introduce different digitally enhanced printed materials creating a foundation for new digital information services linked to the published content. In addition, readers would also need supportive devices with optical input appropriate for extracting the codes embedded into the digitally enhanced printed materials and for providing multimedia feedback.

We use the acronym newsputer [14] (an abbreviation from newspaper and computer) to denote the augmented functionality and wealth of services that become possible with such digitally enhanced printed media. The newsputer concept is about a pervasive environment where:

1. Diversity of print publications would carry embedded digital enhancements in the form of barcodes, digital watermarks, carpet codes, and so on.

**FIGURE 9.1**
Printed materials with embedded QR-codes.

2. A wide selection of devices such as mobile phones, PDAs, compact audio and video players, and so forth would be available for interacting with such digitally enhanced printed media.

Rather than exploring different technologies and methods for document enhancing, in the following sections we will focus on the design and development of a more general, code independent, newsputer software environment for composition, management, deployment and utilization of digitally enhanced printed materials. Our target is to create an experimental software environment that we could use for tests and evaluations in the course of our newsputer-related research project. The developed newsputer software, however, will also play an important role in the envisaged newsputer infrastructure.

## 9.4 The Newsputer Software Environment

Our newsputer software environment is meant to support both publishers and readers and should therefore provide at least two separately adjustable frames of access to the related digitally enhanced document content. The publisher's frame of access is mainly dedicated to the digital code attachment and management process and is maintained by the Code Attachment System (CAS). The reader's frame of access on the other hand enables users to get instant feedback with targeted digital content and is maintained by the Reader Support System (RSS). In addition to the two main frames of access we also provide a maintainer's frame of access handled by the Maintainer Support System (MSS).

### 9.4.1 Code Attachment System (CAS)

Within the publisher's frame of access, CAS exposes the following functionality:

1. Importing and initial setup of digital page content for printed pages
2. Definition of code related domains and matching to document layouts by
   - Employing code data templates
   - Manual adjustment of code domains
   - Link attachment and editing

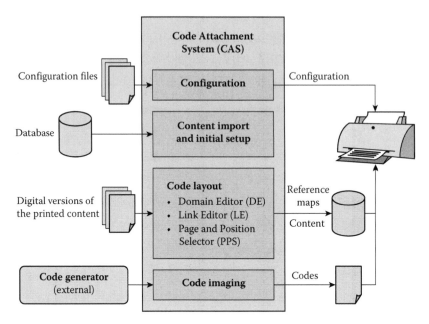

**FIGURE 9.2**
A schematic diagram of the CAS structure and interconnections.

3. Embedding of digital codes in the printable document content by
   - Generation of appropriate digital code images
   - Layering and blending the images with the document content

A schematic illustration of the CAS functionality and interconnections is shown in Figure 9.2.

### 9.4.2 Reader Support System (RSS)

The RSS, maintaining the reader's frame of access, offers the following user-oriented functionality:

1. Identifying and acquiring online or off-line access to the digital content corresponding to the currently used printed document
2. Accessing registered document reference maps, previously created with CAS
3. Recognizing and invoking control functions and maintaining a corresponding global state

A schematic diagram of the RSS functionality and interconnections is shown in Figure 9.3.

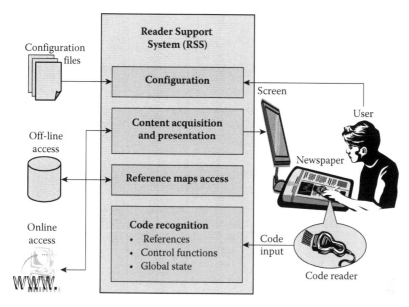

**FIGURE 9.3**
A schematic diagram of the RSS structure and interconnections.

**TABLE 9.1**

Sample Control Functions

| Name | Value |
| --- | --- |
| On/Off | Turns on or off the control functions |
| Get | Obtains additional content related information |
| Help | Provides system operation support |
| Search | Searches for related information on the Internet |
| Music | Plays associated audio |
| Immediate/delayed | Changes the response mode |

The control functions and the global state associated with them have been introduced in order to allow users to employ an overlaid operation. That is, users can invoke different actions by touching a single spot on the document surface. We employ that for an overloading mechanism that allows us to associate multiple actions with one and the same domain of the document surface. Some sample control functions are shown and briefly explained in Table 9.1.

## 9.4.3 Maintainer Support System (MSS)

Finally we will give a brief explanation of the maintainer's frame of access and the MSS. Since printed materials come in great diversity of designs, layouts, volumes, and other features, to accommodate them, CAS has to be highly configurable. Similarly, since users and their needs are quite diverse,

**TABLE 9.2**

CAS-Related Configuration Parameters

| Name | Value |
|------|-------|
| Page width | 2,000 |
| Page height | 1,400 |
| DPI | 72 |
| Type | Newspaper |
| Number of pages | 20 |
| Reusing target | 1 week ago |

**TABLE 9.3**

RSS-Related Configuration Parameters

| Name | Value |
|------|-------|
| Age | 25 |
| Country | Japan |
| Primary language | Japanese |
| Secondary language | English |
| Default search engine | Google |
| User defined functions | none |

the RSS would also need extensive reconfigurations. Instead of handling the configuration independently within each of the systems, we have opted for a unifying approach. Indeed, the employed MSS provides a maintainer's frame of access compliant with the needs of the support staff and allowing full system reconfiguration. Different sets of configuration parameters are supported for the CAS and the RSS systems.

In Table 9.2, for example, where a subset of the CAS configuration parameters is shown, most of the values are associated with the specific properties of the concerned document. The RSS related configuration parameters shown in Table 9.3 in contrast are more user-oriented and are employed for defining specific customer profiles.

## 9.5 Reference Implementation

For our experiments we have prepared reference implementations of the CAS and MSS components of the software environment in JAVA. Currently there are stand-alone JAVA applications, but we are also considering CAS and MSS applet versions that would be accessible through any JAVA-enabled browser.

The CAS component is structured into three functional parts namely a domain editor (DE), a link editor (LE), and a page and position selector (PPS).

**FIGURE 9.4**
A snapshot of the DE window.

A sample snapshot of the DE window is shown in Figure 9.4. The left part of the window shows a partial view of the currently selected document content. Users can define rectangular domains by directly pointing, clicking, and dragging with the mouse cursor on the screen. The DE automatically tracks and manages the following parameters for each of the defined domains:

- X-coordinate of the domain origin
- Y-coordinate of the domain origin
- Width of the domain
- Height of the domain
- Links associated with the domain

The first four parameters are directly derived from the mouse-based user actions while the last one can only be manipulated through the LE interface as explained in the following paragraphs.

Textual information about the currently active and the currently selected domains is shown on the right side of the DE window. Users can point and click on any of the textual fields to open it for direct typing or editing. All domain origin and size changes in the right part of the window are reflected in the domain graphical representation at the left part of the window and vice versa.

Clicking on the "link" field opens the LE window as shown in Figure 9.5a. In this window both links to additional information and special control functions can be specified. The link field in the top of the window is direct typing and history browsing enabled. The direct typing functionality is mostly used for entering new, unrelated links and for copying and pasting of existing links. The history browsing functionality on the other hand supports context-dependent

(a)                                                        (b)

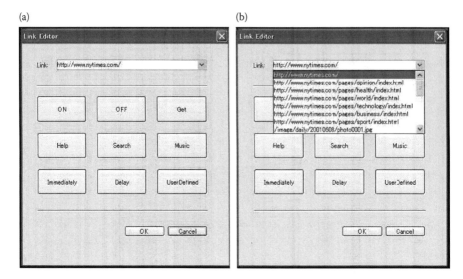

**FIGURE 9.5**
Snapshots of the LE window.

link selection and manipulations (Figure 9.5b), and is thus helpful in optimizing the process of registering overlapping and similar links.

In addition to the standard textual links, the LE also supports functional links that perform special control functions. In contrast to textual links, the functional links are not typed or directly manipulated but rather generated by the configurable functional buttons in the central part of the LE window. Each generated functional link corresponds to a printed functional button, embedded in the document content.

The DE and LE functionality discussed so far only allows processing of partial document views that are limited to the available screen and window sizes. However, publishers need to handle voluminous printed materials such as books with hundreds of pages as well as newspapers and posters with pages of large sizes. To address the publisher's needs we provide the PPS interface shown in Figure 9.6. Through it users could flip pages in forward and backward directions or directly select them by page numbers. With each page we associate an active area, shown in gray in Figure 9.6, which can be freely moved and resized. The active area of the currently selected page is shown in the DE window and is made available for domain editing.

The MSS component has been implemented as a stand-alone configuration editor for direct manipulation of XML files. While experienced system maintainers could open and edit the configuration XML files with text-based editors, the developed MSS component provides a more user friendly GUI. As a reference we show some sample snapshots of the MSS window in Figure 9.7. The MSS also parses and automatically corrects any inconsistencies found in the input XML configuration files.

**FIGURE 9.6**
A snapshot of the PPS window.

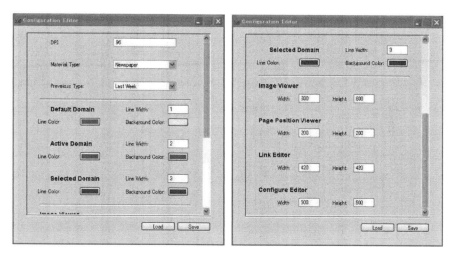

**FIGURE 9.7**
Snapshots of the MSS window.

## 9.6 Experiments

We have employed the experimental software environment discussed in the previous sections for implementing the CLUSPI-enhanced newspaper shown in Figure 9.8.

(a)                                                    (b)

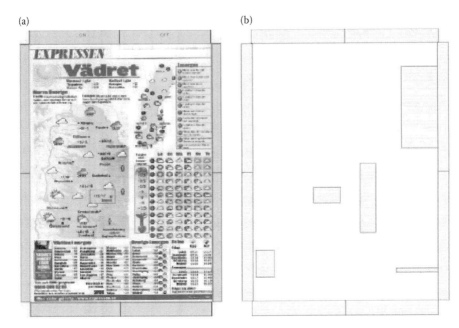

**FIGURE 9.8**
A CLUSPI-enhanced newspaper.

The five domains denoted by the rectangles in the internal part of Figure 9.8b are directly associated with the newspaper content. The remaining eight domains on the edges are for the control functions. The newspaper content (Figure 9.8a) and the domain map (Figure 9.8b) are actually overlaid and printed together as a single document. This way the standard document content and the CLUSPI carpet code are seamlessly blended into a digitally enhanced printed document with newsputer functionality.

Since the CLUSPI layer creates no visual disturbance, readers could use the resulting printed document as a standard newspaper. By employing a CLUSPI enabled optical input device, readers could also take advantage of the added newsputer functionality. For our experiments we have used a USB camera connected to a notebook computer. The readers point and click with the camera on the digitally enhanced printed document and obtain instant feedback with updated additional content. To illustrate this newsputer functionality we use the domain defined around the city of Umea on the map of Sweden, denoted by a rectangle in Figure 9.8a. When pointing and clicking the domain as shown in Figure 9.9a, the reader get an instant, up-to-date local weather report for the city on the notebook screen as shown in Figure 9.9b.

(a) (b)

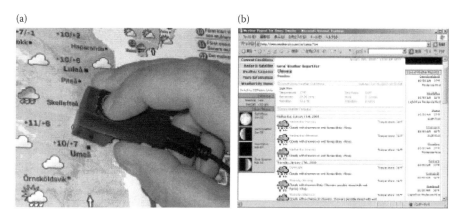

**FIGURE 9.9**
Invoking the newsputer functionality.

## 9.7 Active Knowledge and Semantic Surfaces

So far we have introduced the concept of newsputers and have shown how newspapers, books, magazines, and other printed materials can be integrated into computer-based pervasive infrastructures. We have also explained how enhanced environments supporting such integration can be created, and have provided experimental results demonstrating the practicality of the approach.

Of course, to transform this promising approach into a mature technology, it would be necessary to enrich the features of the environment and, what is especially important, to consider some other, more diverse applications of newsputers. In this line of thought, newsputer-based multimedia and multiple views of events, objects, and phenomena could be employed in self-explanatory materials and process components for education, intelligent advertising, gaming, entertainment, and so on.

Self-explanatory software components and human-computer interfaces [15,16] could support software engineers, ordinary users, and even people with disabilities [17,18] in overcoming obstacles and difficult problems. Such components and interfaces employ a plurality of views in different multimedia formats to provide explicit knowledge about the corresponding object and processes and to make them self-explanatory. Multimedia views are often associated with:

- Moments of time
- Positions in space
- Diversity of materials
- Functional attributes

- Structural representations
- Hierarchy and abstraction levels
- Environmental neighborhood relations

For a self-explanatory object, all of its associated multimedia views need to be organized and presented in such a way that the semantic richness of the object-related data and knowledge would be clearly pronounced. This means that the right group of corresponding multimedia frames has to be extracted and presented through the right channel at the right point of time. The explanatory process can be initiated either by conscious user actions or through some indirect means based on observations of user behavior. If the later method is used, self-explanatory objects could disburse context-dependent multimedia information even in the absence of specific user requests and could thus be considered as active knowledge entities.

Interactions with self-explanatory objects and implementation of active knowledge components could be achieved through digital encoding of object and environment surfaces. Since coverage of diverse surfaces would be needed, we suggest employment of carpet codes that could be applied to flat and curved surfaces. Such codes carry both surface position, and object identification information, and can lead to total identification solutions, if properly applied. In this way walls in different places outside and inside buildings, along with attached pictures and other environmental objects, could be augmented with semantics about "what," "where," "how," and "why" through the embedded carpet encoding. This could become a new basis for supporting ubiquitous computing through transformation of self-explanatory features into self-awareness features. In other words, the rich semantics of real world objects would be exposed through the ubiquitous computing devices employed as human interface components and capable of transparent identification and tracking of surface embedded carpet codes.

Such interfaces and corresponding knowledge are for enhancing people abilities to overcome the barriers of geography, time, and individual intelligence. This is illustrated in Figure 9.10 where recognition of environment embedded codes on semantic surfaces provides the right context for active knowledge objects carrying out communications and videoconferencing with fixed and mobile equipment.

The semantic surface approach could also activate hidden reserves of human intelligence. Many everyday human activities require extensive rote memorization and involve high levels of abstraction and mental simulations dictated by the constant need for adaptation of people to existing technologies, and so forth. This often leads to various troubles and health disorders and, in some cases, even to more serious illnesses. The human brain is a very adaptive biological computer but it is questionable as to what extent it should be forced to follow every advance in current technologies

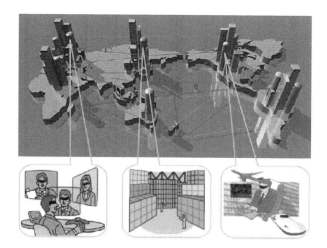

**FIGURE 9.10**
Semantics surfaces in ubiquitous computing environments.

and data–knowledge representation. We believe that the active knowledge semantic surface approach could alleviate the need for rote memorization, could minimize unnecessary mental simulation, and could thus reduce the need of adaptation to a great extent. Hopefully by reducing technological burden on human minds we could open more space for creativity, stimulate positive thinking, and create conditions for deeper understanding and increased performance without mental stress.

The proposed method offers novel opportunities for accessing IT resources and for information analysis and understanding. Self-explanatory interfaces would be aware of what a person knows and could thus adjust to individual mental and physical abilities as well as to the time, the place, and the type of activity that happens. In this way advanced adaptive features of domain-oriented interfaces could also be implemented and, for example, specific market segments such as education, tourism as well as different professional and age groups could be targeted.

In addition to printed materials, semantic surfaces are also applicable at more global levels. At city and street levels, for example, they can be employed for determining positions of cars and people and could be of particular help for people with disabilities. Precise tracking of positions and orientations of robots and special machines, like fire department vehicles and ambulance cars, could be of great importance when disasters strike. Position tracking could also be used as a basis for extracting and supplying rich, context-dependent information about places, related tourist and history data, available services, shops, local businesses, and so forth. At building and room levels precise position tracking may provide guidance and orientation and could be crucial for determining the escape routes in case of fire and other disasters. The proposed method provides higher resolution than global positioning systems and, since the codes are embedded in the surrounding environment, it can be customized to better fit the needs of the local communities.

## 9.8 Conclusion and Future Work

The work reported in this chapter could be extended with more focused searches for associations, timestamps, and specific links to information through semantic web operations. This will lead to an integrated information space that brings together and seamlessly interlinks the physical world with the information world. The active knowledge for semantic surfaces layered on physical objects and backgrounds then become physical windows into the semantic web and provide for instant, context-dependent access to its vast information resources. In other words, the position-based knowledge acquisition and pervasive carpet encoding will open new, promising ways for creating and searching information resources, and for developing effective educational materials and corresponding educational processes. They will be a self-explanatory and self-awareness basis for a great variety of technologies related to concepts of ubiquitous computing. To implement that, however, some standards on code assignments and corresponding code recognition devices would need to be introduced for codes of different scales.

## References

1. Crawford, S. 1983, October. The origin and development of a concept: The information society. *Bull. Med. Libr. Assoc. 71* (4): 380–85.
2. Candela, L., Castelli, D., Pagano, P., and Thanos, C. 2007, March–April. Setting the foundations of digital libraries: The DELOS manifesto. *D-Lib Magazine* 13 (3/4).
3. NewsStand Inc., http://www.newsstand.com/
4. Doermann, D., Liang, J., and Li, H. 2003, August 3–6. Progress in camera-based document image analysis. In Proc. of the Seventh Int. Conf. on Document Analysis and Recognition (ICDAR2003), Edinburgh, Scotland.
5. Shih, T., Chang, W.-C., Wang, T.-H., Lin, H. W., Chang, H.-P., Huang, K.-H., Sie, Y.-L., Tzou, M.-T., and Yang, J.-T. 2005, July 5–8. The hard SCORM LMS: Reading SCORM courseware on hardcopy textbooks. In Proc. of the Fifth IEEE Int. Conf. on Advanced Learning Technologies (ICALT'05), Kaohsiung, Taiwan.
6. Ferreira, S., Garin, V., and Gosselin, B. 2005, August 29. A text detection technique applied in the framework of a mobile camera-based application. In Proc. of Camera-Based Document Analysis and Recognition (CBDAR2005) Workshop of ICDAR, Seoul, Korea.

7. Koike, H., Sato, Y., Kobayashi, Y., Tobita, H., and Kobayashi, M. 2000, April 1–6. Interactive textbook and interactive venn diagram: Natural and intuitive interfaces on augmented desk system. In Proc. of the SIGCHI Conf. on Human Factors in Computing Systems, Hague, Netherlands.

8. Klemmer, S., Graham, J., Wolff, G., and Landay, J. 2003, April 5–10. Books with voices: Paper transcripts as a tangible interface to oral histories. In Proc. of the SIGCHI Conf. on Human Factors in Computing Systems, Ft. Lauderdale, Florida.

9. QR Code. http://www.qrcode.com/, accessed August 31, 2009.

10. Katayama, A., Yamamuro, M., and Sonehara, N. 2004, October 27–29. Fast watermark detection scheme for camera-equipped cellular phone. In Proc. of the 3rd International Conference on Mobile and Ubiquitous Multimedia (MUM2004), College Park, Maryland.

11. Singer, B. and Norrie, M. 2003, August 26–30. Multi-layered cross-media linking. In Proc. of the Fourteenth ACM Conf. on Hypertext and Hypermedia (HT2003), Nottingham, UK.

12. Kanev, K. and Kimura, S. 2006. Direct point-and-click functionality for printed materials. *The Journal of Three Dimensional Images* 20 (2): 51–59.

13. Kanev, K. and Kimura, S. 2005. Digital information carrier. Japanese Patent No. 3635374.

14. Arundale, J. 1993. Newspapers or newsputers? *Online & CDROM Review* 17 (5): 300–301.

15. Mirenkov, N., Vazhenin, A., Yoshioka, R., Ebihara, T., Hirotomi, and Mirenkova, T. 2001. Self-explanatory components: A new programming paradigm. *International Journal of Software Engineering and Knowledge Engineering–World Scientific* 11 (1): 5–36.

16. Watanobe, Y., Mirenkov, N., Yoshioka, R., and Monakhov, O. 2008. Filmification of methods: A visual language for graph algorithms. *Journal of Visual Languages and Computing* 19 (1): 123–50.

17. Mirenkov, N. 2001. Self-explanatory components: A new way to enhance user's ability. *Journal of Shanghai University (English Edition)* 5 (Suppl.): 115–23.

18. Hirotomi, T. and Mirenkov, N. 2003. Self-explanatory components: A basis for new communicators. *Journal of Visual Languages and Computing* 14 (3): 215–32.

# 10

Minimizing the Stego-Image Quality
Impact of Message Embedding Using
the DM Allocation Method

Chin-Chen Chang, Wei-Liang Tai, and Chia-Chen Lin

## CONTENTS

Message embedding is the art of covert communication in which secret messages are embedded into innocuous looking objects, known as covers, through slight modification. The covers concealing secret data are called stego objects. The goal of message embedding is to embed a secret message so that its very existence in the stego object cannot be revealed. Thus, security is mostly influenced by the embedding efficiency, which is the number of bits embedded per embedding change. In this paper, we will show that a disk modulo method called DM can be used for covert communication without sharing the placement of embedding changes. This allows construction of message embedding methods with improved security. We then describe an efficient message embedding method for minimizing the embedding impact using a DM allocation method with the goal of decreasing the number of embedding changes. Finally, the embedding efficiency of the proposed method is compared with theoretically achievable bounds.

## 10.1 Introduction

Message embedding is the science of covert and undetectable communications, which protects secret information from illegal access by third parties. In contrast with cryptography, which encrypts the transmitted information into meaningless form but can expose the secret information to detection by malicious attackers, message embedding (so-called steganography) hides secret information in a cover so that attackers find it difficult to judge whether the hidden secret information exists or not. Message embedding was originally formulated as the "Prisoners' Problem" by Simmons [1]. Take the following scenario for example. Alice and Bob are prisoners in separate jail cells far from each other who want to devise an escape plan. They are allowed to exchange messages, but their communication is monitored by a warden, Eve, who is always on the lookout for suspicious activity. Note that this scenario involves a passive warden who observes the traffic but cannot interfere with the communication. As a result, the prisoners can succeed in their escape if they can exchange secret messages without arousing the warden's suspicion.

The prisoners resort to message embedding and embed the details of their escape plan into innocuous looking objects. To embed a message, the sender slightly modifies the cover object and then generates the stego object. The main requirement of any message embedding system is statistical undetectability. Thus, the prisoners' goal is to hide the details of the escape plan so that warden Eve cannot tell whether the transmitted objects contain embedded messages or not. The formal definition of this requirement in the information-theoretic model was given by Cachin [2].

According to Kerckhoffs's principle, a cryptosystem should be secure even if everything about the system, except the key, is public knowledge. In other words, the security of a cryptosystem must depend solely on the key. Thus, Alice and Bob must use a key, which is a secret shared between them, to design their message embedding process. In general, the key is used to devise the selection rule for identifying a subset of the cover objects that could contain embedded messages. The placement of embedding changes in the cover object is called the selection channel. To minimize the detectability of embedded messages, the selection channel must be revealed as little as possible during communication since this knowledge can aid the warden [3]. However, an obvious problem at this point is that neither the warden nor the recipient know the selection channel and are thus unable to read the embedded message.

The study of such nonshared selection channels is equivalent to writing in a memory with defective cells [4,5]. The defective memory that is modeled as a discrete memory-less channel is a special case of the informed Gel'fand-Pinsker channel [6]. Heegard and El Gamal [7] studied storage capacity based on the Gel'fand-Pinsker theory for computer memory with defects. The message embedding technique in the passive warden's scenario focuses on embedding for the noise-free case. We also consider message embedding in grayscale images with squared error distortion in the present communication.

For example, we use simple least significant bit (LSB) embedding for digital images in which the set of least significant bits (LSBs) of all pixels is the array of binary cells. The LSBs of secret message-carrying pixels (the selection channel) are regarded as functioning cells, while the LSBs of unused pixels correspond to stuck cells. The challenge is to embed secret messages with nonshared selection channels so that the recipient, who has no information about the stuck cells (the selection channel), can still correctly extract the embedded messages.

The most important attribute influencing steganographic security of message embedding schemes is embedding efficiency [8,9], which is defined as the average number of secret bits embedded per embedding change. In steganography, embedding efficiency is used to quantify how effectively a given message embedding scheme embeds secret messages. In general, fewer changes during the embedding process mean a smaller chance that the embedding modifications will be detected. However, the number of embedding changes is not the only factor influencing security. In fact, for two message embedding schemes using the same embedding operation, the one that introduces fewer embedding changes will be harder to detect and thus will provide greater security. As a result, it is desirable to increase embedding efficiency in order to reduce the possibility of detection by a third party.

The disk modulo (DM) method [10,11] finds an approach to uniformly distribute files on multiple disks while maximizing parallel disk I/O access. The DM uses the Hamming weight of a binary string (modulo operation) to determine an appropriate disk for optimal match queries. To improve security, we propose message embedding using a DM allocation method that enables communication with nonshared selection channels. Moreover, we also show that the DM allocation method can be applied to improve the embedding efficiency of message embedding schemes.

To make this paper self-contained, in Section 10.2, we introduce the terminologies and the basic concepts of steganography necessary to explain the embedding method. We also state known bounds on achievable embedding efficiency in the same section. In Section 10.3, we review some traditional ±1 message embedding techniques. The proposed method using the DM allocation method is explained in Section 10.4. Performance analyses appear in Section 10.5 and the performance is compared to theoretically achievable bounds.

---

## 10.2 Message Embedding

We assume that the grayscale cover image $C = \{c_1, c_2, \ldots, c_n\}$ is a vector of integers in the range $[0,255]$ where $n$ is the number of pixels. Let $M = \{m_1, m_2, \ldots, m_w\}$ be the embedded messages with a probability $1/w$ and independent of $C$.

We use the embedding function Emb to modify the values of selected pixels so that the stego-image $S = \{s_1, s_2, ..., s_n\}$ conveys the desired message using the extraction function Ext:

$$S = \text{Emb}(C, M, K),$$

$$M = \text{Ext}(\text{Emb}(C, M, K), K),$$

where $K$ is the stego key shared between the sender and the recipient.

Note that the embedded message can always be extracted from the stego-image $S$ without error; that is, the message $M'$ can be extracted by a recipient so that

$$P(M' \neq M) = 0.$$

The stego-image must always be close to the cover image; that is, the expected distortion

$$D_{\text{exp}} = E[d(C, S)] = \sum_{i=1}^{w} d(C, \text{Emb}(C, M, K)) \times \text{Pr}(M = m_i),$$

should be as small as possible and

$$d(C, S) = \frac{1}{n} \sum_{i=1}^{n} D(c_i - s_i)$$

is the distortion between $C$ and $S$. Here, the distortion measure considered is the squared error: $D(c_i - s_i) = (c_i - s_i)^2$. The value $\log_2 |M|$ is called the embedding capacity (in bits) and

$$R = \frac{1}{n} \log_2 |M|$$

is called the embedding rate (so-called relative payload, in bpp).

We further defined $E = R/D_{\text{exp}}$ as embedding efficiency. Fridrich and Soukal [12] gave the following upper bound on embedding efficiency $E$ for the embedding rate $R$ for the LSB embedding scheme:

$$E \leq \frac{R}{H^{-1}(R)}, \quad 0 \leq R \leq 1,$$

where $H(x) = -x \log_2 x - (1 - x) \log_2 (1 - x)$ is the binary entropy function, and $H^{-1}(x)$ is the inverse function of $H(x)$.

Willems and van Dijk [8,13] defined a rate-distortion function to evaluate the performance of message embedding schemes. They show that it is not useful to consider schemes that have $|c_i - s_i| > 1$ for some component $i$. According to them, the squared error cannot be larger than one; hence, this measure for message embedding is called $\pm 1$ steganography. They give the rate-distortion function as the upper bound on the embedding rate of $\pm 1$ steganography subject to the constraint of expected distortion $D_{exp}$

$$r(D_{exp}) = \begin{cases} H(D_{exp}) + D_{exp}, & \text{for} \quad 0 \le D_{exp} \le 2/3 \\ \log_2 3, & \text{for} \quad D_{exp} > 2/3 \end{cases}.$$

The rate-distortion function tells us the large embedding rate, given a certain distortion level. A plot of $r(D_{exp})$ is shown in Figure 10.1. We can see that to achieve an embedding rate of 1, we need an average distortion of at least 0.22.

To evaluate embedding efficiency, we rewrite the rate-distortion function as an upper bound on embedding efficiency $E$ with a given embedding rate $R$

$$E = \frac{R}{r^{-1}(D_{exp})}, \quad 0 \le R \le \log_2 3,$$

where $r^{-1}$ is the inverse function of $r$. Figure 10.2 illustrates a theoretically achievable bound on the embedding efficiency for $\pm 1$ steganography.

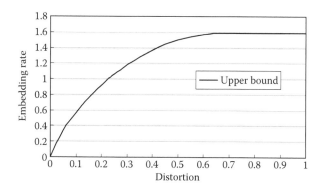

**FIGURE 10.1**
Rate-distortion function for squared error and $\pm 1$ embedding changes.

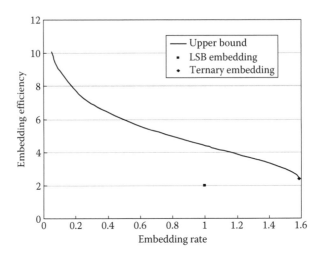

**FIGURE 10.2**
Theoretically achievable bound on embedding efficiency.

## 10.3 Conventional ±1 Steganography

Traditional embedding schemes for implementing ±1 steganography are introduced here to demonstrate the performance improvement offered by the proposed scheme. The LSB embedding is a simple algorithm for ±1 steganography, in which secret messages are conveyed in the LSB of the color values of selected pixels. Then, ternary embedding is applied to improve embedding efficiency. This section also presents performance analyses of conventional embedding methods for comparison with the proposed scheme.

### 10.3.1 Least-Significant-Bits Embedding

Suppose that a pixel's grayscale value $c_i$ is represented as a binary vector $b_7, \ldots, b_1, b_0$ so that

$$c_i = \sum_{h=0}^{7} b_h \times 2^h.$$

Secret messages are embedded only in the least significant bits (LSBs) of this binary representation. To implement ±1 steganography, we assume that each message bit $m_i$ is embedded in the 1 LSB of each pixel so that

$$s_i = m_i + \sum_{h=1}^{h} b_h \times 2^h,$$

where $s_i$ is the stego value. Thus, LSB embedding can be operated with a maximum error $D = 1$. The following examples show what happens in LSB embedding.

We obtain the expected distortion $D_{exp} = (0 + 1 + 1 + 0)/4 = 1/2$ and an embedding rate $R = 1/1 = 1$. Using the rate-distortion pair $(R, D_{exp}) = (1, 1/2)$, we achieve $r(D_{exp}) = 2D_{exp}$. Embedding efficiency is thus

$$E = \frac{R}{D_{exp}} = \frac{r(D_{exp})}{D_{exp}} = 2.$$

These results together with an embedding efficiency upper bound are illustrated in Figure 10.2. To obtain better embedding ability, ternary embedding was devised to improve efficiency.

## 10.3.2 Ternary Embedding

In ±1 steganography, there are three possibilities for each pixel: change it by ±1 or leave it unmodified. Thus, we assume that a pixel's grayscale value $c_i$ is said to be in group $g$ if $c_i$ modifies $3 = g$, where $g = 0, 1, 2$, and message symbol $m_i \in \{0, 1, 2\}$ is to be embedded in each pixel $c_i$. If $c_i$ is in group $m_i$, no modification is made. Otherwise, $c_i$ should be changed to $s_i$ in group $m_i$ so that $|s_i - c_i|$ is minimal. The recipient can determine the message symbol $mi$ by simply looking at the group of $s_i$. The following table describes this ternary embedding more clearly.

We can see that ternary embedding can be operated with maximum error $D = 1$. The expected distortion is $D_{exp} = (2 \times 3)/9 = 2/3$ and the corresponding embedding rate is $R = \log_2 3/1 \approx 1.58$. Using the resulting rate-distortion pair $(R, D_{exp}) = (\log_2 3, 2/3)$, the embedding efficiency can be computed as:

$$E = \frac{R}{D_{exp}} = \frac{\log_2 3}{\frac{2}{3}} \approx 2.38.$$

Combining the examples shown in Tables 10.1 and 10.2, and the comparison results demonstrated in Table 10.2, we can see that ternary embedding

**TABLE 10.1**

Examples of LSB Embedding

| $c_i$ | $m_i = 0$ | $m_i = 1$ |
|---|---|---|
| $128 = 10000000$ | $s_i = 128$ | $s_i = 129$ |
| | $D = 0$ | $D = 1$ |
| $129 = 10000001$ | $s_i = 128$ | $s_i = 129$ |
| | $D = 1$ | $D = 0$ |

**TABLE 10.2**

Example of Ternary Embedding

| $c_i$ | $m_i = 0$ | $m_i = 1$ | $m_i = 2$ |
|-------|-----------|-----------|-----------|
| 12 | $s_i = 12$ <br> $D = 0$ | $s_i = 13$ <br> $D = 1$ | $s_i = 11$ <br> $D = 1$ |
| 13 | $s_i = 12$ <br> $D = 1$ | $s_i = 13$ <br> $D = 0$ | $s_i = 14$ <br> $D = 1$ |
| 14 | $s_i = 15$ <br> $D = 1$ | $s_i = 13$ <br> $D = 1$ | $s_i = 14$ <br> $D = 0$ |

is more efficient than LSB embedding. After studying some message embedding schemes, we propose a ±1 message embedding scheme using a DM allocation method to minimize the impact of embedding.

## 10.4 Proposed Scheme

According to Willems and van Dijk [13], the impact of embedding becomes detectable whenever the maximum allowable error is larger than one; thus, we limit ourselves to ±1 embedding changes. The goal of message embedding is to design schemes that have a high embedding rate but a low embedding change rate. Thus, a DM allocation method is applied to the proposed scheme to decrease the embedding change rate. It also enables improved security in communication by eliminating the need to share the selection channel. Let us now review a few basic concepts from the DM allocation method that are relevant to explain the proposed scheme.

### 10.4.1 DM Allocation Method

Throughout this paper, we use some standard concepts and results from DM allocation technology. A file cannot reside in the memory in a large database; thus, all records are divided into buckets and stored on disks. The task is to allocate files among accessible disks to maximize disk access concurrency and therefore minimize response time. The problem of disk allocation for Cartesian product files on multiple disk systems was first considered by Du and Sobolewski [10]. They proposed an allocation method that assigns all buckets of a Cartesian product file to an $ND$-disk system and show that it is strictly optimal. Before describing the DM allocation method, it is necessary to state relevant definitions and assumptions.

Let $F$ be an $N$-attribute binary Cartesian product file. Each bucket [$b_1, b_2, ..., b_N$] is assigned to a disk unit

$$\left(\sum_{j=1}^{N} b_j \times p_j\right) \mod ND,$$

where each $p_j$ and $ND$ are relatively prime and $j = 1$ to $N$. To consider an $N$-attribute multiple-ary file, an efficient allocation method proposed by Chen et al. [11] was defined as follows. Each bucket $B = [b_1, b_2, ..., b_N]$ is assigned to a disk unit

$$DM(B) = (h_1(b_1) + h_2(b_2) + ... + h_N(b_N)) \mod ND,$$

where $h_i$ is a hash function. It is shown to be strictly optimal for all possible partial match queries when the number of disks is greater than 3. Let us consider an example. Let $F$ be a 3-attribute 3-ary file and $ND = 14$. Table 10.3 gives the hash function. The distribution for assigning all buckets to ND disks is shown in Table 10.4. Having explained the background, we now outline the principle of the proposed method.

**TABLE 10.3**

Hash Function for a 3-Attribute File Defined

| b | 0 | 1 | 2 |
|---|---|---|---|
| h1(b) | 0 | 1 | 2 |
| h2(b) | 0 | 3 | 6 |
| h3(b) | 0 | 4 | 9 |

*Source:* Chen, C. Y., Lin, H. F., Chang, C. C., and Lee, R. C. T. *IEEE Transactions on Knowledge and Data Engineering*, 9, 148–60, 1997.

**TABLE 10.4**

Distribution for Assigning all Buckets to *ND* Disks

| Bucket | Assigned Disk | Bucket | Assigned Disk | Bucket | Assigned Disk |
|---|---|---|---|---|---|
| [0, 0, 0] | 0 | [1, 0, 0] | 1 | [2, 0, 0] | 2 |
| [0, 0, 1] | 4 | [1, 0, 1] | 5 | [2, 0, 1] | 6 |
| [0, 0, 2] | 9 | [1, 0, 2] | 10 | [2, 0, 2] | 11 |
| [0, 1, 0] | 3 | [1, 1, 0] | 4 | [2, 1, 0] | 5 |
| [0, 1, 1] | 7 | [1, 1, 1] | 8 | [2, 1, 1] | 9 |
| [0, 1, 2] | 12 | [1, 1, 2] | 13 | [2, 1, 2] | 0 |
| [0, 2, 0] | 6 | [1, 2, 0] | 7 | [2, 2, 0] | 8 |
| [0, 2, 1] | 10 | [1, 2, 1] | 11 | [2, 2, 1] | 12 |
| [0, 2, 2] | 1 | [1, 2, 2] | 2 | [2, 2, 2] | 3 |

## 10.4.2 Embedding Process

For a grayscale cover image C, permute and divide all pixels of the cover image into many groups, each containing N pixels, according to a secret key. The number of pixel groups is denoted as NP. Denote the pixel values in the nth pixel group as $g(n, 1), g(n, 2), ..., g(n, N), n = 1, 2, ..., NP$. Suppose that message ND-ary symbol $m \in \{0, 1, ..., ND - 1\}$ is to be embedded in each pixel group.

1. For the nth pixel group, assign the pixel value $g(n, i)$ by using the following symbol assignment function

$$t(n, i) = Q(g(n, i)) = g(n, i) \bmod 3, 1 \leq i \leq N$$

and obtain the corresponding 3-ary symbols $t(n, 1), t(n, 2), ..., t(n, N)$.

2. Assign each pixel group $x = \{t(n, 1), t(n, 2), ..., t(n, N)\}$ to a disk unit by using DM allocation function DM(x).

3. If DM(x) = m, no modification is made.

4. If DM(x) ≠ m, find the closest stego-vector $y = \{t'(n, 1), t'(n, 2), ..., t'(n, N)\}$ so that DM(y) = m and the Hamming distance H(x, y) is as small as possible where

$$H(\mathbf{x}, \mathbf{y}) = \sum_{i=1}^{N} \left( t(n, i) \oplus t'(n, i) \right).$$

Assume a single embedding change, the symbol $t(n, j)$ gives the change, and $g(n, j)$ is changed into $g'(n, j)$ so that $Q(g'(n, j)) = t'(n, j)$ and $|g'(n, j) - g(n, j)|$ is minimal.

## 10.4.3 Extraction Process

For a grayscale stego-image S, permute and divide all pixels of the stego-image into many groups, each containing N pixels, according to a secret key. The number of pixel groups is denoted as NP. Denote the pixel values in the nth pixel group as $g'(n, 1), g'(n, 2), ..., g'(n, N), n = 1, 2, ..., NP$.

1. For the nth pixel group, assign the pixel value $g'(n, i)$ by using the following symbol assignment function

$$t'(n, i) = Q(g'(n, i)) = g'(n, i) \bmod 3, 1 \leq i \leq N$$

and obtain the corresponding 3-ary symbols $t'(n, 1), t'(n, 2), ..., t'(n, N)$.

2. Assign each pixel group $y = \{t'(n, 1), t'(n, 2), ..., t'(n, N)\}$ to a disk unit by using DM allocation function DM(y) to extract message $m = DM(y)$.

### 10.4.4 Example of Embedding and Extraction

Suppose that the *ND*-ary message symbol *m* is embedded into the first pixel group, where $g(1, 1) = 121$, $g(1, 2) = 122$, $g(1, 3) = 120$, $ND = 14$, and $m = 11$. The sender uses the symbol assignment function $Q$ to generate a 3-ary vector $\mathbf{x} = \{t(1, 1), t(1, 2), t(1, 3)\} = \{1, 2, 0\}$. As Table 10.4 shows, vector $\mathbf{x}$ is assigned to disk unit 7 by using the DM allocation function; that is, $DM(\mathbf{x}) = 7$. Since $DM(\mathbf{x}) \neq 11$, the stego-vector $y = \{1, 2, 1\}$ is found so that $DM(y) = 11$ and the Hamming distance between $\mathbf{x}$ and $\mathbf{y}$ is the smallest: $H(\mathbf{x}, \mathbf{y}) = 1$. The symbol $t(1, 3)$ gives the change, and $g(1, 3) = 120$ is changed into $g'(1, 3) = 121$ since $Q(121) = 1 = t'(1, 3)$ and $|121 - 120|$ is minimal.

Given the first stego pixel group where the pixel values are $g'(1, 1) = 121$, $g'(1, 2) = 122$, and $g'(1, 3) = 121$, the recipient generates a 3-ary vector $y = \{1, 2, 1\}$ and simply calculates the DM allocation function to extract the message symbol $DM(y) = 11$. In this case, we embed a 14-ary message symbol in three pixels with a distortion $D = 1$. Since the impact of embedding is limited to $\pm 1$ embedding changes, it becomes undetectable. Moreover, the recipient can read the correct messages but does not need to know the selection channel because the placement of the embedding changes is not communicated directly. As a result, the embedding ability is more efficient than with either LSB embedding or ternary embedding.

## 10.5 Performance Analysis

One of the goals of this paper is to maximize embedding capacity while keeping distortion as small as possible. In this section, we show how the parameters of the DM allocation method influence the proposed scheme. This section also discusses the more complicated issue of the role of group sizes and disk units as well as the choice of DM allocation function. Finally, we briefly discuss the embedding efficiency of the proposed scheme in comparison with LSB embedding and ternary embedding.

In the proposed method, an *ND*-ary message symbol is embedded into a pixel group that contains *N* pixels. Thus, the embedding rate is

$$R = \frac{1}{N} \log_2 |ND|.$$

We have performed a number of experiments to see how the embedding rate and distortion change for different group size *N*'s and disks size *ND*'s. A surprising result was that we get the upper bound on embedding efficiency whenever each $3^q$-ary message symbol is embedded into each pixel group of size $(3^q - 1)/2$, where *q* is an integer. Table 10.5 gives an example of how disk

**TABLE 10.5**

Maximum Error for Different Disks Sizes $ND$ $(N, ND)$

| $(N, ND)$ | Maximum Error $D_{\max}$ |
|---|---|
| $(4, 2)$ | 1 |
| $(4, 3)$ | 1 |
| ... | 1 |
| $(4, 9)$ | 1 |
| $(4, 10)$ | $1 + 1 + 1 + 1 = 4 = N$ |
| $(4, 11)$ | $N$ |

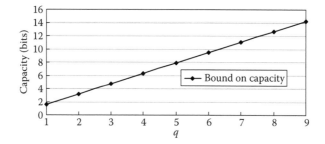

**FIGURE 10.3**

Maximum capacity in bits with $D_{\max} = 1$ for various $q$.

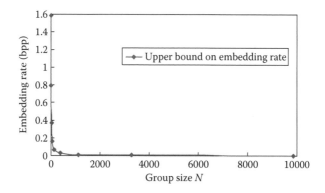

**FIGURE 10.4**

Upper bound on embedding rate with $D_{\max} = 1$ for group sizes $N$.

sizes $ND$ influence embedding distortion. We used groups of $N = 4$ with disk sizes $ND$ ranging from 2 to 11. We can see that the maximum embedding rate with group size $N = 4$ and $D_{\max} = 1$ is $\log_2 9/4$. Figure 10.3 shows the maximum capacity in bits with $D_{\max} = 1$ for various $q$. It is also very apparent that the capacity increases very fast with group sizes $N$.

The upper embedding rate bound with $D_{max} = 1$ for group sizes $N$ is given in Figure 10.4. We can see a very high variability in embedding rate for different group sizes $N$. In general, groups with a larger size have a smaller embedding rate. We looked at the influence of group size, and our experiments showed that groups of one pixel gave us the same embedding rate as with ternary embedding. We discovered there is a close relationship between group sizes and disk size. The influence of group sizes $N$ and disk sizes $ND$ is clear with the DM allocation function. The embedding rate rapidly decreases with group size, as can be seen in Figure 10.4. The next step is to outline a methodology for building a theoretical upper bound on the embedding efficiency for the proposed method.

We now state a theoretically achievable bound on embedding efficiency. Since there are $3^q$ ways for making one or fewer changes in $N$ pixels; that is $D_{max} \le 1$, the proposed scheme has the capacity in bits

$$Cap = \log_2 |ND| \le \log_2 3^q, \text{ for } q \in Z.$$

It can be rewritten as a theoretically asymptotic upper bound on embedding efficiency for a given embedding rate $R = Cap/N$:

$$E \le \frac{\log_2 3^q}{ND_{exp}}.$$

To determine the maximum average distortion, we consider the expected distortion of the upper bound. Here, we embed a $3^q$-ary message symbol in each pixel group of size $(3^q - 1)/2$ by performing, at most, one embedding change; that is, $D_{max} = 1$. Thus, we have the expected distortion

$$D_{exp} = \frac{2}{(2)\left(\dfrac{3^q - 1}{2}\right) + 1} = \frac{2}{3^q},$$

and an embedding rate of

$$R = \frac{\log_2 3^q}{\dfrac{3^q - 1}{2}} = \frac{2\log_2 3^q}{3^q - 1}.$$

Using the resulting rate-distortion pair $(R, D_{exp}) = (2\log_2 3^q/(3^q - 1), 2/3^q)$, we obtain an upper bound on the embedding efficiency

$$E = \frac{R}{D_{exp}} = \frac{3^q \log_2 3^q}{3^q - 1}, \text{ for } q \in \mathbf{Z},$$

which is defined as the average number of embedded bits that can be embedded per change.

We call an embedding using the pair $(N, ND)$ with group size $N$, embedding rate $R$, and the expected distortion $D_{exp}$ optimal if all other embedding with the same group size $N$ have rate $R' \leq R$ or distortion $D'_{exp} \geq D_{exp}$. Table 10.6 gives the optimal parameters $(N, ND)$ of the DM allocation function. Group sizes that are too small will generate capacity that is too small. Although larger group sizes generally achieve a larger capacity, the number of pixels will cause a decrease in embedding rate. We have also observed that ternary embedding is a special case of the proposed method with the parameters $(1, 1)$ of the DM allocation function.

Figure 10.5 plots our theoretically achievable upper bound on the rate-distortion pair $(R, D_{exp}) = (2 \log_2 3^q/(3^q - 1), 2/3^q)$ together with the upper bound on the rate-distortion function. Although for a given distortion our rate-distortion result is smaller than the bound on the rate-distortion function, it can be seen that the difference is very small for small distortion levels. Moreover, we can approach the upper bound of the rate-distortion function when our rate-distortion pair is $(R, D_{exp}) = (\log_2 3, 0.67)$.

The goal of message embedding is to design schemes with a high embedding efficiency; that is, a high embedding rate but low change density. Thus, we compare the performance of the proposed method, LSB embedding and ternary embedding. The comparison results, given in Figure 10.6, show that the proposed method is more efficient than the LSB embedding, and is furthermore close to the bound in embedding efficiency given by Willems and van Dijk [13]. From Figure 10.6, we can see that the embedding rate decreases with increasing embedding efficiency. Note that when $(N, ND) = (1, 3)$, both the proposed method and ternary embedding provide the same family of schemes, which can embed $\log_2 3$ bits of messages into one pixel with 2/3

**TABLE 10.6**

Optimal Parameters $(N, ND)$ of the DM Allocation Function

| $N$  | 1 | 4 | 13 | 40 | 121 | 364 | 1,093 | 3,280 | 9,841 |
|------|---|---|----|----|-----|-----|-------|-------|-------|
| $ND$ | 3 | 9 | 27 | 81 | 243 | 729 | 2,187 | 6,561 | 19,683 |

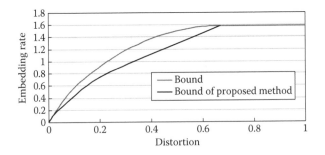

**FIGURE 10.5**
Theoretically achievable upper bound on rate-distortion function.

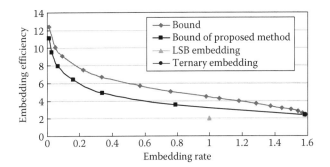

**FIGURE 10.6**
Performance comparison between the proposed method, LSB embedding, and ternary embedding.

changes on average. We also observe that an increase of $N$ results in a low change density. Low change density preserves the quality of the stego-image and allows the embedding to remain imperceptible to bad actors, resulting in improved steganographic security. As a result, the proposed method is best among these conventional ±1 message embedding techniques.

Finally, we close this section with a note on how the proposed method improves security. In the proposed method, the impact of embedding becomes undetectable since it is limited to ±1 embedding changes. Further, the DM allocation function enables communication without sharing the placement of embedding changes. The warden is unable to read the secret messages because the placement of the embedding changes is not communicated directly. Thus, the proposed method minimizes the detectability of the hidden data and can approach the theoretically upper bound of embedding efficiency.

## 10.6 Conclusions

In this paper, we propose a novel message embedding method that uses the concept of DM allocation method. We limit ourselves to so-called ±1 embedding changes, which requires that the sender modifies each pixel by at most one, the smallest possible modification. We also give a theoretically achievable bound on embedding efficiency for a given embedding rate. In the proposed method, $\log_2 3^q$ bits of messages can be embedded in $(3^q - 1)/2$ pixels by performing at most one embedding change. Therefore, our communication setup provides better embedding efficiency compared with traditional ±1 message embedding methods.

Moreover, the DM allocation method-based message embedding enables communication without the need to share the placement of the embedding changes, thus minimizing the detectability of the hidden messages. As a result, security is improved compared to schemes with the public selection channel because the embedded areas remain unavailable to both the recipient and any illegitimate attacker.

## References

1. G. J. Simmons. 1984, August. The prisoners' problem and the subliminal channel. Proceedings of CRYPTO 83: Advances in Cryptology, Santa Barbara, CA, 51–67.
2. C. Cachin. 1998, April. An information-theoretic model for steganography. Proceedings of the Second International Workshop on Information Hiding, Lecture Notes in Computer Science, vol. 1525, 306–18.
3. A. Westfeld and R. Böhme. 2004, May. Exploiting preserved statistics for steganalysis. Proceedings of the 6th International Workshop on Information Hiding, Lecture Notes in Computer Science, vol. 3200, Toronto, Canada, 82–96.
4. A. V. Kuznetsov and B. S. Tsybakov. 1974. Coding in a memory with defective cells. *Problemy Peredachi Informatsii* 10 (2): 52–60.
5. J. Fridrich, M. Goljan, P. Lisoněk, and D. Soukal. 2005, October. Writing on wet paper. *IEEE Transactions on Signal Processing* 53 (10): 3923–35.
6. S. I. Gel'fand and M. S. Pinsker. 1980. Coding for channel with random parameters. *Problems of Control and Information Theory* 9 (1): 19–31.
7. C. Heegard and A. El Gamal. 1983, September. On the capacity of computer memory with defects. *IEEE Transactions on Information Theory* 29 (5): 731–39.
8. M. van Dijk and F. Willems. 2001, May. Embedding information in grayscale images. Proceedings of 22nd Symposium on Information and Communication Theory in the Benelux, Enschede, The Netherlands, 147–54.
9. J. Fridrich, P. Lisoněk, and D. Soukal. 2007, July. On steganographic embedding efficiency. Proceedings of the 8th International Workshop on Information Hiding, Lecture Notes in Computer Science, vol. 4437, Alexandria, VA, 282–96.
10. H. C. Du and J. S. Sobolewski. 1982, March. Disk allocation for Cartesian product files on multiple-disk systems. *ACM Transactions on Database Systems* 7 (1): 82–101.
11. C. Y. Chen, H. F. Lin, C. C. Chang, and R. C. T. Lee. 1997. Optimal bucket allocation design of k-ary MKH files for partial match retrieval. *IEEE Transactions on Knowledge and Data Engineering* 9 (1): 148–60.
12. J. Fridrich and D. Soukal. 2006, September. Matrix embedding for large payloads. *IEEE Transactions on Information Forensics and Security* 1 (3): 390–95.
13. F. M. J. Willems and M. van Dijk. 2005, March. Capacity and codes for embedding information in gray-scale signals. *IEEE Transactions on Information Theory* 51 (3): 1209–14.

# 11

## Indexing and Retrieval of Ubiquitous Visual Information

Lei Ye, Jianhua Ma, and Runhe Huang

**CONTENTS**

It is a challenge to index and retrieve visual information generated from ubiquitous multimodal sensors based on its perceptual relevance. This chapter describes the key techniques and some fundamental problems in indexing and retrieving visual information including visual information description, visual similarity, and feature aggregation. First, standard moving picture experts group (MPEG-7) visual features and their similarity matching methods are introduced. Second, experimental evaluation of some basic visual features is presented to help readers with practical knowledge of their performance in visual information indexing and retrieval. Then, two key problems of visual similarity and feature aggregation are discussed for construction of comprehensive indexing and retrieval systems. The system performances resulted from different aggregation strategies are demonstrated. Finally, a practical content-based image retrieval system, Web Image Search, is presented with its structure and user process in the application of techniques are presented in this chapter.

## 11.1 Introduction

Ubiquitous or pervasive computing defines a new paradigm for the twenty-first century as a result of ubiquity in computing beyond the mainframe, personal computers, and advances of the media sensing, computing, and usage along with the developments and applications of communication and networking. Ubiquitous multimodal sensors capture visual information in the form of moving and still pictures. In many applications, visual information needs to be organized according to relevance or compared with examples in a networked database to provide service to users. In either case, it is necessary to index the visual information based on its perceptual content and extract multimedia documents.

> *Scenario 1:* In a ubiquitous environment, often networked with the Internet, certain images are required to be searched over the Internet. For example, a user may take a picture of a landmark location and then search for similar pictures taken by other people that would be available on the Internet, anywhere and everywhere.
>
> *Scenario 2:* All images or videos captured by wearable cameras are kept in a personal multimedia database called lifelog image/video, which is ubiquitous since images or videos are taken everywhere. Indexing, retrieval, and summarization of the lifelong images and videos will help the user to relate his or her past experience.
>
> *Scenario 3:* The ubiquitous videos are captured by surveillance cameras in multiple locations (e.g., shops, hospitals, streets, etc.). When some incident occurs, the surveillance camera key video frames could be retrieved to identify certain activities.

This chapter will describe the key techniques for indexing and searching for visual information including visual information description, visual similarity, and feature aggregation.

## 11.2  Visual Information Description

To index and search visual information, the images and videos need to be characterized using quantified visual features. There are a number of visual features proposed in the literature [5,11]. In this section, we will focus on the visual features standardized by the MPEG-7 [1,2].

MPEG-7, formally known as Multimedia Content Description Interface, provides interoperability among systems and applications in audiovisual content descriptions that help to identify, index, and retrieve audiovisual information. See Section 11.3 for the visual information description of MPEG-7 Visual. It lists a set of visual descriptors for specific visual features such as color, texture, shape, and motion. The main descriptors are briefly explained in the following. More comprehensive specification can be found in ISO/IEC information technology [1,2].

### 11.2.1  Color Feature Descriptors

One of the most important visual perceptions for humans is color. Color has proved to be one of the most effective and efficient features for similarity indexing and retrieval and therefore plays a critical role in visual similarity of content. Several feature descriptors based on color are specified in MPEG-7.

#### *11.2.1.1  Dominant Color Descriptor (DCD)*

This descriptor specifies a set of dominant colors. The dominant color feature is extracted from a set of pixel color values in the RGB color space and quantizes the color vectors in the image based on the Generalized Lloyd Algorithm (GLA). The color vectors are the centroids of color clusters in the CIE LUV color space, called ColorBin and the number of bins is denoted as Bin-Num. The dominant colors are then represented in the RGB color space, which is converted back from color centroids in the CIR LUV color space. Along with the color vectors, percentage and color variance are specified for pixels corresponding to the dominant colors. In addition, spatial coherency is specified in the descriptor as a single value computed by the weighted sum of predominant color spatial coherency, proportional to the number of pixels corresponding to each dominant color. The predominant color spatial coherency describes pixels corresponding to the dominant color that appear to be a solid color. The extraction algorithm is described in detail in ISO/IEC information technology [2].

The dominant color descriptor (DCD), as the name suggests, captures the dominant colors in pictures.

The MPEG-7 standard defines two forms of description representations. One is defined in Description Definition Language (DDL) and the other in binary form. To provide readers with some idea of how features are represented, the syntaxes of the two representations are presented here for the first descriptor introduced in this chapter. Others are omitted and can be found in ISO/IEC information technology [1].

- **DDL Representation Syntax**

```
<complexType name="DominantColorType" final="#all">
  <complexContent>
    <extension base="mpeg7:VisualDType">
      <sequence>
        <element name="ColorSpace" type="mpeg7:ColorSpaceType"
minOccurs="0"/>
        <element name="ColorQuantization" type="mpeg7:ColorQua
ntizationType" minOccurs="0"/>
        <element name="SpatialCoherency"
type="mpeg7:unsigned5"/>
        <element name="Value" minOccurs="1" maxOccurs="8">
        <complexType>
          <sequence>
            <element name="Percentage" type="mpeg7:unsigned5"/>
            <element name="Index">
              <simpleType>
                <restriction>
                  <simpleType>
                    <list itemType="mpeg7:unsigned12"/>
                  </simpleType>
                  <length value="3"/>
                </restriction>
              </simpleType>
            </element>
            <element name="ColorVariance" minOccurs="0">
              <simpleType>
                <restriction>
                  <simpleType>
                  <list itemType="mpeg7:unsigned1"/>
                  </simpleType>
                  <length value="3"/>
                </restriction>
              </simpleType>
            </element>
          </sequence>
        </complexType>
      </sequence>
    </extension>
  </complexContent>
</complexType>
```

- **Binary Representation Syntax**

  The binary representation syntax is presented in Table 11.1.

  To evaluate the feature similarity, a matching method for each descriptor is described by MPEG-7. The matching method using the DCD is expressed as follows.

$$\text{Difference}(D_1, D_2) = W_1 \times SC_{\text{Diff}} \times DC_{\text{Diff}} + W_2 \times DC_{\text{Diff}}, \qquad (11.1)$$

where $SC_{\text{Diff}} = |\text{SpatialCoherency}_1 - \text{SpatialCoherency}_2|$, and $D_1$ and $D_2$ are the *DCD* of two images, respectively; $DC_{\text{Diff}}$ is the difference between two sets of dominant colors; $W_1$ and $W_2$ are weights of the first and second terms, respectively. $DC_{\text{Diff}}$ can be computed by the following distance function.

The similarity between two DCDs, $F_1$ and $F_2$, can be measured by the following distance function

$$D(F_1, F_2) = \sqrt{\sum_{i=1}^{N_1} p_{1i}^2 + \sum_{j=1}^{N_2} p_{2j}^2 - \sum_{i-1}^{N_1} \sum_{j=1}^{N_2} 2a_{1i,2j} p_{1i} p_{2j}}, \qquad (11.2)$$

**TABLE 11.1**

Dominant Color Binary Representation Syntax

| DominantColor { | Number of bits | Mnemonic |
|---|---|---|
| Size | 3 | uimsbf |
| Color Space Present | 1 | bslbf |
| if(ColorSpacePresent){ | | |
| Color Space | Color Space | Color SpaceType |
| } | | |
| Color Quantization Present | 1 | bslbf |
| if(Color Quantization Present){ | | |
| Color Quantization | Color Quantization | Color QuantizationType |
| } | | |
| Variance Present | 1 | bslbf |
| Spatial Coherency | 5 | uimsbf |
| for( k=0; k<Size; k++ ) { | | |
| Percentage | 5 | uimsbf |
| for( m=0; m<3; m++ ) { | | |
| Index | 1–12 | uimsbf |
| if(VariancePresent) { | | |
| Color Variance | 1 | uimsbf |
| } | | |
| } | | |
| } | | |
| } | | |

where $p_{ki}$ is percentage and $a_{k,l}$ is the similarity coefficient between two colors $c_k$ and $c_l$,

$$a_{a,l} = \begin{cases} 1 - d_{k,l}/d_{\max} & d_{k,l} \le T_d \\ 0 & d_{k,l} > T_d \end{cases}' \tag{11.3}$$

where $d_{k,l}$ is the Euclidean distance between two colors

$$d_{k,l} = \| c_k - c_l \|. \tag{11.4}$$

$T_a$ is the maximum distance for two colors to be considered similar and $d_{\max} = \alpha T_d$.

If color variances are present, the following similarity measure is used.

$$D_v(F_1, F_2) = \sqrt{\sum_{i=1}^{N_1} \sum_{j=1}^{N_2} p_{1i} p_{1j} f_{1i1j} + \sum_{i=1}^{N_1} \sum_{j=1}^{N_2} p_{2i} p_{2j} f_{2i2j} - \sum_{i=1}^{N_1} \sum_{j=1}^{N_2} 2 p_{1i} p_{2j} f_{1i2j}}, \tag{11.5}$$

where

$$f_{xiyj} = \frac{1}{2\pi \sqrt{v_{xiyjl} v_{xiyju} v_{xiyjv}}} \exp\left[ -\left( \frac{c_{xiyjl}}{v_{xiyjl}} + \frac{c_{xiyju}}{v_{xiyju}} + \frac{c_{xiyjv}}{v_{xiyjv}} \right) \right], \tag{11.6}$$

and

$$c_{xiyjl} = (c_{xil} - c_{yjl})^2 \tag{11.7}$$

$$v_{xiyjl} = (v_{xil} - v_{yjl})^2. \tag{11.8}$$

This matching method is a good example of applying different similarity measures when different data are available in the descriptor.

### 11.2.1.2 Scalable Color Descriptor (SCD)

This descriptor is a color histogram in hue, saturation and vale (HSV) color space that is encoded by a Haar transform. The histogram in HSV color space is uniformly quantized into 256 bins and the histogram values are then nonlinearly quantized. The 4-bit values then are transformed by a Haar transform.

The SCD captures the color distribution in images.

The similarity matching for the SCD can be performed in both the Haar transform and histogram domains. The $L_1$ norm is recommended for matching in both domains.

### 11.2.1.3 Color Layout Descriptor (CLD)

This descriptor specifies a spatial distribution of colors in $YC_bC_r$ color space. It is extracted from the $8 \times 8$ array of local representative colors, which are defined by the *DCT* coefficients of color components in one of the 64 partitioned blocks of images or video frames.

The CLD captures the spatial distribution of colors.

The similarity matching for the CLD can be measured by the following distance function.

$$
\begin{aligned}
D = &\sqrt{\sum_{i=0}^{Max\{Number\ of\ YCoef-1\}} \lambda_{Yi}(YCoef\ f_1[i] - YCoef\ f_2[i])^2} \\
&+ \sqrt{\sum_{i=0}^{Max\{Number\ of\ YCoef-1\}} \lambda_{C_bi}(C_bCoef\ f_1[i] - C_bCoef\ f_2[i])^2} \qquad (11.9) \\
&+ \sqrt{\sum_{i=0}^{Max\{Number\ of\ YCoef-1\}} \lambda_{C_ri}(C_rCoef\ f_1[i] - C_rCoef\ f_2[i])^2},
\end{aligned}
$$

where the $\lambda$'s are weights for each coefficient.

### 11.2.1.4 Color Structure Descriptor (CSD)

This descriptor specifies both color and the structure of color in hue, max, min, diff (HMMD) color space. It characterizes the relative frequency of $8 \times 8$ structuring elements that contain image pixels with a particular color scanning the image, instead of characterizing the relative frequency of individual image pixels with a particular color, which is what color descriptors such as SCD does. This descriptor is capable of distinguishing between two images of identical amounts of a given color where the structure of the groups of pixels having that color is different. The CSD containing 256 bins is extracted directly from the image based on a 256-cell quantization of the HMMD color space. Those containing 128, 64, and 32 bins are extracted based on unification of the bins of the 256-bin descriptor. The descriptor elements are organized in an M element array of 8-bit integer values. The bins of an M-bin descriptor are associated bijectively to the M quantized colors of the M-cell color space. For the M = 256 case, the bin value represents the number of structuring elements in the image that contain one or more pixels with a color.

The CSD captures the local structure of color.

The similarity matching for CSD can be measured by $L_1$ norm.

## 11.2.2 Texture Feature Descriptors

Many natural and artificial pictures contain strong visual patterns, called texture, as a result of multiple colors or intensities in the image. Several feature descriptors based on texture are specified in MPEG-7.

### 11.2.2.1 Homogeneous Texture Descriptor (HTD)

This descriptor specifies the region texture using the energy and energy deviation in a set of frequency channels. It is extracted from the partitioned frequency space in the polar frequency domain with equal angles of 30° in the angular direction and with an octave division in the radial direction. The frequency layout of 30 feature channels is shown in Figure 11.1. The 2-D Gabor function is applied in each feature channel. The energy $e_i$ of a feature channel is defined as the log-scaled sum of the square of the Gabor-filtered Fourier transform coefficients of an image. The energy deviation $d_i$ of a feature channel is defined as the log-scaled standard deviation of the square of the Gabor-filtered Fourier transform coefficients of an image.

The homogeneous texture descriptor captures the texture of homogeneous properties.

The similarity matching for homogeneous texture descriptor can be measured by the following function

$$D = \sum_k \left| \frac{TD_1(k) - TD_2(k)}{\alpha(k)} \right|, \tag{11.10}$$

where $TD$'s are descriptors of two images and the normalization value $\alpha(k)$ is recommended as the standard deviation of the descriptors of all candidate images.

The homogeneous texture descriptor is sensitive to intensity, rotation, and scaling of images. The following matching methods provide invariant matching.

- **Intensity-Invariant Matching**
  The average intensity of the image in the feature descriptor is eliminated for similarity matching.

- **Rotation-Invariant Matching**
  The query Image feature vector is shifted in angular directions, $TD_{query}|_{m\varphi}$, and the similarity is computed as the minimum among all shifted versions as

$$D(TD_{query}, TD_{database}) = \min\{D(TD_{query}, TD_{database}, m\phi) \mid m = 0, \dots, 5\}, \tag{11.11}$$

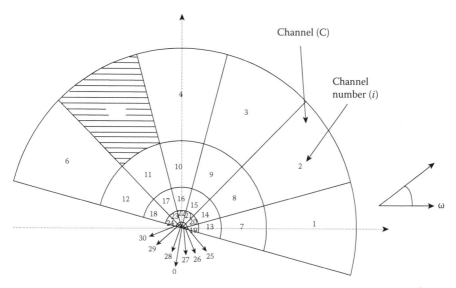

**FIGURE 11.1**
Frequency layout. (From ISO/IEC TR 15938-8:2002 Information technology.)

where

$$D(TD_{\text{query}}, TD_{\text{database}}, m\phi) = D(TD_{\text{query}} |_{m\phi}, TD_{\text{database}}), \qquad (11.12)$$

and $\phi = 30°$.

- **Scale-Invariant Matching**
  The query image is scaled with $N$ different zooming factors, $TD_{\text{query}} |_{n}$, and the similarity is computed as the minimum among all scaled versions as

$$D(TD_{\text{query}}, TD_{\text{database}}) = \min\{D(TD_{\text{query}}, TD_{\text{database}}, n) \,|\, n = 0, \ldots, N\}, \quad (11.13)$$

where

$$D(TD_{\text{query}}, TD_{\text{database}}, n) = D(TD_{\text{query}} |_{n}, TD_{\text{database}}). \qquad (11.14)$$

This matching method is a good example of invariant matching for variant features.

### 11.2.2.2 Texture Browsing Descriptor (TBD)

This descriptor specifies the perceptual characterization of a texture, which is similar to a human characterization, in terms of regularity, coarseness, and directionality. It is extracted from the frequency layout as shown in Figure 11.1 at different scales decomposed using Gabor wavelet. The dominant direction is based on directional histograms computed from $4 \times 6$ filtered images that are used to construct four directional histograms. Each histogram essentially captures the first-order statistics of the distribution of edge pixels over $K = 6$ directions. The scale is computed based on contrast of projections along two dominant orientations. The regularity is computed by assigning credits to those rejections that pass a consistency check as the periodicity of the basic texture elements, called textons.

The texture browsing descriptor (TBD) is not a typical similarity matching descriptor. It is used for browsing images. One can select any of the components and browse along that dimension.

### 11.2.2.3 Edge Histogram Descriptor (EHD)

The EHD specifies the spatial distribution of five types of edges. It extracted small square image blocks in each of the 16 nonoverlapping subimages. The directional edge information is obtained from mean values for four sub-blocks of image blocks. If the maximum of the five directional edge strengths is greater than a thresholding value, the block is decided to have the corresponding edge type.

The EHD captures spatial distribution of edges in images that are nonhomogeneous texture.

The similarity matching for EHD can be measured by the following function

$$D = \sum_{i=0}^{79} |\,\text{Local\_Edge}_1[i] - \text{Local\_Edge}_2[i]\,|$$

$$+\, 5 \times \sum_{i=0}^{4} |\,\text{Global\_Edge}_1[i] - \text{Global\_Edge}_2[i]\,| \qquad (11.15)$$

$$+\, \sum_{i=0}^{64} |\,\text{Semi\_Global\_Edge}_1[i] - \text{Semi\_Global\_Edge}_2[i]\,|,$$

where the Global_Edge histogram and Semi_Global edge histogram are generated from the 80 local edge histograms in the descriptor. The Global_Edge histogram is obtained by accumulating and normalizing the bin values of the corresponding edge type of the local edge histogram while the histogram is obtained by the grouping subsets of the local edge histogram as shown in Figure 11.2.

**FIGURE 11.2**
Subimages for Semi_Global histogram. (From ISO/IEC TR 15938-8:2002 Information technology.)

This matching method is a good example of the distance calculation of the derived secondary features from the primary feature available in a descriptor.

### 11.2.3 Shape Feature Descriptors

Shape is a prominent feature of image objects perceived by humans and provides powerful visual characters of images for similarity matching.

The current application of these feature descriptors includes binary images, trademarks, and presegmented images. The application of shape feature descriptors is limited by the availability of shape information in images, which is hard to obtain automatically from ubiquitous images and videos. The more detailed description of these descriptors can be found in ISO/IEC information technology [1].

The MPEG-7 standard defines several feature descriptors based on shape.

- **Region Shape Descriptor**
  The region shape descriptor specifies any type of shape ranges of an object in images from a simple shape like a triangle or circle to complex ones like a donut or a trademark. It uses a set of ART (Angular Radial Transform) coefficients and robust to noises that may be introduced in the process of segmentation. It allows minor nonrigid deformations, split objects, and even an object with separated regions.

- **Contour Shape Descriptor**
  The contour shape descriptor specifies a closed contour of a 2-D object or region in images. It is based on the Curvature Scale Space (CSS) representation of the contour extracted from a list of contour points of an object or region. It can distinguish between shapes that have similar region-shape properties but different contour-shape properties and robust to noise, nonrigid deformations, and distortions in the contour due to perspective transformations.

- **Shape 3-D Descriptor**
  The Shape 3-D descriptor specifies the shape of 3-D mesh models. It uses a function of two principal curvatures. However, it is strongly dependent on the accuracy of estimates of the principal curvatures.

### 11.2.4 Motion Feature Descriptors

Although color, texture, and shape feature descriptors can be used in indexing video frames, motion features in videos can provide more visual information about motion characteristics in video sequences. The more detailed descriptions of motion features and feature extraction are beyond the limitations of this chapter and can be found in ISO/IEC information technology [1].

The MPEG-7 standard defines several feature descriptors based on motion.

- **Camera Motion Descriptor**
  The camera motion descriptor specifies 3-D camera motion parameters that automatically are extracted or generated by capture devices. It supports the following basic camera operations: fixed, panning, tracking, tilting, booming, zooming, dolling, and rolling. The subshots with all frames characterized by a particular type of camera motion determine the building blocks for this camera motion descriptor.

- **Motion Trajectory Descriptor**
  The motion trajectory descriptor specifies the motion trajectory of a moving region, defined as the spatio-temporal replacement of one of its representative points such as its centroid. It can be used to check if objects enter sensitive areas for surveillance videos.

- **Parmetric Motion Descriptor**
  The parmetric motion descriptor specifies the motion of objects in video sequences, as well as global motion. The descriptor characterizes the evolution of arbitrarily shaped regions over time in terms of a 2-D geometric transform. The descriptor describes translational, ration-scaling, affine, perspective, and quadratic models of 2-D motion from one frame to the next.

- **Motion Activity Descriptor**
  The motion activity descriptor specifies the motion activities including intensity, direction, spatial distribution, spatial localization, and

temporal distribution of activity. It is based on $16 \times 16$ macroblock motion vectors and captures the intensity of motion and pace of action in a video segment.

## 11.2.5 Experimental Evaluation

Each visual feature, as described in the previous subsection, characterizes one aspect of visual information of images and videos. They have various powers for indexing and retrieval of images and videos in various application domains. Their statistical properties and expressiveness for perceived similarity are investigated. A statistical analysis reveals the properties and qualities of the descriptors while comparative physical and perceived similarity analysis shows their expressiveness for indexing and retrieving visual information that is consistent to the human perception. Some interesting results for basic visual features are presented in the following. These provide guidance on future selection of specific applications.

### 11.2.5.1 Statistical Analysis

Four statistical methods, mean and variance of description elements, distribution of elements, cluster analysis, and factor analysis are used to reveal the redundancy and sensitivity of MPEG-7 visual feature descriptors [6].

- **Redundancy Analysis**
  Generally, all MPEG-7 descriptors are highly redundant. The relationship of description elements to redundancy-free varies from 4:1 to 7:1. color layout, color structure, and SCDs are independent of each other and other descriptors. The DCD is absolutely independent of all other color descriptors and shows no similarities to texture descriptors. The elements of the homogeneous text descriptor are highly self-similar and redundant and the EHD consists of clusters of redundant elements while the TBD is independent of all others.

- **Sensitivity Analysis**
  All color descriptors are robust to variations in the quality of the content, however color layout, color structure, and scalable color (except dominant color), descriptors perform poorly on artificial object with few colors and very badly on monochrome content. The EHD performs excellently while the homogeneous texture descriptor works poorly for color images and relatively poorly for texture regions. The text browsing descriptor produces partially ambiguous results. The region shape descriptor performs well for any type of image content.

The same study has shown that most descriptors have peaks and holes in the distribution of descriptor elements that can be compressed.

### 11.2.5.2 Perceptive Analysis

A comparative study on physical similarity measured by descriptors and perceived similarity by human subjects is important to show the expressiveness, or effectiveness, of various descriptors for indexing and retrieval applications. Experiments have been carried out for MPEG-7 color and texture descriptors [13].

The groundtruth sets of the queries are generated from the human subjects for each query. Then the MEPG-7 descriptors are applied to retrieve the images and the performance is measured by precision and recall, which will be described in Section 11.3. Figure 11.3 shows the results for three different queries named Bush, Opera House, and Party. It shows that different descriptors perform differently for different queries. For the Bush query, the color structure works best and the EHD works significantly better than color layout and homogeneous texture descriptors while for the Opera House query, the color layout and edge histogram descriptors do equally better than the color structure and homogeneous texture descriptors. However, for the Party query, all descriptors perform equally poor for almost all of the recall range. There is no one descriptor that can perform satisfactorily for all queries. These observations have led to the research on methods to aggregate the individual feature similarities and to design the aggregation strategies based on individual queries that will be discussed in Section 3.

### 11.2.5.3 Effectiveness for Indexing

To evaluate the effectiveness of MPEG-7 descriptors for indexing applications, experiments have also been conducted to describe the evolutional changes in image time sequences [12]. It shows the applicability of MPEG-7 for measuring indexing images with gradual changes. Figure 11.4 shows the visual changes of a rotting banana over time and Figure 11.5 shows the normalized distances of these changes measured by MPEG-7 color and texture descriptors using their respective distance functions presented in Section 11.2 with the first image as the reference. The results demonstrate that all four descriptors, dominant color, color layout, scalable color, and homogeneous texture are able to index the changing images correctly with their monotonic increasing distances. It is also found that the distance functions of the color layout and SCDs are more consistent to the degree of the perceived visual changes, noting that the small visual change of the banana from time point 1 to time point 2 are measured by these two descriptors with small distances rather than by the other two.

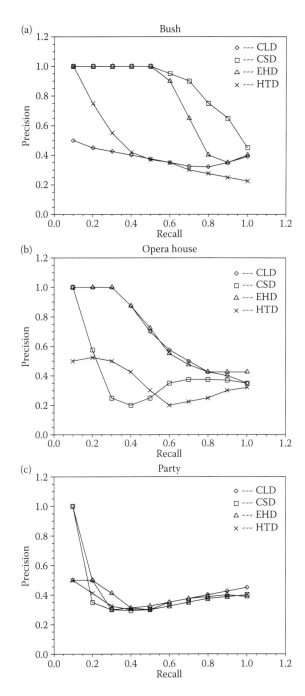

**FIGURE 11.3**
Retrieval results of MPEG-7 descriptors. (From Zhong, Y., Master's thesis, University of Wollongong, 2007. With permission.)

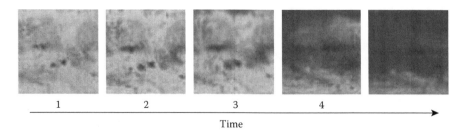

**FIGURE 11.4**
Visual changes of a banana over time. (From Ye, L., Cao, L., Ogunbona, P., and Li, W., *Visual Content Processing and Representation*, Springer, Berlin-Heidelberg, 189–97, 2006. With permission.)

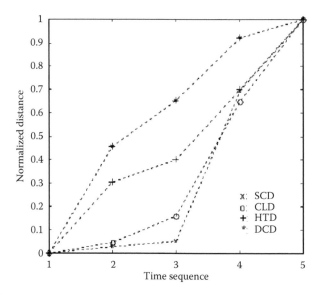

**FIGURE 11.5**
Normalized distances of visual changes measured by MPEG-7 descriptors. (From Ye, L., Cao, L., Ogunbona, P., and Li, W., *Visual Content Processing and Representation*, Springer, Berlin-Heidelberg, 189–97, 2006. With permission.)

## 11.3 Visual Similarity and Feature Aggregation

Visual similarity of images and videos has been believed to be perceived using multiple visual features, though this process is little known. The evaluation results in Section 11.2 have suggested that no single color descriptor can adequately quantify the whole visual content of images and videos perceived collectively with multiple visual features and no single descriptor can perform satisfactorily for any and all queries. Individual color descriptors characterize one certain aspect of visual information. Visual content ought to be characterized by multiple features to better match the perceived similarity. The MPEG-7 standards specify individual visual features and do not provide solutions to this problem, which is open for further research and innovations.

### 11.3.1 Relevance and Visual Similarity

Let $Q$ be the query image set; $C_Q$ the relevant subset of $Q$ in the image collection $C$. An image retrieval system $\Leftrightarrow$ is a mapping from $Q$ to $\Leftrightarrow_Q$, as

$$\Re : Q \to C_Q \subset C. \tag{11.16}$$

Therefore, in set theoretic term, an image retrieval system $\Leftrightarrow$ is realized as an operation to determine the retrieved images in the collection, that is $C_Q$, as

$$C_Q = \{I_i \in C : P_Q(I_i) \text{ is true}\}, \tag{11.17}$$

where $P_Q$ is a predicate that is true for the image $I_i$ satisfying the predicate given the query $Q$. An image is relevant to the query if the image meets the information needs of the user as represented by the query. The ultimate goal of the content-based image retrieval (CBIR) research is to define retrieval strategies that the predicate $P_Q$ is true when the image is relevant to the query $Q$ and false when the image is irrelevant to the query $Q$. This is not a trivial problem. Unfortunately, so far no such perfect retrieval strategies to realize the predicate have been discovered for any practical systems that are able to guarantee that the retrieved image set $C_Q$ contains all and only relevant images for a query. What the CBIR research strives to achieve is to formulate the retrieval strategies so that $C_Q$ contains as many relevant images and few irrelevant ones as possible. Therefore, it is necessary to evaluate the performance of CBIR systems. The most commonly used measure is the precision and recall. If $T \subset C$ is the subset of relevant images in $C$ and the $|T|$ the number of images in $T$, where $|\cdot|$ is the counting measure, then

$$\text{precision} = \frac{|T \cap C_Q|}{|C_Q|} \tag{11.18}$$

and

$$\text{recall} = \frac{|T \cap C_Q|}{|T|}, \tag{11.19}$$

where $\cap$ is a set intersection. It is worthwhile to point out that the relevant image subset may not be known in some cases, such as Web image retrieval.

The fundamental problem of a probabilistic image retrieval system is what the probability is that an image is relevant with respect to the query example.

The event $r$ denotes that the image $I$ is relevant when $P_Q(I)$ is true. The question is then answered by estimating the probability of relevance $P(r|I, Q)$. According to the Probability Ranking Principle, the optimality is achieved by ranking the images in order of their probability of relevance. One of the difficulties of this model is that the underlying assumptions often do not hold in practical applications, for example, the probability of relevance is not correctly known in practice, for instance, the probability of relevance is estimated based on the basis of whatever data is not always accurate.

The other model realizes a predicate using similarity. (See Lew [8] for a comprehensive review on nonmetric similarities.) The relevance is estimated by the similarity. Each image is represented as multiple feature vectors in high-dimensional feature spaces and the similarity of images in each feature space is estimated by a distance in the space. One of the difficulties of this model is that it is not obvious how to measure the visual similarity of images from feature vectors, for instance, the common metrics that measure most of our physical spaces are not consistent to the perceived similarity of image content.

The similarity is then defined either by the probability of relevance $P(r|I, Q)$ or the distance of feature vectors $D(d_1, d_2, \dots, d_L)$.

In both cases, the problem is transformed to a ranking problem by either the probability of relevance or the similarity. The top ranked images are outputs of the retrieval. The Equation 11.17 can be expressed as

$$C_Q = \{I_i \in C : S(I,Q) > t_s\}, \tag{11.20}$$

where $t_s$ is a threshold.

In the case where the relevant image set is unknown, another applicable performance measure is the retrieval rate defined as number of relevant images in top $n$ images

$$RR = \frac{\text{number of relevant images in top } n \text{ images}}{n}. \tag{11.21}$$

How to assert the question $\{I_i \in C : r_Q(I_i)$ is true$\}$ is a nontrivial problem.

The practical approach in similarity-based image retrieval is to define the similarity as the visual similarity of resultant images to the example image or images. In turn, the visual similarity is measured by a dissimilarity metric of visual features in practice, which is also conveniently called a distance between two images, although it is not a distance in mathematical terms in many cases as presented in Section 11.2.

The dissimilarity of two images is hence measured by a function of individual feature distances as

$$D(d_1, d_2, \ldots, d_L) \tag{11.22}$$

where $d_l : l = 1, 2, \ldots, L$ are distances of $L$ adopted visual features between two images. It is worthwhile to point out that the vector formed by feature distances $\boldsymbol{d} = (d_1, d_2, \ldots, d_L)$ is a vector in the vector space $R^L$ where $R$ is the set of real numbers.

It is an open problem what the function $D(\ldots)$ that is consistent to the perceived similarity of image content should be. Efforts have been made to discover a suitable function for various CBIR systems, which can be categorized as feature combination (also known as early fusion) and feature aggregation (also known as late fusion). Similarity is not necessarily a metric.

A feature combination approach is heavily based on statistical component analysis techniques that combine all elements of individual feature vectors into one large combined vector and employ a metric in the combined vector space, like the Euclidean distance. Then the dimension of the combined feature vector is reduced in most cases using discriminant analysis techniques. This approach is also known as dimension deduction and feature selection. However, it ignores the perceptive significance of elements of individual feature vectors and the possible physical meanings of individual feature elements. Some common metrics, such as Euclidean, $p$-norm, for example, are used to estimate the similarity in the derived vector space, which equivalently yields a dissimilarity function.

The feature aggregation approach, on the other hand, treats the problem as a classifier combination of multiple classifiers defined from the individual feature dissimilarity metrics, which is formulated as a multicriteria decision problem.

The common methods in multicriteria decision can be categorized as follows.

- Linear combination: The use of linear functions like SUM and PROD for the combination of feature similarity.
- Nonlinear combination: The use of rank-based classifiers like majority voting.
- Statistical combination: The use of methods like the Dempster–Shafer technique and Bayesian combination.

Figure 11.6 depicts the feature aggregation in CBIR systems. In Figure 11.6, $x_i$ and $q_i$ are feature vectors of an image in the collection and the query image; $s_i$ is the feature similarity and $s$ is the aggregated similarity of multiple features;

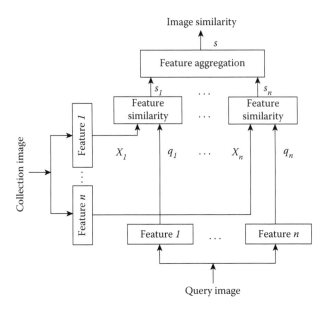

**FIGURE 11.6**
Feature aggregation in content-based image retrieval.

and $i = (1, \dots, n)$ is the $i$th visual feature. The overall aggregated similarity of multiple visual features is the ultimate criterion for image ranking; therefore, it determines the overall performance of the retrieval. In the literature, physical similarities in retrieval systems are defined either as a probability that one image is similar to another or a metric of visual features of images. It does not matter that the similarity is defined as a probability or a metric; it must answer the question of how the aggregated similarity of multiple visual features can be measured and measured consistently as humans do.

Various classifier combination schemes are used that equivalently yield a dissimilarity function $D(\dots)$. Though combination schemes are evaluated against their statistical performance, the question remains what combination scheme would meet the query requirements.

### 11.3.2 Linear Combination of Visual Features

The intuitive and straightforward solution is to aggregate multiple features using a linear combination of individual feature distances. To factor the relative significance of various features, the weighted combination is a popular suggestion [4,7], as expressed in the following.

$$D(d_1, d_2, \dots, d_L) = \sum_{l=1}^{L} w_l d_l \qquad (11.23)$$

Efforts have been taken to find good weights for particular visual features. As a matter of fact, the problem of aggregating the features has been transformed into the problem of finding suitable weights. As suggested in the experiments presented in Subsection 11.2.5.2, visual features perform differently on different queries. A promising solution would be to derive the variable weights from the query for each query.

From one query image, it is difficult to determine the query concept, which is what users look for by presenting the example. The similarity-based retrieval using one example simply finds the close match of the means of all features of images. The weights are either equal (no weighting) or preselected. Query by one example cannot realistically lead to scalable, satisfactory query performance [3]. Inspired by the weighting technique in relevance feedback techniques [10], the weights are normally set to be inversely proportional to the variances of the feature distances of multiple example images. The principle of this strategy is to assign higher weights to features that are close to each other in examples and considered characterizing the more significant features of the query.

Figure 11.7 shows a comparison of unweighted and weighted linear aggregation of features that are proposed in Ren [9]. Figure 11.7a and Figure 11.7b are the query examples and the groundtruth image set, respectively, which are the relevant images available in the image collection. Figures 11.7c and 11.7d are the results with equal weights (unweighted) and the proposed weights, respectively, where the groundtruth images are highlighted with dark bars. The weighted system is able to retrieve all seven groundtruth images in the top 10 output images while the unweighted system can only find five groundtruth images in the top 20 output images as shown in the figures.

The difficulty of the linear combination of features is not only the difficulty to determine the weights but also the loss of the visual significance in summing distances of individual visual feature.

## 11.3.3 Classifier Combination

The common combinatorial operations are summarized as follows. Let $C_1, C_2, \ldots, C_n$ denote the individual features, $s_{Ci}(x)$ be the similarity defined for the $i$th feature of the image $x$ and $s(x)$ is the aggregated overall similarity.

- **MAX (Maximum)**
  The maximum feature similarity is chosen as the overall image similarity. This operation ensures that two images are considered similar if all features are similar.

$$s_{MAX}(x) = \max \{s_{C1}(x), \ldots, s_{Cn}(x)\}. \tag{11.24}$$

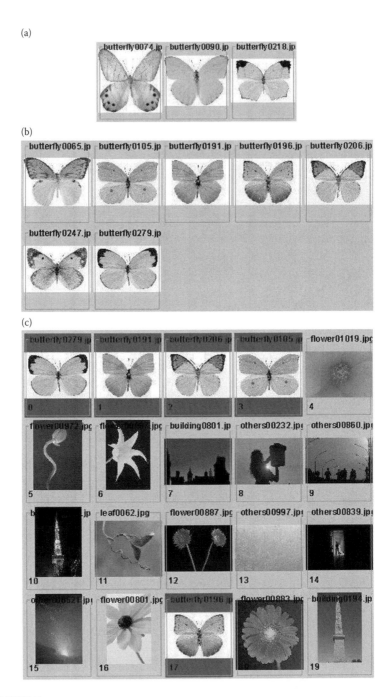

**FIGURE 11.7**

A Comparison of weighted and unweighted linear feature aggregation (a) Query images (b) Groundtruth image set (c) Retrieval results without weights (d) Retrieval results with proposed weights. (From Ren, F., Master's thesis, University of Wollongong, 2006. With permission.)

(d)

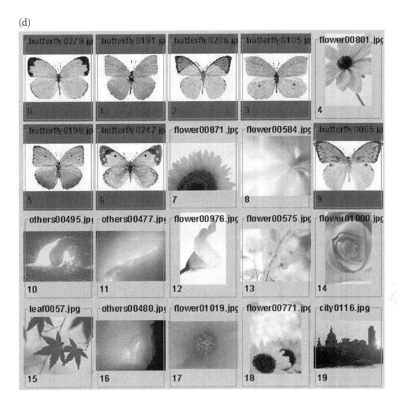

**FIGURE 11.7** (Continued)

- **MIN (Minimum)**
  The minimum feature similarity is chosen as the overall image similarity. This operation ensures that two images are considered similar if any features are similar.

$$s_{MIN}(x) = \max \{s_{C1}(x), \ldots, s_{Cn}(x)\}. \tag{11.25}$$

- **SUM (Summation-Weighted SUM)**
  The summation function adds the original-weighted feature similarities to obtain the overall image similarity.

$$s_{SUM}(x) = \sum_{i=1}^{n} w_i s_{Ci}(x), \tag{11.26}$$

where $w_i$ represents the weighting for the $i$th feature, which incorporates the significance of features in the feature aggregation, if available. This is the linear aggregation presented in the previous subsection, which is listed here for convenience of discussion.

- **PROD (Product-Weighted PROD)**
  The original-weighted feature similarities are multiplied to obtain the overall image similarity.

$$s_{\text{PROD}}(x) = \prod_{i=1}^{n} w_i s_{Ci}(x). \tag{11.27}$$

- **AVG (Average-Median)**
  The overall image similarity is the value of the summation of the feature similarities divided by the number of features.

$$s_{\text{AVG}}(x) = \frac{\displaystyle\sum_{i=1}^{n} w_i s_{Ci}(x)}{n}. \tag{11.28}$$

- **Average Vote-Weighted Average Vote**
  Feature similarity is measured as a vote that is a number between 0 and 1. The overall image similarity is the sum of all original-weighted votes divided by the number of features.

$$s_{\underline{\text{AVG}}}(x) = \frac{\displaystyle\sum_{i=1}^{n} w_i s_{Ci}(x)}{n}. \tag{11.29}$$

- **BORDA**
  The collection images are ranked according to their feature similarities. Each ranking gives a collection image a number of points, with the winner receiving the highest number of points. The overall image similarity is the sum of the number of points assigned to each collection image.

- **MV (Majority Voting)**
  Collection images are assigned binary votes, relevant or irrelevant, according to their feature similarities. The overall image similarity is also represented as relevant or irrelevant in accordance with the summation of votes.

- **Bayesian Combination**

- **Dempster-Shafer Approach**

## 11.4 A Case of Application: Web Image Search

Images are available ubiquitously on the Internet. It is a challenge for users to find what they need. Most commercial web search services, such as Google, Yahoo! and MSN Live, and so on, provide some web image search facility. These are mostly based on text retrieval techniques using keywords supported by image metadata like image file names and associated web texts. However, it is difficult to use words to describe the perceived visual content of images or to search visually for similar images with keyword-based search techniques. This section describes a web image search system based on image visual content and the user process to search similar images from example images.

Figure 11.8 depicts a structure of a Web image search system based on image visual content. Images are first retrieved using keywords from the Internet. The keyword searched images may include images with very different visual content. The user then selects some desirable images as examples to articulate the query. The examples are analyzed to derive feature weights to signify the importance of the visual features that manifested in the selected examples.

Figure 11.9 shows the user interface of the system. Figure 11.9a is the output images from the keyword search that are presented to users to select the desirable images. Three images are selected as examples to search for similar images as shown in the figure. Figure 11.9b shows the similar images retrieved from the Internet to the three selected example images.

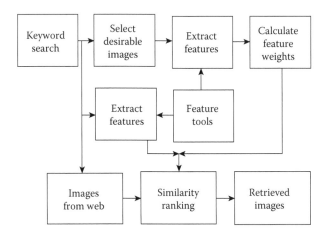

**FIGURE 11.8**
Web image search based on visual content.

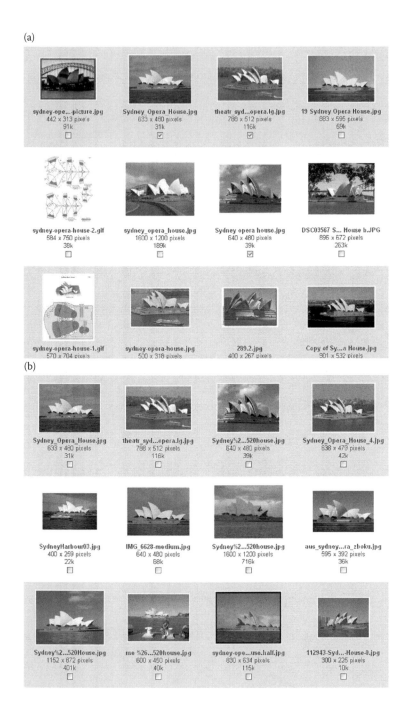

**FIGURE 11.9**
User process of web image search (a) Output images from keyword–based search (b) Output images from content–based search.

## 11.5 Conclusions

This chapter described the basic techniques and some fundamental problems for indexing and retrieving ubiquitous visual information. Section 11.2 introduced MEPG-7 visual descriptors and their similarity matching methods, with the focus on some basic visual features. Some experimental results were presented to demonstrate the properties of various features that help readers with some practical knowledge of their performance in potential applications. Section 11.3 discussed two key problems of visual similarity of images and visual feature aggregation techniques including linear weighting and classifier combination. Experimental results were presented to show how the aggregation strategy can improve the performance of retrieval. In the last section, a practical system and user process was presented to show how a system is constructed from techniques in this chapter.

## References

1. ISO/IEC 15938-3:2002 Information technology: Multimedia content description interface. Part 3: Visual.
2. ISO/IEC TR 15938-8:2002 Information technology: Multimedia content description interface. Part 8: Extraction and use of MPEG-7 descriptions.
3. T. E. Bjoerge, and E. Y. Chang. 2004, June. Why one example is not enough for an image query. In IEEE International Conference on Multimedia and Expo., vol. 1, 27–30, 253–56.
4. G. Das, S. Ray, and C. Wilson. 2006. Feature re-weighting in content-based image retrieval. In *Image and Video Retrieval, Proceedings,* ed. H. Sundaram et al., vol. 4071, 193–200.
5. R. Datta, J. Li, and J. Z. Wang. 2005, November. Content-based image retrieval: Approaches and trends of the new age. In Proceedings of the 7th ACM SIGMM International Workshop on Multimedia Information Retrieval, no. 253262, Singapore.
6. H. Eidenberger. 2004. Statistical analysis of content-based mpeg-7 descriptors for image retrieval. *Multimedia Systems* 10:84–97.
7. Q. Iqbal and J. K. Aggarwal. 2003, September. Feature integration, multi-image queries and relevance feedback in image retrieval. In the 6th International Conference on Visual Information Systems (VISUAL 2003), 467–74, Miami, FL.
8. Michael S. Lew, ed. 2001. *Principles of Visual Information Retrieval.* Springer-Verlag London Berlin.
9. Fenghui Ren. 2006. Multi-image query content-based image retrieval. Master's thesis, University of Wollongong.
10. Y. Rui, T. S. Huang, M. Ortega, and S. Mehrotra. 1998, September. Relevance feedback: A power tool for interactive content-based image retrieval. *IEEE Transactions on Circuits and Systems for Video Technology* 8 (5): 644–55.

11. A. W. M. Smeulders, M. Worring, S. Santini, A. Gupta, and R. Jain. 2000, December. Content-based image retrieval at the end of the early years. *IEEE Transactions on Pattern Analysis and Machine Intelligence* 22 (12): 1349–80.

12. L. Ye, L. Cao, P. Ogunbona, and W. Li. 2006. *Visual content processing and representation*, vol. 3893/2006 of lecture notes in computer science, chapter description of evolutional changes in image time sequences using MPEG-7 visual descriptors, 189–97. Berlin-Heidelberg: Springer.

13. Y. Zhong. 2007. A weighting scheme for content-based image retrieval. Master's thesis, University of Wollongong.

# Part III

# Ubi-Media Applications

# 12

## A Peer-to-Peer Digital Human Life Memory Store for Sharing Serendipitous Moments

Azizan Ismail, Madjid Merabti,
David Llewellyn-Jones, and Sud Sudirman

### CONTENTS

Every person has their own serendipitous moments; joyful moments that they hope can be kept in their mind forever. Unfortunately our memory can sometimes fail to retrieve all the details of when, where, and why something happened. Today, with advances in technology, we are able to capture our serendipitous moments as digital images, videos, audio, text, and other forms of data, making use of the huge capacities of computer storage available to us. To capitalize on this, we describe a system that combines the technologies of digital human life memory stores with peer-to-peer networking. As well as allowing the storage, annotation, and indexing of digital human life memories, such a system can also support the sharing of memories and personal serendipitous moments between trusted peer group members. Such a system can not only correlate the memories of a single user, but can also find relationships between memories, such as serendipitous moments, shared with multiple users across the peer-to-peer network. Based on these ideas we propose a flexible and scalable system for storing serendipitous moments in a human life memory framework using a peer-to-peer network that allows them to be shared between friends without the need for a central server.

## 12.1 Introduction

### 12.1.1 Background of Research

People are capturing and storing an ever-increasing amount of digital memories, with new types of information constantly being added such as GPS location data, heart-rate monitor recordings, TV viewing logs, and so on. The need to keep, manage, and share our personal memories in a digital way has became important through the necessity of remembering so many things such as the names of people, places, phone numbers, e-mail addresses, and general knowledge, and also to allow us to share our personal experience, improve communication between people, personal reflection and analysis, review conflicts, and to identify relevant information and evidence in our mind. Without a digital means to store this knowledge there is a possibility that we might forget or lose important data forever.

However, keeping everything may have negative side effects, such as information overload and less effective data searching and browsing. Therefore, there are technological solutions that can be used for updating and managing the relevant and critical data for future use. The rapidly developing technologies for recording multimedia (such as digital cameras for images and video), ubiquitous sensors, and the availability of low-cost storage can make the accumulation and retention of a lifetime of digital memories possible. The aim of our research is to help people share and store life moments and memories in a digital human life memory across their entire lifetime.

Serendipity is a word that originated from an obscure Persian fairytale "The Three Princes of Serendip" in the tenth century. This Arabian tale concerns three princes; as they traveled they were always making discoveries, through accident and sagacity. One of the most famous examples of serendipity is Christopher Columbus's discovery of America when he was looking for India. On the other hand, according to the Encarta English Dictionary [1], serendipity is a "natural gift for making discoveries quite by accident," and according to Dictionary.com [2], serendipity means "good fortune or luck." This word also became Britain's favorite word according to The Word, London's Festival of Literature 2000, which commissioned a nationwide survey, reported by BBC online news on September 18, 2000 [3].

The need to capture serendipitous moments and to save and share them with others in digital ways has become important through the necessity of remembering so many things such as names of people, places, and events and to identify relevant information and evidence in our minds. Without a digital means to store joyful moments there is a possibility that we might forget or lose them forever.

The novelty of our work therefore lies in the consideration of memory relationships, both within an individual's serendipitous moments and across the serendipitous moments of multiple individuals, as a primary consideration of our life memory database system. We believe the use of a peer-to-peer (P2P) network structure as a means to reflect such relationships to be unique, yet with genuine benefits in regards to the identification of relevant memories and sharing of memories between individuals.

### 12.1.2 Problem Statement

Many researchers refer to LifeStreams [4], MyLifeBits [5–8], Total Recall [9] and Haystack [10] as successful personal digital management systems (human memories) because they considered the problems of different types of media. Each of these projects has tried to establish a good personal database management system. Problems relating to the management of personal digital data (especially serendipitous moments), the sharing of them with others, the way data should be presented, and the methods for allowing understanding of such information and content are becoming increasingly complex. In the context of personal serendipity moments, the complexity of digital data has increased significantly in the recent past with the introduction of affordable digital cameras (still image and video), portable audio recorders and players, and smart mobile phones capable of supporting, capturing, and storing information as text (e-mails, instant messages), images, video, GPS, and sound.

We have encountered the following four challenges in the design of a system to store personal serendipity moments:

- Capturing the semantics of the information that may be spread across different media
- Supporting efficient and effective interactions between users and the information through P2P networking
- Providing media content analysis to reflect users' identities
- Providing privacy policies, security, and access control to personal life digital data

## 12.2 Literature Review and Theoretical Framework

### 12.2.1 Managing Serendipitous Moments in Human Life Memory

As explained in the U.K. Grand Challenge in Computing 2003 [11], there is a real need for the development of techniques for storing personal serendipitous moments and other lifetime memories in a manner that is robust to changes in hardware, operating systems, and indexing strategies. The computer and programs that operate on the data will change frequently over a human lifetime, but the data must outlast the systems that analyze it.

In fact, the idea to keep an individual or private file was envisioned by Vannevar Bush in Memex [12] as long ago as 1945. His vision was for a device "in which an individual stores all his books, records, and communication, and which is mechanized so that it may be consulted with exceeding speed and flexibility." The Managing Human Life Memories project moves beyond Bush's vision to store personal storage because it also allows a user to share their lifetime memories with family and trusted friends.

The work most similar to ours is perhaps that of MyLifeBits [7]. MyLifeBits is a database of resources (media) and links, but it is a stand-alone system and therefore their media content cannot be shared with others. They proposed a new system for storing digital media collections using an improved user interface, allowing the easy manual annotation of images and also providing for audio annotations to be created. They used a SQL Server with an Index Server supplying full-text search and used sensor cameras to continually record personal experiences.

There are several other projects that deal with personal memory. Cheng et al. in Total Recall focus more on the privacy and security issues of personal data storage [9]. The Total Recall system continually records personal experiences (memories) by using personal sensors (a special camera and microphone). Adar et al. in their Haystack project emphasize the relationship between an individual and their environment [10]. When an individual interacts with the environment, Haystack gathers data about those interactions and uses this metadata to further personalize the retrieval process. Lifestreams [4] is a personal store that abandons hierarchy in favor of time-based visualization. The project supports a form of saved query to filter what is viewed.

Another project, Singh et al., proposed a prototype called eVITAe (for electronic vitae) [13]. The eVITAe prototype is intended to chronicle the life of its user through various media collected by the user over their lifetime. Czerwinski et al. list five main challenges in managing digital memories: identifying valuable personal records, interpreting and correlating data from multiple sources, viewing and interacting with records, security and privacy of personal data, and data sharing [14]. Much research into lifetime memories has been undertaken but none of the existing approaches have considered the sharing of memories in a peer to peer network environment.

The novelty of our work therefore lies in the consideration of memory relationships, both within an individual's serendipitous moments and across the lifetime memories of multiple individuals, as a primary consideration of our memory database system. We believe the use of a P2P network structure as a means to reflect such relationships to be unique, yet with genuine benefits in regards to the identification of relevant memories and sharing of memories between individuals.

There are four major areas that form the core for our work. They are: the way in which multimedia data can be retrieved, media content analysis for personal lifetime evaluation, the manner in which multimedia data can be accessed through P2P networks, and the privacy and security of such data.

### 12.2.2 Multimedia Data Retrieval

Information flows into the real human memory from the outside world through an individual's senses of sight, hearing, smell, taste, and touch. The web site entitled The Brain From Top to Bottom [15], which has links to the Canadian Institute of Health Research, states that memory is simply a way we store and recall things we have sensed. Nerve network patterns store memories. We recall a memory only when we activate that network of interconnected neurons. In addition our memory is divided into two sections: short-term memory and long-term memory. We store information from our six sensory capabilities in areas located throughout the cortex. Some of this data then moves into short-term memory. Finally, some of that information goes into our long-term storage in various parts of the cortex, much of it returning to the sensory cortex areas where we originally received it. Only the data that catches our attention (such as a police car behind us) or because we will need it soon (a telephone number) goes into our short-term memory. We hold short-term data for maybe 30 seconds. Short-term storage is small; it holds about seven independent items at one time, such as carry numbers when calculating arithmetic.

In this research, we try to transfer the concept of human memory into a multimedia database system. Everything we capture constitutes part of our lifetime experience, which we can store in the system. The raw data which we need to edit, we store in a temporary database for a maximum of two months. After two months, a pop-up alert will suggest to the user to remove existing data onto external devices (such as CDs or DVDs).

More than 100 articles on multimedia data retrieval research—especially in the area if content-based retrieval—have been published. From the point of view of this project, multimedia data retrieval is important because everything we store in our personal digital memory we may need to retrieve back for future use. We use various findings from previous research especially in the area of automatic image annotating, semantic retrieval for video, text and image retrieval, and music (e.g., MP3) retrieval.

The first issue is how to meet the user's needs when they want to retrieve back their thousands of previous photos or images. Jeon et al. [16] proposed automatic image annotation and retrieval using a Cross-Media Relevance Model (CMRM). Nontext media (images, video, and audio) may have little value if not annotated with additional text. Although through normal text annotation for images, the process would not be easy and it becomes difficult to fulfill complex queries. Through automatic image annotation, we can easily retrieve a particular image. There are two ways the CMRM can be used. First the blobs corresponding to each test image were used to generate words and associated probabilities from the joint distribution of blobs and words, which corresponds to a document-based expansion. Each test image can be annotated with a vector probability for all of the words in the vocabulary. This is referred to as the Probabilistic Annotation-Based Cross Media Relevance Model (PACMRM). This model is useful for ranked retrieval, but is less useful for people to look at. Another method is the Fixed Annotation-Based Cross-Media Relevance Model (FACMRM). This is not useful for ranked retrieval but easy for people to use when the number of annotations is small. Second, a query word (or multiple words) is used to generate a set of blob probabilities from the joint distribution of blobs and words, corresponding to query expansion. This vector of blob probabilities is compared with the vector of blobs for each test image using Kullback–Liebler (KL) divergence and the resulting KL distance is used to rank the images. They call this model the Direct-Retrieval Cross-Media Relevance Model (DRCMRM). There is room for improvement of this proposed technique in terms of accuracy and reliability. The existing automatic image annotation techniques usually use common words to associate with several different image regions. As a result, uncommon words have little chance of being used for annotating images, consequently giving inaccurate results to queries. To resolve this, a proposed solution is to raise the number of blobs that are associated with uncommon words. It is also possible to use text anthologies with a combination of image features to make improvements to the current automatic image annotation techniques.

Another issue is how to retrieve audio (music, sound, humming, and voice) from the database. Liu et al. [17] proposed an approach to retrieve MP3 music objects and voice-based objects on their energy distributions. In their method, they define an MP3 phase as the logical unit for indexing MP3 objects. It is then segmented into a sequence of MP3 phase units after the object is inserted into the MP3 music database. They used PCVs (Polyphase-Filter Bank Coefficient Vectors) as discriminators for each MP3 phase. The PCVs of an MP3 slot represents the average energy distribution in the 32 sub-band; therefore a certain pitch error can be tolerated. The PCV of an MP3 slot is also designed to identify any sudden change in pitch or volume within the whole MP3 phase. The MP3 similarity measurement function is used to retrieve the selected MP3 phases. There are several disadvantages of the proposed method: only MP3 audio can be tested and not any other type such as MIDI, AVI, WAV, RealAudio, DTS, or other audio formats, and samples cannot be less than 16 bits (mono) and cannot have a higher frequency than 44.1 kHz (stereo). Jang et al. [18] proposed the content-based music retrieval process named Super MBox. They used the acoustic input data (singing, humming, or a musical instrument playing), which is directly recorded from a PC microphone for a duration of eight seconds, at an 8-bit resolution, in mono and with 11,025 units of sample rate. For their system they used pitch tracking using two methods: first is an autocorrelation function and second is through the use of the Average Magnitude Difference Function (AMDF). Both methods are performed in the time domain and have comparable performance. They have tested Super MBox and verified the proposed optimization scheme for HFM (Hierarchical Filtering Method). The system apparently demonstrates the feasibility of real-time music retrieval with a high success rate. Another issue arises when new formats of music are released. Thus it is important to combine or to improve the sound retrieval so that it can be used for multiple sound formats.

A final issue in multimedia data retrieval is how to retrieve video from the database. Semantic retrieval is one of the hot topics in video retrieval. Hamminchi et al. [19] describe how a tree-based system works (semantic retrieval), their algorithm, and details of the tree embedding problem for MPEG-7. They propose a new framework of semantic searches to solve user problems for querying multimedia data based on a tree embedding approximation algorithm, combining the MPEG-7 standard and an ontology. They use reasoning and embedding tree mechanisms in a complementary way to retrieve MPEG-7 data from a database. In their framework, they define a specific ontology domain and integrate it with the MPEG-7 database during the indexing process. Their proposed algorithm is excellent for MPEG-7 query work using the tree-based retrieval technique. However, their semantic retrieval research is based only on MPEG-7 data and has not been applied directly to other formats.

### 12.2.3 Accessing Multimedia Data through P2P Networks

Many researchers refer to Napster [20], Gnutella [21,22] and JXTASearch [23] when considering peer-to-peer (P2P) networks. Napster was the original P2P application that made the P2P idea popular. Peer-to-peer systems offer an alternative to traditional client-server systems for some application domains. In P2P systems every node (peer) of the system acts as both client and server and provides part of the overall information available from the system. The P2P approach circumvents many of the problems associated with client-server systems but results in considerably more complex searching, node organization, security, and so on. P2P networks can be divided into two classes: structured and unstructured networks [24]. In a structured P2P overlay network the topology is tightly controlled and the content is placed not at random peers but at specified locations that will make subsequent queries more efficient. Structured P2P systems often use Distributed Hash Tables (DHTs) as a substrate. In a Distributed Hash Tree approach such as Chord, Pastry, and Tapestry, a global identification scheme for the peers is exploited in order to decide what part of the search space the peer is associated with [25]. On the other hand, unstructured overlay networks organize peers in a random graph in a flat or hierarchical manner (e.g., using a Super-Peer layer) and use flooding or random walks or expanding-ring Time-To-Live (TTL) search, and so on, for the graph to query content stored by overlay peers. Each peer visited will evaluate the query locally on its own content, and can therefore support complex queries, which can be difficult to achieve in structured networks.

A particular issue involved in the access of multimedia through P2P networks is how to create an efficient way to access the huge personal multimedia database with maximum security. Yang [26] focuses on three systems (symbolic query on symbolic database, monophonic acoustic query on symbolic database, and polyphonic acoustic query on polyphonic acoustic database). He proposes a music indexing framework known as the Music-Audio Characteristic Sequence Indexing System or MACSIS. Two protocols are used with replicated databases separated in two different conditions. For the first one, each query is processed by one response node at a time. For the second one, each query is processed by several nodes simultaneously. On the other hand, the protocol for general P2P networks tends to be divided into two phases (a pre-search phase and a search phase). By using Yang's protocol, a system throughput can be improved and waste reduced. Both (replicated database and generic) protocols will involve query setup, query processing, result generation, and process interruption. The advantage of the proposed protocols is not limited to the music search domain. They can also be applied to many other domains as long as the algorithm remains suited to certain conditions. The disadvantage of the proposed protocols is that they assume all nodes in the P2P network are cooperative, and will follow the specified protocols without deviating from them.

Klampanos et al. [27] focus on full text searching and retrieval of documents and explore information retrieval (IR) approaches in P2P environments. They propose an architecture for IR over large semicollaborating P2P networks. They define a network where peers have to cooperate in order to perform IR without the need to share any detailed information with the rest of the network. Lu [28] explores more on content-based retrieval in P2P networks that use a hybrid P2P architecture and focuses on the resource selection conducted by directory nodes. For content-based retrieval in P2P networks, both the retrieval accuracy and the efficiency of query routing are important. The authors therefore measure the performance of different resource selection and document retrieval algorithms in hybrid P2P networks by the retrieval accuracy and efficiency of query routing. Most of the algorithms that have been used in their experiments require a query matching rule, which defines the number of query terms that need to be matched for a query.

Overall, content-based retrieval has been found to be more accurate and efficient than name-based retrieval in hybrid P2P networks and content-based resource selection is more accurate and more efficient for content-based retrieval in hybrid P2P networks, compared with simple match-based resource selection algorithms. Unfortunately, considering the body of existing work in the area would suggest that the security and privacy issues have been somewhat left behind. There remains a need to combine security and privacy issues into the processes involved in the access of multimedia data through P2P networks.

### 12.2.4 Data Privacy and Security

In this section we discuss previous work focusing on the need for further research in security policies, privacy frameworks for data integration and sharing, and policies for access control based on authorization for general multimedia data and for video databases.

Verdon [29] explains that every software developer needs to create a good security policy and then needs to follow it. A good security policy can protect our personal and confidential data, our software applications, and our databases when attacked by hackers or as a result of customer complaints arising through lawsuits. The author lists several policy types such as corporate security policies, acceptable-use policies, privacy policies, e-mail policies, information system policies, network security policies, data classification policies and so on. A number of suggestions from the author can be considered in relation to human life memory stores. E-mail policies, for example, such as the requirement that a user cannot simply forward an e-mail to a third party that contains another person's e-mail address, can protect other e-mail addresses from being exposed to the wrong people. A significant information system secure operation policy (system access control, firewall policy, physical security, etc.) and a data classification policy are the most critical aspects needed to be considered as the main security policy in my project. The issue here is how to integrate a good privacy policy, security policy, and network security policy into one system.

As highlighted by Clifton et al. [30], data integration and sharing is a big challenge for the database community. The authors propose a privacy framework named Sharing Scientific Research Data, which includes privacy views, privacy policies, and purpose statements, for defining private data and privacy policies in the context of data integration and sharing. The authors develop schema-matching solutions that do not expose the source data or schema. They need to develop the components of match prediction, human verification of matches, and mapping creation before matching algorithms that preserve privacy can be done.

Bechara et al. [31] explain about the problem of multimedia data authorization and access control when multimedia data is accessed through the Internet. This problem becomes more complex when attempting to protect multimedia objects with no textual description in multimedia applications. The authors propose an extended RBAC model as used by Ferraiolo [32] with a user model (describing both the user-related information and the application services), a role model (extending the RBAC model so that it can describe roles on a user-related basis), a rule model (adding object features), a policy model (to consider all of the requirements of current methods), and a link model (allowing the representation of various types of links) to address role specification based on content features such as shape, color, and relation rather than on textual descriptions.

The most important issue in managing human life digital memories is that of keeping personal digital information safe and private without the owner having difficulty accessing or preserving it. It will become increasingly important to develop techniques for storing large amounts of complex data over decades or indeed centuries, in a manner that is robust to changes in hardware, operating systems, and indexing strategies. The computer and programs that operate on the data will change frequently over a human lifetime, but the data must outlast the systems that analyze it. Questions will be asked of the data that were not predicted when the data was indexed, so the indexing strategies must change over time.

## 12.3 Proposed Architecture

### 12.3.1 Data Capture

For our architecture we use active capture methods to capture our lifetime experiences, especially personal serendipitous moments. Active capture means people capturing photos or video on their own initiative. Active capture is very effective for allowing someone to recall an event back in the future. It can be more personally meaningful compared to passive images captured using an automatic sensor camera [33].

It is not unusual for more recent models of digital cameras to include features that can considerably improve the capture of digital memories, such as face and smile detection and antiblur modes. They can increasingly be paired with GPS devices to establish photo locations. These can be used to record where shots were taken, allowing them to be displayed where they belong on a world map based on latitude and longitude. For our prototype we store images at $3264 \times 2448$ resolution; a 1.0 gigabyte memory stick can store up to approximately 341 such photos in raw form. We use a small, light camera making it convenient to carry around the neck and easy to operate for the capture of photographs and video. A serendipitous moment is not a well-planned situation and it can happen at any time and any place. Therefore it is important that the camera can be easily left in standby mode, to allow pictures or video of serendipitous moments to be shot at any time or place.

## 12.3.2 Data Storage

In our system, everything we capture is stored in a database on a personal computer hard disk that we call the memory. The database can store content and metadata for a variety of item types, including contacts, documents, email, events, photos, music, and video. There is an option for users either to edit their raw data to make it more interesting and compress them into an appropriate format or to immediately store photos or video directly to the database. A user must add annotations of a specific name, feeling, or expression to their media data. This is important for future use of that media data, as without annotation the data will be difficult to search by retrieving or querying from the database. For example, suppose I captured a serendipitous moment at my sister's wedding several years ago. I may have difficulty in remembering when that photo was captured if there are no clues to the event. We can therefore annotate that image with a date (e.g., "November 1990" this may be applied automatically) and a feeling or expression (e.g., "Happy"). As we store memories in our brain, we also attach them to other related memories, such as "unforgettable moment in life," and thus concepts with older memories. In the permanent database, we tag annotated data with a GPS location and where the event happened (e.g., "ABC Club"), the event name (e.g., "Noraini's Wedding 1999") and its relationships with other data. We can then retrieve the concept at a later date by following some of the pointers that trace the various meaning codes and decoding the stored information to regain meaning. Media data with proper annotation can be short-listed into a group when the user retrieves data by date, event, expression, and so on. Returning to our example, if I want to retrieve my serendipitous moment, I can put "happy" or "joyful moment" in the retrieval text box of the photograph, then the system will list all the photos with that annotation and also other related media connected with other pointers so that one hint may allow me to recover the whole meaning. The user can then utilize instance links to relate all of the digital data (photo, video, audio, GPS location, and images) when retrieving all information related to the memory of interest.

## 12.3.3 Data Sharing

In our system, we share our media content through peer-to-peer networking without using a central server.

We will now consider a scenario to demonstrate how our system works in sharing serendipitous moments. In July 2007 I took photographs at my friend's wedding. On that occasion, I bumped into my other friends Jundi and Ida. Though the serendipity lies not in seeing them at the wedding party, but rather how we met in the first place. Ten years ago, I was also the cameraman for Jundi and Ida and now they already have two children. At this point in time, Ida had a desire to share her memories of her wedding ceremony with her friends. She started to retrieve all of her wedding photos and videos.

By using the system, she can easily retrieve her collections of photographs and videos by name, event, date, file type, size, expression, and location. When she retrieved photos with expression mode and chose "Happy," she would realize that in addition to sharing her serendipitous moments during her wedding ceremony, she can also share serendipitous moments during birthday parties and other events.

When the database is ready, Ida can make her serendipitous moments available to share within the peer group. Ida can also invite other peer members to share their serendipitous moments not just from wedding and birthday ceremonies but any serendipitous moment that may exist in their lifetime memories. As a result, many peer members contribute to share their serendipitous moments by sharing photos and videos through the system. In addition, when a user captures their serendipitous moment or their lifetime activity and pairs with the GPS receiver, the user can see the location and movement on a map by date. The user can retrieve peer groups and their own serendipitous moments and ask the system to generate a report on how many of them are similar.

Back to the scenario, Ida can retrieve a personal photo or video map by month or year using GPS locations and create location metadata. She can also download shared files from her peer group and map movement by month or year in a split window and activate the function that traces two different individuals' movements. When one of the peer members shares their metadata location with Ida, she may discover that at certain dates or locations they were present at the same place, even if they were not aware of it at the time. As a result, they can share their stories behind the scenes. The advantage of the system is that a user can track back to where they were at a specific date, establish who they met with, what happened, and correlate this movement between two or more people even if it happened many years earlier.

### 12.3.4 System Design

We consider the use of JXTA P2P networking technology for the system since JXTA [34] provides an open set of P2P protocols that enable any device on the network to communicate, collaborate, and share resources. The JXTA peers create a virtual, ad hoc network on top of existing networks, hiding their underlying complexity. In the JXTA virtual network, any peer can interact with other peers, regardless of location, type of device, or operating environment, even when some peers and resources are located behind firewalls or are on different network transports. The JXTA technology supports multi-platform operation and avoids the constraints of hierarchical client-server architectures.

For the purposes of our architecture, the most important characteristics are that JXTA technology can run on any device, including cell phones, PDAs, two-way pagers, electronic sensors, desktop computers, and servers. It is based on proven technologies and standards such as HTTP, TCP/IP, and

XML and JXTA technology is not dependent on any particular programming language, networking platform, or system platform and can work with any combination of these. Using peer groups, we can establish a set of peers with naming within the group and mechanisms to create policies for creation and deletion, membership, advertising, and discovery of other peer groups and peer nodes, communication, security, and content sharing.

Returning again to our scenario, Ida is working on her personal computer at home and has several peer members within her group using similar devices. By using JXTA peer-to-peer networking, she can also share her serendipitous moment with other peer members who are using different platforms or devices. For example, Ida can communicate with Azizan using a different operating system and even with Ahmad using a mobile phone. Figure 12.1 shows how Ida, Azizan, and Ahmad can share their serendipity content through the use of a JXTA Virtual Network that can support multiple peer group platforms.

### 12.3.5 JXTA Content Manager Service

The Content Manager Service (CMS) allows JXTA applications to share and retrieve content within a peer group. Each item of shared content is represented by a unique content ID and a content advertisement that provides meta-information about the content, such as its name, length, mime type,

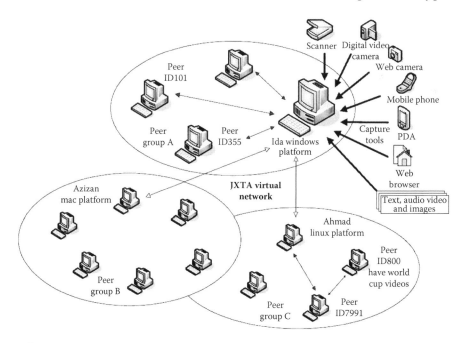

**FIGURE 12.1**
The JXTA peer-to-peer platform for sharing serendipitous moments.

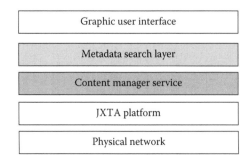

**FIGURE 12.2**
Content manager service architecture.

and description. The CMS also provides a protocol based on JXTA pipes for transferring content between peers. Unlike some other P2P systems, peers running CMS are not required to use HTTP in order to exchange content.

The overall architecture of our metadata-life memories sharing system is shown in Figure 12.2. In the architecture, the JXTA platform provides basic P2P communication mechanism, including peer discovery, pipe, and advertisement. The content manager service built upon the JXTA platform, implements content publication and transmission. The metadata search layer, as could be seen from the figure, is implemented as an extended service based on the existing content manager service, providing rich metadata search mechanism and the general purpose query language interface. Users manipulate the content using the graphics user interface implemented through Java Swing API. Seeking to provide a general P2P infrastructure across the boundaries of programming languages, platforms, and network protocols, the JXTA platform implements several core protocols based on common concepts like peer identifier, peer group, message, advertisement, and pipe. Figure 12.3 shows the architecture of the JXTA peer.

On the basis of the JXTA core layer, several Sun JXTA services and diverse JXTA community services have been implemented. Among them is the CMS, a file sharing and retrieving service that allows JXTA applications to share and exchange content within the scope of peer groups. Topmost is the real-world JXTA applications. They rely on the JXTA protocols and services to meet different needs pertaining to P2P networking. This is where our resource-sharing application lies. In a CMS, each piece of content is represented by a globally unique identifier generated by a 128-bit MD5 checksum algorithm from the content. The content ID is utilized when comparing and requesting content. Through the use of the content advertisement, an XML file describing services and messages, each peer can publish its content along

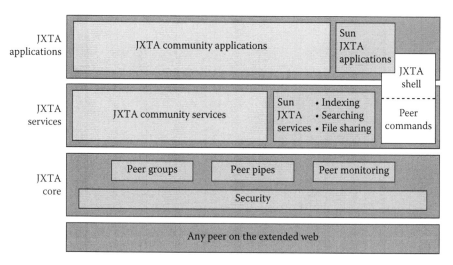

**FIGURE 12.3**
JXTA architecture.

with content ID, file name, and some basic metadata, such as name, length, MIME type, and so forth.

## 12.4 Implementation

### 12.4.1 Face and Smile Detection

The system will first divide the images from photo databases into two groups: photos with human images and photos without human images. This process will be implemented with face detection software where the system detects human faces from the images and groups all of the images with human faces into one group and places the remaining images without human faces into another group. Secondly, all of the images with human faces will be sorted using a smile detection process whereby all smiling faces from the images will be sorted into a subgroup to help identify happy serendipitous moments. Figure 12.4 shows the different markers for each of the photo images after the face and smile detection process has been applied.

### 12.4.2 User Interface

The design of the interface has been chosen to support user interest and intention with heterogeneous multimedia data. The user can explore and relate to their serendipitous moments and life activities with friends. The

(a)                                        (b)

(c)                                        (d)

**FIGURE 12.4**
Three types of images: (a and b) photoswith human images, (c) photo without human images, and (d) photo with human images and smiling faces (using the smile detection process).

use of an event model allows information to be presented to users in a manner that is independent of media and data source. In the interface various views of the data (such as those across time, events, or other attributes) are tightly linked to each other, so that interaction in terms of any one of them are instantaneously reflected in all the views (for example, selecting an event by time leads it to be highlighted in the spatial view). This is essential for maintaining context and helping the user build their own lifetime experience database.

The user can also display a satellite image and zoom to various levels. Figure 12.5 shows a split window for two individuals' photos including GPS locations. Figure 12.6 shows the location logs for two different people with photos and GPS locations marked on a map where a tracing location was activated. Figure 12.7 shows the location at increased magnification, using a satellite image for the photo locations using GPS data.

Currently we have developed a web-based interface to allow the user to manually load media to the system. First the user needs to log in to the system after which they can start uploading all or a selected subset of media to the system. In our design, each specific piece of media content (such as

**FIGURE 12.5**
Correlated memory interface: photos from two peer group members are displayed next to each other.

photographs, video, audio, text, or e-mails) is stored in specialized tables in a relational implementation.

Another table is created to store events, along with related information and different attributes such as event name, location, start and end time, expression, and links to other media. Relationships between various events are also captured here. Date and time is attributed to the media content automatically ensuring it can be easily sorted by the system. Photo and video properties (latitude and longitude, date taken and modified, file name, size, etc.) can be displayed by clicking the right mouse button. The user can display the location of their photo or video by activating the online link button to Google's maps.

The system gives an option to the user to annotate their media content by completing the text fields for names, events, and comments, and by choosing from a drop down list to set a suitable expression (happy, sad, neutral, etc.). Figure 12.8 shows the record insertion form for direct access to the user database. The system provides direct manipulation techniques to facilitate natural user–system interactions. In this paradigm, users can directly perform different kinds of operations on items (events) of interest. Furthermore, combining the query and presentation space alleviates the cognitive load from the user's perspective, unlike traditional query environments. The user can display the time a photo was taken with a person's GPS recorder location to create

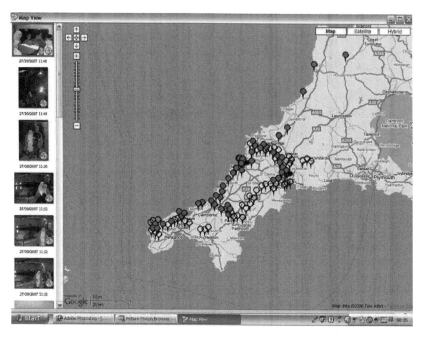

**FIGURE 12.6**
Map interface: red dots for Ida's GPS points and blue dots for another peer member's GPS points.

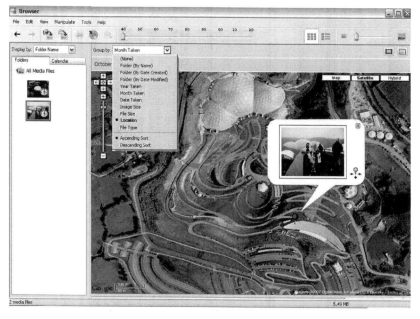

**FIGURE 12.7**
Displaying photo locations using Google maps.

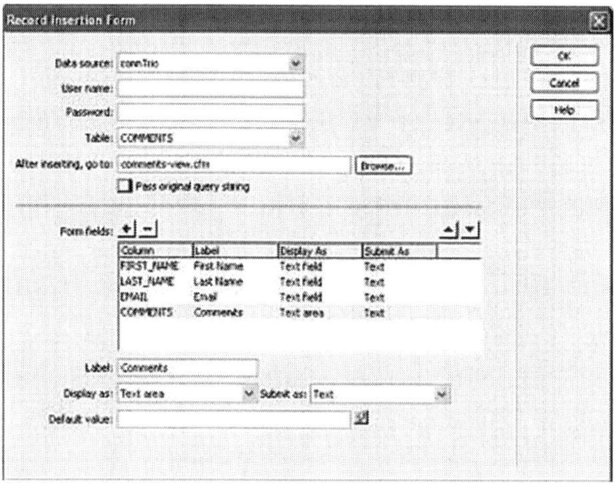

**FIGURE 12.8**
Database record insertion form.

location metadata. Photos can also be linked to calendar events to indicate a photo of the event, effectively turning the calendar into a photo diary.

### 12.4.3 System Report

The system supports exploration and allows users to define information at varying resolutions of time, location, and event. With a database, one can also create reports to understand the contents of the user's serendipity moment (lifetime) store and how annotation, filtering, extraction, viewing relations, and details on demand have been provided in order to help users look in depth at their lifetime database. The reports can be provided in the form of log files, charts, graphs, or text lists. The system generates charts for the collected data properties in case a user wants to see the pattern of their serendipitous moments and lifetime experiences, and this can be particularly useful in helping them to compare with friends, discover new serendipitous moments, evaluate, change, or improve their lives.

### 12.4.4 Security

The system is divided into two sections: one acting as a personal database system (updating personal media content online or off-line) and the other one for sharing media content within a peer group. A user is required to log in using a username and password to access the personal database area. Users also need to log in to the system (web based) as a peer group to share media content. Figure 12.9 shows this separation of data access through the use of a username and password.

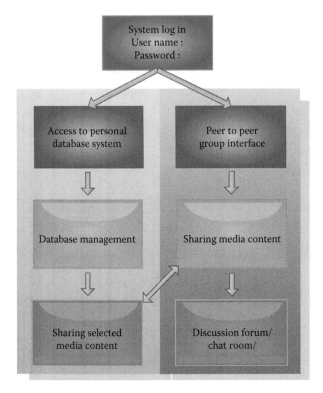

**FIGURE 12.9**
Security and data flow for the system.

## 12.5 Conclusion and Future Work

A framework for sharing and storing serendipitous moments in human life memories as described here can become a reality through the use of our proposed system using a P2P networking environment. The novelty of our work therefore lies in the consideration of lifetime memory management and analysis, memory relationships, both within an individual's lifetime memories and across the lifetime memories of multiple individuals, as a primary consideration of our lifetime memory database system. We believe the use of a P2P network structure as a means to reflect such relationships to be unique, yet with genuine benefits in regard to the identification of relevant memories and sharing of memories between individuals. Finally, we need to do further research and development especially in the media content analysis, automatic annotation, face and location detection processes, and different visual presentations to archive our research objectives.

# References

1. MSN Encarta: Online encyclopedia, dictionary, atlas, and homework. 2007. http://encarta.msn.com
2. http://Dictionary.com 2007.
3. UK's favourite word is a surprise. 2007. http://news.bbc.co.uk/1/hi/uk/930319.stm
4. E. Freeman and D. Gelernter. 1996. LifeStreams: A storage model for personal data. *ACM Sigmond Record* 25 (1): 80–86.
5. J. Gemmell, R. Lueder, and G. Bell. 2003. The MyLifeBits lifetime store. In Proceedings of the 2003 ACM SIGMM Workshop on Experiential Telepresence ETP '03.
6. J. Gemmell, G. Bell, and R. Lueder. 2006. MyLifeBits: A personal database for everything. *Communications of the ACM* 49 (1): 88–95.
7. J. Gemmell, G. Bell, R. Lueder, S. Drucker, and C. Wong. 2002. MyLifeBits: Fulfilling the Memex vision. In Proceedings of the 10th ACM International Conference on Multimedia MULTIMEDIA '02.
8. J. Gemmell, L. Williams, K. Wood, R. Lueder, and G. Bell. 2004. Passive capture and ensuing issues for a personal lifetime store. In Proceedings of the 1st ACM Workshop on Continuous Archival and Retrieval of Personal Experiences CARPE'04.
9. W. C. Cheng, D. G. Kay, and L. Golubchik. 2004. Total recall: Are privacy changes inevitable? In Proceedings of the 1st ACM Workshop on Continuous Archival and Retrieval of Personal Experiences CARPE'04.
10. E. Adar, D. Kargar, and L. A. Stein. 1999. Haystack: Per-user information environments. In Proceedings of the 8th International Conference on Information and Knowledge Management CIKM'99.
11. A. Fitzgibbon and E. Reiter. 2003. Memories for life: Managing information over a human lifetime. In Grand Challenges in Computing Research, and Grand Challenges in Computing Education, Newcastle.
12. V. Bush. 1945. As we may think. *The Atlantic Monthly,* 176:202–8.
13. R. Singh, R. Knickmeyer, P. Gupta, and R. Jain. 2004. Designing experiential environments for management of personal multimedia. In Proceedings of the 12th Annual ACM International Conference on Multimedia MULTIMEDIA'04.
14. M. Czerwinski, D. W. Gage, J. Gemmell, C. C. Marshall, M. A. Pérez-Quiñones, M. M. Skeels and T. Catarci. 2006. Digital memories in an era of ubiquitous computing and abundant storage. *Communications of the ACM* 49 (1): 44–50.
15. The brain from top to bottom: How memory works. 2007. http://thebrain.mcgill.ca
16. J. Jeon, V. Lavrenko, and R. Manmatha. 2003. Automatic image annotation and retrieval using cross-media relevance models. In Proceedings of the 26th Annual International ACM SIGIR Conference on Research and Development in Information Retrieval SIGIR'03.
17. C.-C. Liu and P.-J. Tsai. 2001. Content-based retrieval of MP3 music objects. In Proceedings of the 10th International Conference on Information and Knowledge Management CIKM'01.

18. J.-S. R. Jang and H.-R. Lee. 2001, October. Hierarchical filtering method for content-based music retrieval via acoustic input. In Proceedings of the 9th ACM International Conference on Multimedia MULTIMEDIA'01.

19. S. Hammiche, S. Benbernou, M.-S. Hacid, and A. Vakali. 2004. Semantic retrieval of multimedia data. In Proceedings of the 2nd ACM International Workshop on Multimedia Databases MMDB'04.

20. S. Saroiu, K. P. Gummadi, and S. D. Gribble. 2002. Measuring and analyzing the characteristics of Napster and Gnutella hosts, 2006, 2003. www.mpi-sws.mpg.de/~gummadi/papers/mm.pdf.

21. The Gnutella protocol specification v0.4. 2006. http://www.stanford.edu/class/cs2446/gnutella_protocol_0.4.pdf

22. The annotated Gnutella protocol specification v0.4. 2006. http://rfc-gnutella.sourceforge.net/developer/stable/index.html

23. S. Botros, S. Waterhouse. August 2001. "Search in JXTA and other distributed networks," First International Conference on Peer-to-Peer Computing, pp. 30–35.

24. D. Tsoumakos and N. Roussopoulos. 2006. Analysis and comparison of P2P search methods. In Proceedings of the 1st International Conference on Scalable Information Systems InfoScale'06.

25. E. K. Lua, J. Crowcroft, M. Pias, R. Sharma, and S. Lim. 2005. A survey and comparison of peer-to-peer overlay network schemes. *IEEE Communications Survey and Tutorial* 7 (2):72–93.

26. C. Yang. 2003. Peer-to-peer architecture for content-based music retrieval on acoustic data. In Proceedings of the 12th International Conference on World Wide Web WWW'03.

27. I. A. Klampanos and J. M. Jose. 2004. An architecture for information retrieval over semi-collaborating peer-to-peer networks. In Proceedings of the ACM symposium on Applied Computing SAC'04.

28. J. Lu and J. Callan. 2003. Content-based retrieval in hybrid peer-to-peer networks. In Proceedings of the 12th International Conference on Information and Knowledge Management CIKM'03.

29. D.Verdon. 2006, July–August. Security policies and the software developer. *IEEE Security and Privacy* 4 (4):42–49.

30. C. Clifton, M. Kantarcio, A. H. Doan, G. Schadow, J. Vaidya, A. Elmagarmid, and D. Suciu. 2004. Privacy-preserving data integration and sharing. In Proceedings of the 9th ACM SIGMOD Workshop on Research Issues in Data Mining and Knowledge Discovery DMKD'04.

31. B. A. Bouna and R. Chbeir. 2006. Multimedia-based authorization and access control policy specification. In Proceedings of the 3rd ACM Workshop on Secure Web Services SWS'06.

32. D. F. Ferraiolo, J. F. Barkley, and D. R. Kuhn. 1999. A role-based access control model and reference implementation within a corporate intranet. *ACM Transactions on Information and System Security (TISSEC)* 1 (2):12–41.

33. A. F. Abigail, J. Sellen, M. Aitken, Ss Hodges, C. Rother, and K. Wood. 2007. Do life-logging technologies support memory for the past?: An experimental study using sensecam. In Proceedings of the SIGCHI Conference on Human Factors in Computing Systems CHI'07.

34. Xiang, X., Y. Shi, and L. Guo. 2004. "Rich metadata searches using the JXTA content manager service." in 18th International Conference on Advanced Information Networking and Applications AINA 2004 (1)pp. 624–629.

# 13

## Real-Time Recording and Updating of Personal Journeys via Ubiquitous Computing Technologies

**Han-Bin Chang, Hsuan-Pu Chang, Hui-Huang Hsu,
Louis R. Chao, and Timothy K. Shih**

### CONTENTS

Managing personal travels is a desirable, yet tedious work. In this chapter, we introduce an intelligent travel book management system. With this system, the user can easily handle multimedia information during a journey, including recording the real experience and updating the travel route according to instant-fetched shared information from other users. The multimedia information includes photos, audios, videos, and notes taken during the journey. All the information can contain location information. Agent and mobile computing technologies are used to implement the system. The user would be able to use the system to record and update his or her personal journey in a ubiquitous manner through a mobile device, like a personal digital assistant (PDA) or a smart phone. Thus ubiquitous management of traveling records and plans becomes feasible.

## 13.1 Introduction

The intelligent travel book management system is designed for helping visitors to manage and edit multimedia resources generated from their journeys. Since users always focus on enjoying impressive buildings, natural scenery, and outdoor attractions, using a mobile device works best to handle multimedia information. On a tour, visitors always take pictures using their cameras, video cameras, or other mobile devices. In this case, users prefer making simple operations rather than complex ones since editing such multimedia resources and uploading them to webpage servers is not the most important task when they are sightseeing. With such requirements, we developed an intelligent travel book management system to help users easily keep and manage their multimedia resources obtained on a journey. Several major components are included in this system: the client application, the agent, the server platform, and the interfaces for management.

The mobile client application is a first-line component to users. On a journey, visitors can carry a PDA to capture images or videos. Like other applications run on a PDA, a program called intelligent travel book wizard will help users to make easy operations to those multimedia resources. Users can use this wizard to process captured images and add some comments if necessary. Combining with the Global Positioning System (GPS) installed on the PDA, the wizard will transfer these images or video files, personal annotations, and coordinates of the event automatically. So visitors can focus on experiencing wonderful scenes on their journey without worrying about the procedures on managing photos and videos. Images, videos, and comments will be packaged and transferred to a server automatically. Location information provided by a GPS component is also uploaded to the server side. Such resources will be analyzed in advance to help visitors edit their own travel books after they finish their journey.

Agent technology plays an important role in the developed intelligent travel book management system. Information captured from users is uploaded to the server side. Such information is composed of image, video, audio, and textual comments. It is not easy or convenient if the user does not have such a system. In contrast, users can edit their travel books via interactive interfaces with the system. If the authors need more information and materials from the Internet, they need to browse web pages using searching engines. It is not convenient for visitors to collect such information from the Internet to augment contents of their own travel book. The intelligent travel book agent will collect information and materials from the Internet automatically. But, how do agents collect appropriate information to help users to edit their own travel books? According to the location coordinates provided by GPS component in the users' PDA, the agent can get information around the location of the users. Furthermore, the agents can obtain textual data like names of scenes or buildings and use such data to collect textual and multimedia resources from the Internet. Such materials collected by agents will be put into a resource pool. Users can edit their own travel book and access these materials after they logged into the travel book management server.

The server platform is designed for storing and analyzing information collected from users. On a journey, visitors capture images or videos and add annotations. The intelligent travel book wizard will upload these resources automatically to the server side. There are also other resources and materials collected by agents to be uploaded to the server. By such materials, users can edit their own travel books and publish them on the Internet. The management server not only stores information from the wizard application and agents but also would be responsible for presenting travel books in the form of web pages.

The final component is the management interface. The interfaces are a subcomponent for providing means to users so they can access resources stored in the management server. In the developed system, a simple interface is provided to users in the form of web pages. The users log in into the management server using their accounts and passwords, and they can access image, video, audio files, and textual comments recorded on their journey. Other materials collected from agents are also displayed in the resources section. Users can also browse the usage history to observe resources that were used in the past.

The rest of this chapter is organized as follows: adopted methodologies for gathering information around visitors are discussed in Section 13.2, the architecture of the system and detailed functionalities are described in Section 13.3; the implementation of each component mentioned in Section 13.3 are presented in Section 13.4, comments and suggestions from users are listed in Section 13.5; related research and developed systems are mentioned in Section 13.6; and finally, the conclusion and contribution will be discussed in Section 13.7.

## 13.2 Methodologies

In this section, we will introduce adopted approaches for collecting and filtering meaningful information to help users to create their personal travel books. Two major objectives are considered in the developed system: landmark selection and similar travel route searches.

### 13.2.1 Landmark Selection

Landmark selection is an important function to help visitors to collect related resources for creating travel books. With several procedures, landmarks will be generated and transferred to an agent for further processing.

A landmark selection engine is also designed for such a purpose. A detailed description about implementation of this system will be discussed in the next section. On a journey, visitors always take photos of the same region. By functions provided by a GPS on the users' mobile devices, we can obtain some landmarks. These landmarks may help users to find related information through the travel book agent. Not all landmarks around visitors can be referenced for gathering information that users are interested in. For an example, visitors arrive in a region with restaurants, souvenir stores, and parking lots. They may be interested in information other than parking lots. There may be common landmarks in the same region. In order to get an adequate string of landmarks where visitors are located, we developed an algorithm to find keywords filtered from all landmarks in the region for further search via the travel book agent. The algorithm and variable definitions are

- $C_l$: A list composed of candidate landmarks. All landmarks around visitors will be assigned to this list.
- $R_l$: A list composed of representative landmarks. By procedures of this algorithm, several landmarks will be filtered and stored in this list.
- $G(R)$: With a given radius R, the GPS libraries will return a string of landmarks nearby.
- $K_w$: A list composed of keywords from the intersection of candidate landmarks $C_l$.
- $N_k$: A list composed of frequent keywords. In this algorithm, the frequency is assigned as an integer.
- $F_n(K_w)$: A subfunction to indicate index of $N_k$.
- $T_k$: An integer threshold to distinguish if a keyword is necessary for filtering representative landmarks or not.
- $F_k(N_k)$: If a keyword's frequency is larger than the threshold $T_k$, the function $F_k(N_k)$ is used to indicate the keyword in the candidate landmark list $C_l$.
- $K_{wf}$: A list composed of the final results of keywords, by computing $N_k$ and $T_k$.

```
Selecting_representative_landmarks()
    Loop l₁ : i ← 1 ~ n
        C₁[i] ← G(R)
    Repeat l₁ until G(R) = ψ
    Loop l₂ : i ← 1 ~ n
        Loop l₃ : j ← 1 ~ n
            If i ≠ j and If C₁[i] ∩ C₁[j] ≠ ψ
                K_w[n] ← C₁[i] ∩ C₁[j]
```

```
        N_k[F_n(K_w[n])]  ←  N[F_n(K_w[n])] + 1
          Repeat l_3
        Repeat l_2
        Loop l_4 : i ← 1 ~ n
          If N_k[i] > T_k
            K_wf[j]  ←  K_w[F_k(N_k[i]) ]
            j ← j+1
          Repeat l_4
        Loop l_5: i ← 1 ~ n
          Loop l_6 : j ← 1 ~ n
            If C_1[i] k_wf[j] ≠ ψ
      R_1[k]← C_1[i]
      k ← k+1
        Repeat l_6
        Repeat l_5
        End
```

According to this algorithm, we can easily obtain representative land-marks. The first procedure in this algorithm is to collect strings of candidate landmarks from the GPS. The procedure will not end until all landmarks near the visitors have been transferred in a given radius. Two loops are used for calculating frequencies of keywords in candidate landmarks. By making intersections of landmarks, the keywords can be easily obtained. With a given threshold of frequency, unnecessary keywords will be filtered, remaining ones can be used for finding representative landmarks. The final part of this algorithm is using the selected keywords to find landmarks that are related to users. By comparing obtained keywords and candidate landmarks, strings of landmarks composed of substrings with low frequency will be filtered.

In the algorithm, frequently appearing keywords in landmarks will be selected. Two kinds of results have different meanings to visitors. The first kind of keywords indicates where users are. Such keywords help users to obtain information of this area via an agent. The second kind of keywords represents objects like stores or restaurants. If there are buildings with the same purpose located in the area, keywords of these objects will also be considered as representative landmarks. The two types of landmarks are obtained by calculating frequencies. It is probably not enough for agents to gather all the information visitors need, users' comments are also a kind of information that can be used for discovering objects or landmarks besides results generated from this algorithm. This is not included in the algorithm but will be discussed in the next section.

## 13.2.2 Similar Travel Route Search

In the previous section, we introduced an approach to find related informa-tion by filtering keywords from landmarks near visitors. On a journey, visitors probably travel to several locations. In the selecting_representative_landmark approach, only landmarks in the same area are processed. Since visitors may travel in different areas, the relationship between such areas is also essential information to help users find other travel books on the Internet.

By accessing resources from users' mobile devices, we can obtain multimedia, spatial, and temporal information. Multimedia information is the basic unit in the intelligent travel book management system. After visitors captured multimedia resources, spatial and temporal information are also generated. By this information, we can approximately summarize the travel route of visitors. A simple example of a travel route is shown in the following:

| | |
|---|---|
| Location 1: Tamshui MRT station | 08:00 1/13/2008 |
| Location 2: Tamshui Old Street | 08:20 1/13/2008 |
| Location 3: TamKang University | 09:00 1/13/2008 |
| Location 4: Tamshui Fisherman's Wharf | 10:00 1/13/2008 |

In the previous section, representative landmarks, and keywords are generated by an agent for gathering related information. The users can access information of these four locations individually after they have finished their journey.

With spatial and temporal information, users can find more precise resources on the Internet. They may discover travel books with similar travel routes and make references in their next travel route arrangement.

To find a similar travel route is not a hard task for agents both on the Internet and the server side. The travel book agent searches related information by landmarks and keywords. Travel books published on the Internet with one or more target keywords will be returned to the management server.

Discovering travel blogs by using only keywords may be not very efficient. Another approach is searching travel books in the management server. Since users in the server use the same functions to produce travel books with specified tags, comparison of travel routes can be easily made on the server side. Besides landmarks or keywords, traveling time and sequence can also be used for discovering similar travel routes.

## 13.3 The System Architecture

As mentioned in the introduction, the major components of the intelligent travel book management system are the wizard, the agent, and the management server. Figure 13.1 shows the procedure for information collected from users or the Internet and transferred to the management server. The wizard is a program that runs on a user's PDA or other mobile device. The wizard

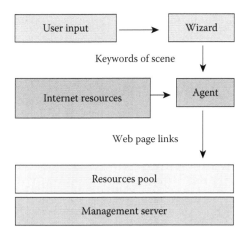

**FIGURE 13.1**
Procedure of gathering and repackaging information from wizard to server. (From Chang, H. B., Chang, H.-P., and Hung, J. C., *J. Software*, 3, 8, 42–48, 2008. With permission.)

can collect information and transfer resources for editing travel books to the management server. The agent on the other hand is responsible for analyzing the information received from the wizard and collecting related resources from the Internet. After the operations of the wizard and the agent, multimedia resources from users and the Internet will be stored on the server side and users can create their own travel books using these integrated resources. We will go through the details of all three components in the following.

### 13.3.1 The Intelligent Travel Book Wizard

The intelligent travel book wizard is designed for collecting user data and providing adequate resources by the agent. The wizard can access hardware resources of the mobile devices. The relationship of the wizard and the PDA hardware resources is shown in Figure 13.2. By using adequate program libraries provided by the operation system running on the PDA, the developer of wizard can easily get information and multimedia resources from the PDA. There are several kinds of user data collected by this wizard: images, audios, videos, temporal, and spatial information.

Images, audios, and videos are common multimedia resources generated by users. Visitors may capture such multimedia resources using their PDAs during their journeys. Such resources can be used for editing travel books for visitors in the future. Handling such resources is the first task of the wizard. The wizard will upload these resources to the server automatically. These multimedia resources are basic units that will be uploaded to the resource pool in the management server. Most of these multimedia resources will not be processed with additional procedures; users can access theses resources in the resource pool easily.

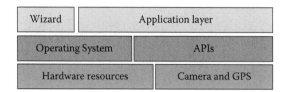

**FIGURE 13.2**
Architecture of intelligent travel book wizard.

There is other information from the above-mentioned multimedia resources; it is the temporal and spatial information. After the user captured a picture via a mobile device, the wizard can get the time stamp when the picture is taken. The time stamp helps the user to manage these multimedia resources. They can browse and access these resources according to the time of each picture. For an example, pictures could be classified into the first, second, and final day of a journey. For pictures that belong to the same day, there would also be temporal relationships between them. By the temporal relationship, the wizard can have a simple way to classify these pictures if the time difference is over a threshold (i.e., the pictures could be taken at different locations).

The final information collected from users is spatial information. This kind of information is the most important part of the users' information. The wizard gets the spatial information by accessing GPS component in a PDA with GPS, the wizard can get coordinates like latitude and longitude. The information captured from GPS is not the final version that will be transferred to an agent for further usage. The wizard needs to perform some procedures to transform such information to useful data.

One objective of the wizard is to locate visitors. By using the GPS component in a PDA, the wizard can get simple coordinates of the users. For further analysis by the agent, the wizard needs to transfer simple coordinates to a string of landmarks. With the advanced functions provided by the GPS, the wizard will get a different kind of information in the form of a string. The final task of the wizard is to find landmarks near users and transfer them to agents for advanced reference web page searching.

Landmark is the basic unit we used for gathering related information. The wizard needs to find all possible landmarks near users first. It can be easily achieved by the GPS component. Landmarks, buildings, and other meaningful objects will be shown to users if an adequate range is assigned. With the coordinates and a predecided surrounding area, several objects around visitors will be stored in the cache of the PDA. The landmarks will be used as keywords for an agent to gather reference web pages from the Internet. Not all landmarks are useful for users to create their travel books. A landmark selection engine is used for filtering unnecessary objects. Figure 13.3 illustrates a dataflow about the filtering process of landmark selection engines. An example might have three candidate landmarks. But

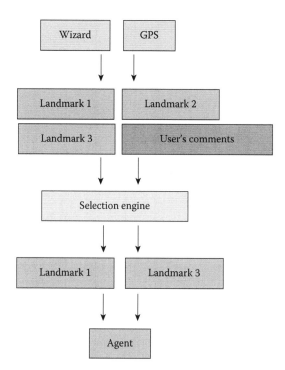

**FIGURE 13.3**
Using selection engine to filter unnecessary landmarks.

not all landmarks would be used for searching web pages on the Internet because it is too inefficient. By processing strings of landmarks and user comments, mark 1 and mark 2 would not be the users' most interesting objects. So they are ignored. The rest will then be sent to the agent for collecting information.

The concept of a landmark selection engine is simple. The input source of the selection engine is a set of strings generated by the GPS in a PDA. The selection engine will find the maximum similar substring in all input landmark strings. There could be several similar names of landmarks in the same region of the visitor. We chose some terms appearing in those strings of landmarks as keywords for further searching by the agent. The approximate procedures are listed as follows:

1. Storing all landmarks around visitors for selecting representative keywords

2. Comparing all input landmark strings and users' comments for generating keywords

3. Passing those selected keywords to an agent for searching related web pages on the Internet

In the second procedure of the developed method, the users' comments are also considered as clues for finding representative keywords. The users' comments reflect direct feelings and thoughts of visitors. Using the comments and string of landmarks nearby will help us to find meaningful objects for visitors.

Although there are some keywords generated by the landmarks selection engine; sometimes, these keywords cannot cover all objects that users are interested. The function provided by the wizard is aimed at helping users to save time from searching reference data on the Internet. Users can still find resources they want in the travel book editing phase if they are not satisfied with these web pages searched by the agent.

The information collected by the wizard also provides another type of record for users. The wizard can obtain geographic data from the GPS and transfer these data to the landmark selection engine and agent. When visitors are on their journeys, the information provided by the GPS will be uploaded in real time. The precise position of visitors changes frequently on a journey. A route of the journey is obtained by connecting every landmark the user passed. This information is also very useful for the users to edit their travel blogs.

### 13.3.2 The Intelligent Travel Book Agent

The agent is designed to help visitors to collect related information from the Internet. It is a program stored in the management server. After the wizard transmitted keywords generated by the selection engine, these keywords will be transferred to servers. These keywords are the most important information for the agent to search resources.

The searching procedures are similar to ordinary searches on the Internet. Taking Google as an example, users just need to feed keywords and set some additional properties and then the searched results will be listed on a Web page. Integrating such a function to the agent is not a hard task; we can use hyperlinks compliant with the HTTP protocol to replace the searching behavior by a simple procedure in the agent. For example, if users want to search information on the Tamkang University, they can key in the keywords in the search area. Searching information of Tamkang University can be represented by a hyperlink such as "http://www.google.com/search?hl=en&q=TamKang+University." The wizard just needs to follow the form and change keywords, a search request will be made to the search engine. The search results are also compliant with the HTTP protocol. The wizard just needs to parse the result web pages and discover the HTTP address of these web pages to get reference links.

There are many hyperlinks in the searching result window. By filtering those unrelated resources, keywords are used to ignore Web pages with names that are there as advertisements or services provided by search engines. By parsing and filtering pages returned by search engines, hyperlinks and comments will be gathered and transferred to the management server.

Besides the landmark selection engine, there are many other kinds of related resources distributed on the Internet. Visitors can capture images and add comments using mobile devices when they are on a journey. The keywords can also be used for searching nontextual resources. Users may be interested in other photos taken at the scene and use those photos to make comparisons with their own in the travel book. Searching related pictures on the Internet is very similar to Web pages. Some search engines provide interfaces to find pictures with keywords. What the agent needs to do is feed keywords to the search engine and parse the resulting Web pages to download image files. In this chapter, complex methods and technologies for retrieving image files are not discussed. The objective of the agent is just sending keywords to the search engine without mentioning what search algorithm is used. The case of a video retrieve is more complicated. Searching related video files for editing travel books is not discussed here.

### 13.3.3 The Intelligent Travel Book Management Server

The intelligent travel note management server is the final part of the developed application. Multimedia materials captured by the user are first transferred to the management server. Such materials are generated by visitors and do not need to be processed in advance. They are stored in the resource pool and users can access them directly while establishing a travel note. Another kind of material is collected from agents. Before users begin to create travel notes, they can not access such multimedia resources.

The management server provides an easy interface to users to access related resources collected from the Internet. After users end a journey and want to begin creating travel notes, they need to log in to the server first. There are not only multimedia resources but also a usage history in the server. In the process of creating travel notes, users can easily download related resources from the server and rearrange them in travel notes.

Some multimedia resources and information collected from users are stored in the intelligent travel note management server. For advanced queries, a simple data structure for these multimedia resources is adopted. Representative tags are listed below:

- Resource name
- Longitude
- Latitude
- Time
- Date
- Media type
- Descriptions

The first tag, resource name, is a basic item for the management server to handle. Users can package a multimedia resource and upload it to the server with a specified name. This name is for users to easily manage their contents. On the server side, a unique ID will be generated automatically.

Longitude and latitude attributes are collected from the GPS module of a user's mobile device. These two attributes are important tags in the multimedia resource information structure. After a user has produced a multimedia resource, the value of longitude and latitude will be recorded automatically. Other users can find these resources in the database of the management server according to their locations. If users want to access multimedia contents related to the location they are at, the travel note wizard (mentioned in the previous section) can also present image, audio, or video to users immediately.

Longitude and latitude attributes provide an easy way to retrieve multimedia objects in specified locations. Users can make a simple query directly to the management server without relying on agents to collect information from the huge network. Although the coverage of retrieval can only be limited to the management server, users can get precise query results quickly. Before users start to make a query to find simple resources at their current location, they need to input a parameter or radius. The meter is adopted as the distance unit of radius. Since latitude and longitude are hard for users, a transferring procedure is necessary to solve this problem. After users input the value of radius, the wizard will transfer the input to management and complete the retrieval process with longitude and latitude attributes. The following algorithm shows a simple verification of target multimedia resources retrieved from the management server.

- $R$: Radius input by users
- $C$: A set of information of users' current location
- $n$: Numbers of multimedia objects in the management server
- $M$: An information set of multimedia objects stored in the management server
- $dist(\ )$: To get distance of two input values
- $e$: A coefficient to transfer meters to value of latitude and longitude
- $Q$: A list to store the query results

```
Selecting target resource with radius(R, C)
  Loop 1 : i ← 1 ~ n
  If dist(M[i].latitude, C.longitude) < R*e and dist(M[i].
  latitude, C.longitude) < R*e
    Q[j] ← M[i]
    j ← j+1
  Repeat 1
  Output Q
End
```

Time and date tags represent temporal information of multimedia objects. Such information can also be used to choose resources. Here is an example. If a tourist visits a place in the morning, he or she may also want to access photos or video taken in the evening. Since tourists usually do not stay at the same location all day, the time attribute of multimedia resources can help users to have a look at scenes in different hours. The date is also used for the same purpose. If users at a specified location in the spring, they may want to view photos of the same scene in winter. Retrieving resources according to time and date tags can fulfill such requirements.

The tags of media type include audio, image, and video. The management server can use this tag to classify multimedia objects into three categories. For every type of object, the client application needs to have an adequate means to present them to users. The final part is description of objects. Using descriptions for retrieving resources is another way for users to make advanced queries. There is little difference from the tags mentioned above. Since descriptions are totally generated by users' input, casual descriptions may lead to bad query results.

Such tags can be used to help users quickly find resources they want. There is little difference from information collected from agents and the resources mentioned above. Information collected from agents presented to users is in a more complete form. They are presented mainly in Web pages. These pages may be captured from Web pages of travel books of other tourists. These multimedia resources can be considered a minimal unit to help users to augment contents of their travel records.

Information of users' history is also stored and analyzed in the management server. Data and records collected from users can help tourists to produce a schedule for their next journey. Besides information presented in users' personal profiles, they can access other traveling records provided by other users. By referencing these data, tourists can find suitable locations for their next journey by comparing data records of other users. In the management server, every procedure of uploading multimedia resources will be considered as a basic record. By analyzing these records, locations visited by users can be stored. A simple approach is adopted for computing the similarity of two users' traveling record in the server.

- $U$: Visited locations by users
- $n$: Number of visited locations of user 1
- $n$: Number of visited locations of user 2
- $S$: Grade of similarity of the two users

```
Getting similarity of two users'
record (U₁, U₂)
   Loop l₁ : i ← 1 ~ n
```

```
Loop l₂: j ← 1 ~ m
   If U₁[i] = U₂[j]
      s ← s+1
      Break
   Repeat l₂
   Repeat l₁
   Output S/n and S/m
End
```

By adopting this approach, users can find the traveling similarity with other tourists in the management server. The procedure of finding the most similar user is a little inefficient since the records of every two users need to be compared once. The number of repeated locations of two users can not indicate the traveling similarity of them because the total visited locations should also be considered in the similarity evaluation. The numbers of repeated scenes will be divided by the number of total scenes visited by tourists. The final results will present the degree of the similarity between users' travel records. With adequate presentation like ranks, users will find similar tourists and can access locations they visited.

There are several services provided by this server. Figure 13.4 shows several components of the management server. The three components in the upper side of this server are the login manager, the travel books manager, and the resources pool. All of them can be considered as front services of this server. The agent program runs on the server, there is no direct interfaces for users to operate it. The travel books manager and the resources pool are interfaces for users to edit travel notes and access multimedia resources in the server. The other three components are on the lower side of Figure 13.4: the Web page server, the database, and the usage history analyzer. All of these components provide services for storing and analyzing information from users.

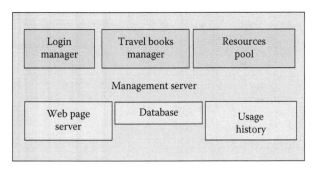

**FIGURE 13.4**
Major components of the management server.

- **The log-in manager**
  The major function of the server is providing adequate interfaces for users to create and edit their travel books. There are two different applications on both the client (the wizard) side and the server (the management server) side. A program for managing users' account is called login wizard that will be installed on the users' mobile devices. The login wizard is a component of the intelligent travel book wizard. It stores a user's account and password in the storage of mobile devices. Users need to login to the management server when they begin a journey. The event that triggers the login attempts is photo taking or video capturing. Assuming that users start the intelligent travel book wizard on their journey, the wizard will login to the management server using the account and password stored in the mobile devices. All of the information collected from login wizard will be processed in the login manager on the server side in order to allow users to login to the server.

  These procedures are made automatically by the wizard. The only thing that visitors need to do is just start the wizard and then the wizard will login to the management server automatically when a picture is taken. The wizard will upload keywords to the agent and multimedia resources to the management server.

- **The travel books manager**
  After the login manager checks the users' identification, users may enter the travel books manager to create and edit their travel notes. The detailed functions and interfaces will be introduced in the implementation section. Users can browse images, videos, comments, and related resources from the Internet. All of these resources are arranged with an adequate layout.

- **The resources pool**
  This is an interface shown in the travel book manager. The resources pool lists all multimedia resources collected by an agent. They are represented in the form of Web page hyperlinks. Titles of Web pages are shown in the resources pool. Users can access them by just clicking these items. The texts, images, videos, and audios in these Web pages can be used as references in users' blogs.

- **Web page server**
  The management server can be a Web page server. Since visitors create and edit their travel notes in the management server, the final goal of travel notes is to share experiences to other users on the Internet. Presently, there are several free services and network

spaces provided by companies or organizations. The behavior of users is not easy to change, forcing them to publish blogs in different platforms may not be a wise decision. On the management server side, travel books created by users are represented with a XML format. Migrating travel notes from the management server to other platforms is an issue that is not discussed here.

- **Database**
  There is a database for storing all of the materials for travel notes. The database is used for storing materials that are not processed like multimedia files from users and agents. Published travel books are also stored in the database. Users can edit or remove them at any time by using an interface of travel books manager.

- **Usage history analyzer**
  The analyzer is designed for recording and analyzing the behavior of users. In the interface of the analyzer, users can access published travel notes and find commonness of all their journeys. There are records of landmarks and routes generated from a user's journeys. By computing and counting the records, a simple report will be produced to help visitors to understand what kind of scenes they love to see. Analyzing the record of routes can also help visitors to improve their schedule arrangement if they want to travel again.

---

## 13.4 Implementation

The developed system utilized Microsoft.NET compact framework to implement the client program on Pocket PC with the Operating System—Windows Mobile 5.0. The test Pocket PC, Dopod 9100, integrates the wireless Internet access, camera, and GPS functions. In our scenario, the user brings the Pocket PC when he is traveling. GPS traces the user's geographic location and updates the information periodically to the server so that the user's route of the trip is able to be constructed and stored in the server at the same time. The user can also take a picture with the device while on vacation, attach relevant audio information as well as captions and time stamps then upload them to the server for further processing.

The pictures taken by the user are temporarily stored in the Pocket PC according to the date during the trip. Figure 13.5 shows the pictures taken for the first day of the trip. The pictures that belong to each day are also

**FIGURE 13.5**
Snapshot of the interface of wizard runs on mobile devices. (From Chang, H. B., Chang, H.-P., and Hung, J. C., *J. Software*, 3, 8, 42–48, 2008. With permission.)

**FIGURE 13.6**
Adding user comments to captured images. (From Chang, H. B., Chang, H.-P., and Hung, J. C., *J. Software*, 3, 8, 42–48, 2008. With permission.)

categorized by location. The user may visit several places within a day and add additional information to describe the pictures. Figure 13.6 illustrates that the user is selecting a picture to display its detail and is also adding text notes and audio description to the picture.

More and more users like to post their travel experiences on a personal blog now. However, users often encounter difficulty in manually uploading and organizing a lot of pictures, or forgetting their itinerary or exact location. Therefore, we have implemented a blog-like Web service that is capable of automatically integrating a variety of information that is relevant to the user's itinerary such as pictures, locations, route, and time. Eventually the information about the trip is retrieved by the Pocket PC and then uploaded to the server for advanced editing.

Instead of a user's time-consuming effort to organize and integrate the information, our service is able to analyze the data and collect additional, related information on the Internet to help the user efficiently finish his or her travel notes. Figure 13.7 illustrates the middle column as the main area for the user to edit his travel experience with text or multimedia files. The left-hand side calendar allows the user to select the date during his travel and display corresponding traveling data that has been stored in the server accordingly. For example, Figure 13.7 shows that the user selects March 12th and the pictures taken on that day will be displayed and categorized according to the different locations. Furthermore, the day's itinerary is mapped out on the right-hand side. Utilizing functions of Google Map, we combine the route formation traced during the user's travel and Google map APIs to draw the day's route with colored lines. The Web sites that are relevant to the user's visited places are also provided as referenced information.

This on-line travel book management service efficiently organizes and integrates all related traveling information. It provides users convenient editing application to record what they really want to remember filtering through relevant information and discarding everything else.

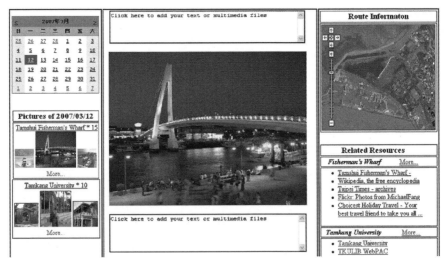

**FIGURE 13.7**
The interfaces of travel books manager.

## 13.5 Users' Experience

Users' experience and suggestions of using the proposed intelligent travel book are summarized as follows:

- The pictures taken during travel are automatically uploaded and categorized according to date and location. Therefore, it greatly saves time to focus on editing the main part of the traveling article.
- The well-arranged photographs and related location information help recall the significant events during a trip.
- The geographic and path tracking data can be valuable information and referenced by others.
- Batteries' lifetime becomes very short on a trip.
- While composing the travel book, the referenced information about the visited places provided by the agent is not all relevant and required.
- The interface of the editing travel note can be improved.

The points suggested by users to improve the system are explained as following; First, the battery of the portable pocket PC is a problem during a travel because many required processes execute continuously such as GPS receiver, camera, and data transmission. Therefore, preparing backup batteries or recharging the battery during traveling is absolutely necessary. Secondly, the geographic and referenced location information is not always accurate because some places visited by users are not famous scenic spots and have no required data on the digital map. Finally, we are working on improving the search results of related location information from the Internet and designing a friendlier interface.

## 13.6 Related Work

Location-based services have been long envisioned as an important element and applied to various applications. Our system provides a different way for individual users to record their travels and differs significantly from traditional travel applications or mobile tourist guides. As discussed in Baus et al. [4], it has been a great concern with the traveler about the place. In addition, an online travel service was dedicated for high quality images [8] and helped travelers to plan their trips. Utilizing rich and accurate metadata of pictures from photo forum sites, a virtual tour system can extract geographic location information and assess the quality of them. Its primary purpose is to help travelers to plan their trip. Similarly, a sophisticated map-based user interface is proposed to allow browsing the location tagged photos [15].

LocoBlog [3] is a space-time photo travel blogging service. Unlike traditional travel blogs, the entries are not only displayed linearly by time, they are also displayed spatially, allowing the viewer to get a greater sense of the journey. Another related research [10] has proposed an approach for integrating context-aware computing to a mobile travel assistant. A compact structure called User Task Model provides an explicit record of activities and intentions.

An interactive system [1] has demonstrated a graphic visualization of travel itinerary with emphasis on time and time zones. The travel system covers search aspects by supporting collaborative World Wide Web browsing and document creation with multiple complementary views of the trip. The system allows collaborative editing via any of the resources.

Web Travel Support Engine [14] was introduced to recommend traveling information to users and considering users' interests by applying the reinforcement theory to analyze the user behavior and study user interests. The agent technology used for the Personal Travel Market [9] has three types of agents, Personal Travel Agent, Travel Broker Agent, and Travel Service Agent, are collaborated to resolve users' travel requirements. The Virtual Travel Agent System [2] is built on the technology of a multiagent information system and the use of the semantic Web that can effectively organize information and service resources. A related system [13] applied on this topic with ontology that is more specific to tourism, package planning, recommendation, scheduling to tourist. An e-tourism scenario [5] has introduced a recommender engine that extents the use of semantics and ontology in the tourism applicative domain.

A research of a mobile game, alien revolt is discussed by De Souza e Silva [7]. This famous java cell phone game is very popular in Brazil. Users equip cell phones with location awareness components and can play games in cites just like they are played on battlefields. In the environment of this game, alien revolt, users can play two kinds of roles, alien hacker and human warriors. With cell phones and GPS components, users can enter various gaming stages in different locations.

The authors also point out that the level of mobile devices in Brazil is not high enough to give high growth to this game. High price and low accessibility do not make Brazil a good environment to develop mobile games. If we want mobile games with location awareness mechanism to be highly accepted by users, the mobile products distribution on the market will have to play an important part.

An interesting issue of location-based games was proposed by Das and Rose [6]. Uncertainty is the major issue in this research. Survival games are very suitable for participants with handheld or wearable sensors and interfaces. Such games are really convenient and exciting. Users need only carry mobile devices and no weaponry to experience battle with no fatigue.

The major uncertainty of this research is about accuracy of the GPS. In the experimental results, there are errors from [4] meters to 104 meters; the average is about 12 meters. Obviously the level of errors is too large to be accepted. It really affects the process of such survival games. The solution in this chapter is using better GPS component with antennas. The average of improved errors is about 4.4 meter. The results are much better than the original ones. In the proposed learning platform in this dissertation, accuracy is not very important as in mobile games. If a higher accuracy is necessary, a higher class of GPS component can be installed in our system.

Some researches focus on providing services to users by GPS components and electronic map. A project of websigns was published by Pradhan, Brignone, Cui, McReynolds, and Smith [12]. The websigns are a kind of hyperlink to physical locations. Users can access relative information of these physical locations by websigns. The kernel components of websigns are virtual beacons and virtual tags. The beacons are based on IR beacons and the tags are based on RFID tags. With these websigns, users can download and browse information from the network. They just need to scan IR beacons or Radio Frequency Identification (RFID) tags, then they can access information on their mobile devices.

Another application of electronics maps was published by Naaman [11]. Yahoo map is a free service to let users examine maps on the network. By integrating with Flickr's services, users can add photos on a Yahoo map. There are also customized profiles so users can edit their own maps with photos.

---

## 13.7 Conclusion and Future

An intelligent travel book management system integrated with the agent and Web technologies is introduced in this chapter. In order to save visitors' time and improve efficiency of creating travel notes, the wizard on the client side, an agent, and a management server are implemented. The wizard runs on the users' mobile devices, collects and filters information from visitors' photos and comments. Geographical information gathered from the GPS component in the mobile devices also provides a means to get precise landmarks on a visitor's journey. The string of landmarks generated by a selection engine will input to an agent, and these related materials on the Internet will be collected for references in the phase of travel notes editing. All resources from users and agents can be accessed by the travel books manager in the server. Visitors can easily create blogs and include the multimedia resources collected on their journey. Other reference web pages and images are also available in the resource pool. By this management system, creating travel notes becomes an easy task and even an entertainment for users.

The methodologies used for retrieving related materials in this chapter are simple. In order to get better search results and performance, other novel approaches of image or video retrieval may be adopted. The arrangement of visitors' routes is an interesting issue, too. The management system only records routes, but does not make suggestions to visitors. Searching information in other visitors' travel notes may be a solution to this problem.

## References

1. Apperley, M., Fletcher, D., Rogers, B., and Thomso, K., "Interactive Visualization of a Travel Itinerary," Proceedings of the Working Conference on Advanced Visual Interfaces, pp. 221–26, Palermo, Italy, May 2000.
2. Balachandran, B. M. and Enkhsaikhan, M., "Development of a Multi-Agent System for Travel Industry Support," Conference on Intelligent Agents, Web Technologies and Internet Commerce, pp. 63, Sydney, Australia, November 2006.
3. Bamford, W., Coulton, P., and Edwards, R., "Space-Time Travel Blogging Using a Mobile Phone," Proceedings of ACM International Conference on Advances in Computer Entertainment Technology, pp. 1–8, Salzburg, Austria, June 2007.
4. Baus, J., Cheverst, K., and Kray, C., "A Survey of Map-Based Mobile Guides," in Liqiu Meny, Alaxander Zipf and Tumasch Reichenbacher (eds.) *Map-Based Mobile Services Theories, Methods and Implementations,* pp. 197–216, Berlin and New York: Springer, 2005.
5. Chang, H. B., Chang, H.-P., and Hung, J. C., "Interactive Traveling Assistant based on Agent Technologies and Mobile Computing." *J. Softwa*re, 3, 8, 42–48, 2008.
6. Corallo, A., Elia, G., Lorenzo, G., and Solazzo, G., "A Semantic Recommender Engine Enabling an Etourism Scenario," Proceedings of the 4th International Conference on Semantic Web, pp. 1092–1101, Galway, Ireland, November 2005.
7. Das, S. K. and Rose, C., "Coping with Uncertainty in Mobile Wireless Networks," 15th International Symposium on Personal, Indoor and Mobile Radio Communications, (PIMRC 2004), vol. 1, pp. 103–8 Barcelona, Spain, September 2004.
8. De Souza e Silva, A., "Alien Revolt (2005–2007): A Case Study of the First Location-Based Mobile Game in Brazil," *IEEE Technology and Society Magazine* 27 (1): 18–28, Spring 2008.
9. Jing, F., Zhang, L., and Ma, W. Y., "VirtualTour: An Online Travel Assistant Based on High Quality Images," Proceedings of the 14th Annual ACM International Conference on Multimedia, pp. 599–602, Santa Barbara, California, October 2006.
10. Jorge, N. S., Donie, O'S., Henri, B., Patrice, C., Clair, M., and Ciara, B., "Experiences in the Use of FIPA Agent Technologies for the Development of a Personal Travel Application," Proceedings of the 4th International Conference on Autonomous Agents, pp. 357–64, Barcelona, Spain, June 2000.

11. Klusch, M., "Toward an Intelligent Mobile Travel Assistant," Proceedings of the 3rd International Joint Conference on Autonomous Agents and Multiagent Systems, pp. 1518–19, New York, 2004.

12. Naaman, M., "Eyes on the World," *IEEE Computer* 39 (10): 108–11, August 2006.

13. Pradhan, S., Brignone, C., Cui, J-H., McReynolds, A., and Smith, M. T., "Websigns: Hyperlinking Physical Locations to the Web," *IEEE Computer* 34 (8): 42–48, August 2001.

14. Shakshuki, E., Ghenniwa, H., and Kamel, M., "A Multi-agent System Architecture for Information Gathering," Proceedings of the 11th International Workshop on Database and Expert Systems Applications, pp. 732–36, London, United Kingdom, 2000.

15. Sukonmanee, P. and Srivihok, A., "Personalisation Travel Support Engine Using Reinforcement Learning," Proceeding of International Conference on Knowledge Management, pp. 287–92, Taipei, Taiwan, December 2004.

16. Toyama, R. L., Roseway, A., and Anandan, P., "Geographic Location Tags on Digital Images," Proceedings of the 11th ACM International Conference on Multimedia, pp. 156–66, Berkeley, California, November 2003.

# 14

## Real-Time Vehicle Navigation Using Vehicular Ad Hoc Networks

Jen-Wen Ding, Fa-Hung Meng, and Ray Yueh-Min Huang

**CONTENTS**

Traffic congestion is a serious problem for many urban cities around the world. This problem, unfortunately, causes many undesirable effects, such as fuel consumption, air pollution, and waste of time. Conventional approaches to alleviating traffic congestion problems include constructing new roads or widening existing roads. However, both approaches are very expensive solutions. With advances in Vehicular Ad hoc NETworks (VANET) and GPS-enabled devices (such as PDAs, smart phones, and car PCs), it is now feasible to make use of the VANET-based vehicle navigation systems to alleviate traffic congestion problems, which is a much more cost effective solution. The VANET system supports two types of communication modes: vehicle-to-roadside (V2R) and vehicle-to-vehicle (V2V). We describe V2R-based and V2V-based vehicle navigation systems in this chapter.

## 14.1 Introduction

Traffic congestion is an annoying but a common phenomenon in many urban cities around the world. Traffic congestion causes many serious problems, such as fuel consumption, air pollution, and waste of time for drivers.

Many reports have indicated that the total cost paid for traffic congestion is tremendous. For example, Schrank and Lomax [7] investigated the traffic congestion problem in 75 U.S. urban areas. The 75 areas used in the analysis include a range of populations from 100,000 to 17 million for the years from 1982 to 2001. They found that the total congestion cost for the 75 areas in 2001 came to $69.5 billion, which was the value of 3.5 billion hours of delay and 5.7 billion gallons of excess fuel consumed.

Traditional ways to alleviate traffic congestion problems are to construct new roads or widen existing roads. Unfortunately, both ways are very expensive solutions. With rapid advances in wireless communication networks and GPS-enabled devises (such as smart phones, PDAs, or car PCs), it is now feasible to use vehicle navigation systems, a more cost effective solution. In the past, many studies have shown that vehicle navigation systems can effectively alleviate traffic congestion problems. For instance, Bose and Ioannou showed that automatically guided vehicles can increase the throughput of roadways to double that of manually driven vehicles [1].

Vehicle navigation systems can be classified into two categories: static routing systems and dynamic routing systems. In static routing systems [11,12], vehicle navigation systems find the shortest path to guide a vehicle from its origin to its destination without considering the real-time traffic conditions on different roads. In implementation, some modifications may be made to change the weights of some roads. For example, some roads are assigned higher or lower weights according to the driver's preference and other factors (such as rush hours or nonrush hours). In dynamic routing systems [2,10,14], the real-time traffic information of different roads will be taken into account when deciding the quickest path. For this reason, we also refer to dynamic vehicle navigation systems as real-time navigation systems. In addition to this type of classification (i.e., static versus dynamic), other viewpoints can be used to classify different vehicle navigation systems. A comprehensive study that classifies different vehicle navigation systems from several viewpoints can be found in Schmitt and Jula [6].

Real-time vehicle navigation systems consist of the following main components: (1) GPS receiver, (2) wireless communication network interface, (3) digital road map, (4) route guidance algorithm, and (5) real-time traffic information collecting algorithm. Real-time navigation systems perform three main functions: (1) collecting local traffic information from different roads, (2) merging the local traffic information to generate wide-area routing information (i.e., route guidance information), and (3) delivering the routing information to the vehicles that need navigation. According to the routing information, guided vehicles can find quicker paths that bypass congested areas to save the driving time and alleviate traffic congestion problems.

To collect real-time traffic information from different roads, we need a proper network technology. However, existing network technologies, such as GPRS/3G and WiFi, are not specially designed to deliver vehicular traffic information. Over the past few years, a new network technology, called vehicular ad-hoc network (VANET), has been quickly developed to deliver vehicular traffic information. VANET is a form of a mobile ad hoc network. It supports two types of communication modes: vehicle-to-roadside (V2R) communication and vehicle-to-vehicle (V2V) communication [4]. A variety of wireless technologies are expected to implement VANET. Over the fast few years, a new VANET technology, Dedicated Short Range Communications (DSRC), has been rapidly developed in Europe, the United States, and Japan [15]. Currently, the DSRC is developed based on IEEE 802.11a and 802.11p protocols [3,4]. The transmission range of DSRC is about 1000 meters (a little longer than a half mile), and the data rate of DSRC ranges from 6 to 27 Mbps.

### 14.1.1 Basic Concept of Real-Time Vehicle Navigation

Real-time vehicle navigation systems are designed to find the quickest path for guided vehicles, instead of the shortest path. This can be explained with reference to Figure 14.1. Figure 14.1 is a snapshot of a road graph, where edges represent roads and nodes represent road junctions. The pair of numbers on each edge represents the travel time and distance of the corresponding road, respectively. For example, it takes eight units of time for a vehicle to traverse from J4 to J6, and the distance from J4 to J6 is three units of length. Due to the changing traffic conditions on all roads, the shortest path is usually not the quickest path. For example, in Figure 14.1, the shortest path from the source node to the destination node is source → J4 → destination. The length of the shortest path is 7 + 4 = 11 units of length, and the travel time is 14 + 4 = 18 units of time. Another path that is quick: source → J3 → J5 → J6 → destination. Although the length of the quickest path is 3 + 2 + 5 + 5 = 15 units of length, and the travel time is 2 + 1 + 3 + 9 = 15 units of time. By taking the quickest path, the driving time can be reduced from 18 to 15 units of time, and the traffic congestion can be alleviated.

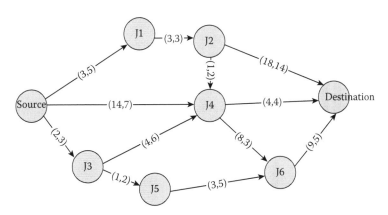

**FIGURE 14.1**
The quickest path versus shortest path.

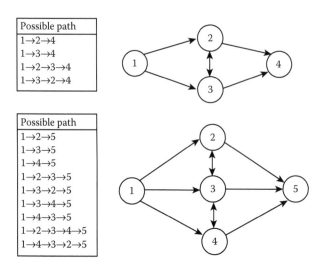

| Possible path |
|---------------|
| 1→2→4 |
| 1→3→4 |
| 1→2→3→4 |
| 1→3→2→4 |

| Possible path |
|---------------|
| 1→2→5 |
| 1→3→5 |
| 1→4→5 |
| 1→2→3→5 |
| 1→3→2→5 |
| 1→3→4→5 |
| 1→4→3→5 |
| 1→2→3→4→5 |
| 1→4→3→2→5 |

**FIGURE 14.2**
Potential routes in road graphs.

In a road graph, there may exist multiple paths between any two given nodes. As shown in Figure 14.2, the number of possible paths increases rapidly with the number of nodes in the graph. Each possible path may be a route candidate since each possible path may take less travel time than the shortest path. Therefore, for a large area, it is challenging to find all quicker paths between any two given nodes according to the real-time traffic information.

### 14.1.2  Three Types of VANET-Based Vehicle Navigation Systems

The VANET-based vehicle navigation systems can be classified into three types of categories: (1) V2R-based, (2) V2V-Based, and (3) hybrid. Figure 14.3 shows the architecture of a V2R-based vehicle navigation systems. V2R-based systems adopt a centralized and hierarchical architecture [8,9,10,14]. In this type of architecture, a large number of roadside units (or wireless sensors) and base stations (BSs) need to be deployed along the side of all roads. The roadside units use V2R communication links to communicate with onboard units of passing vehicles to collect local traffic information, such as the number of cars on a road and the average driving speed on a road. Then, the BSs transmit the collected local traffic information to a traffic information center (TIC). The TIC merges the collected local traffic information to generate a wide-area traffic information. The advantage of V2R-based architecture is that V2R links are reliable since roadside units are fixed and V2R links use one-hop communication. The disadvantage of V2R-based architecture is that it is very costly since an infrastructure consists of a large number of roadside units, BSs, and traffic information centers (TICs).

Figure 14.4 shows the architecture of the V2V-based vehicle navigation systems. V2V-based systems adopt a decentralized and flat architecture

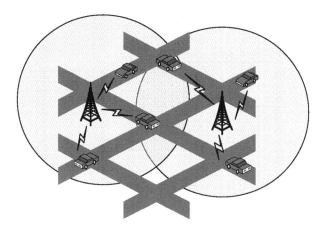

**FIGURE 14.3**
V2R-based vehicle navigation systems.

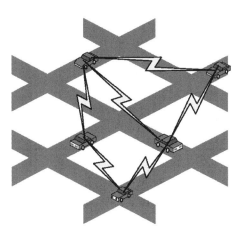

**FIGURE 14.4**
V2V-based vehicle navigation systems.

[2,5,13]. In this type of architecture, there is no need to deploy roadside units and BSs. Also, there is no need for TIC. Real-time traffic information is collected and managed by vehicles themselves using V2V communication links. Traffic information is relayed from one vehicle to another vehicle in a multi-hop fashion. The advantage of V2V-based architecture is that no infrastructure is required, which greatly reduces the overall installation and maintenance cost. However, since vehicles are not stationary and V2V-based architecture employs multi-hop relay, the failure rate of multi-hop relay is higher than one-hop relay. From this point of view, V2R-based systems are more reliable than V2V-based systems.

Since the V2R-based and V2V-based systems have their own advantages and disadvantages, it is nature to combine both approaches. Figure 14.5

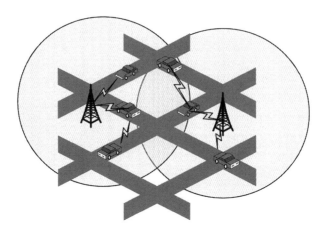

**FIGURE 14.5**
Hybrid architecture of vehicle navigation systems.

shows the hybrid architecture. With the use of V2V communication links, some vehicles that are beyond the transmission range of wireless BSs can transmit data to and from BSs using V2V links. Therefore, the deployment density of BSs can be greatly reduced. On the other hand, with the use of BSs, the length of the multi-hop relay for remote vehicles can be reduced too, thereby increasing the reliability of the V2V links.

## 14.2 Vehicle-to-Roadside-Based Vehicle Navigation Systems

In this section, we describe the main design principles of V2R-based vehicle navigation systems on the basis of the methods proposed [9,14]. A V2R-based navigation system needs an infrastructure to collect real-time traffic information. The infrastructure includes a large number of roadside units (or wireless sensors), BSs, and TICs. The roadside units use V2R links to communicate with onboard units on passing vehicles to collect real-time traffic information. Since the transmission range of roadside units is very limited (e.g., 1000 meters for DSRC), the collected traffic information is first delivered to BSs. The BSs then transmit the collected information to TIC at a remote distance. We refer to the connections from BSs to TIC as backhaul connections. Backhaul connections can be implemented by a variety of wired or wireless technologies, such as frame relay, WiMax, wireless mesh networks, and GPRS/3G. The TIC merges all collected real-time local traffic information, generates wide-area traffic information and routing information, and transmits the routing information to guided vehicles. The routing information may be broadcast to guided vehicles or unicast to guided vehicles in an on-demand fashion. Figure 14.6 depicts the information flow chart of V2R-based navigation systems. Figures 14.7 and 14.8 show two different network architectures of V2R-based navigation systems, which were proposed in [9,13]. In Figure 14.7, wireless sensors are installed along the roadside to collect local traffic information. The roadside sensors transmit collected information to BSs using multi-hop relay. By doing so, the deployment density of BSs can be greatly reduced. In Figure 14.8, DSRC roadside units are responsible for collecting real-time traffic information and transmit it to WiMax BSs. The WiMax BSs are organized as a wireless mesh network. That is, remote BSs can transmit its traffic information to TIC using multi-hop relay. Since the transmission range of WiMax is about 50 km, and the transmission bandwidth of WiMax is about 70 Mbps, the deployment density of BSs can be very small.

TIC is responsible for merging local real-time traffic information, generating routing information, and transmitting routing information to all vehicles. There are two ways to generate routing information: proactive and reactive. In the proactive approach, the routing information is generated periodically. That is, for each source position, the quickest paths from the source position to all potential destination positions are periodically calculated and broadcasted to all vehicles. In implementation, both source

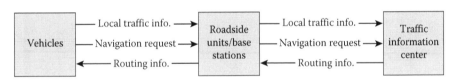

**FIGURE 14.6**
Information flow chart of V2R-based navigation systems.

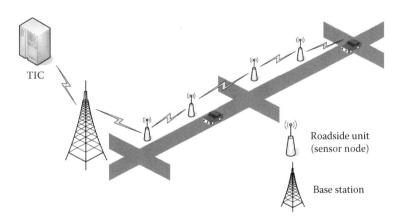

**FIGURE 14.7**
Network architecture used in a wireless mesh network for real-time vehicle guidance.

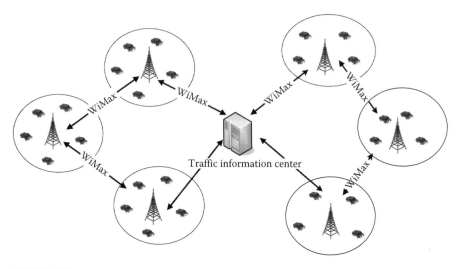

**FIGURE 14.8**
Network architecture used in the design and implementation of a vehicle-to-vehicle based traffic information system.

position and destination position can be represented by the ID of roadside unit. In the reactive approach, the routing information is generated in an on-demand fashion. That is, TIC will only calculate the quickest path for a given source position and a destination position after it receives a navigation request. The on-demand routing information is usually delivered to the requesting guided vehicle via unicast delivery. The main problem with the proactive approach is that a large routing table will be generated because the total number of all pairs of source position and destination position is large. For example, if the number of source positions (or roadside units) is $n = 10^4$, then the routing table will have $n^2 = 10^8$ entries. It is difficult to deliver such a large routing table to all vehicles. To conquer this difficulty, a hierarchical addressing scheme can be employed to reduce the size of the routing table [9]. With hierarchical addressing, a map is divided into several tiers. For example, the digital map in Figure 14.9 is divided into nine tier-3 zones (i.e., A, B, C, …), each of which is divided into nine tier-2 zones (i.e., a, b, c, …). Each tier-2 zone is further divided into nine tier-3 zones. A routing table is generated for each tier-3 zone. Each tier-3 zone is covered by a BS. For example, Table 14.1 shows the routing table for a tier-3 zone "E.e.a". There are only 24 entries in the routing table. However, if there is no hierarchical addressing, the entries in the routing table will be $n^2$, where $n = 9^3$. Another common problem with proactive routing is that the computing complexity is high for finding the quickest path between any two given nodes in a graph. If the Floyd shortest path algorithm is employed, the time complexity will be $\Theta(n^3)$, where $n$ is usually a large number. The main problem with the reactive

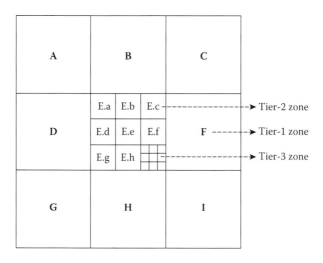

**FIGURE 14.9**
Hierarchical addressing used in a wireless mesh network for real-time vehicle guidance. (Adapted from Wolff, A. and Lee, C. C., Proc. of the 9th International Conference on Applications of Advanced Technology in Transportation, 368–73, 2006. With permission.)

**TABLE 14.1**

Hierarchical Routing Table for Zone E.e.a in Figure 14.9

| Tier 3 | | Tier 2 | | Tier 1 | |
|---|---|---|---|---|---|
| Destination | Next BS | Destination | Next BS | Destination | Next BS |
| A.x.x | | E.a.x | | E.e.b | |
| B.x.x | | E.b.x | | E.e.c | |
| C.x.x | | E.c.x | | E.e.d | |
| D.x.x | | E.d.x | | E.e.e | |
| F.x.x | | E.f.x | | E.e.f | |
| G.x.x | | E.g.x | | E.e.g | |
| H.x.x | | E.h.x | | E.e.h | |
| I.x.x | | E.i.x | | E.e.i | |

*Source:* Adapted from Wolff, A. and Lee, C. C., Proc. of the 9th International Conference on Applications of Advanced Technology in Transportation, 368–73, 2006. With permission.

routing is that an initial delay must be tolerated to wait the response from TIC, which may be long if the TIC is busy.

## 14.3 Vehicle-to-Vehicle-Based Vehicle Navigation Systems

In this section, we look at the main design principles of V2V-based vehicle navigation systems by describing the methods proposed [2, 5]. In the

V2V-based approach, there is no need for an infrastructure. Each vehicle that needs navigation collects real-time traffic information by using V2V communication links, as shown in Figure 14.11. Also, each vehicle that needs navigation will calculate the quickest route by itself according to the collected real-time traffic information.

To reduce the large communication overhead incurred by V2V comm-unication, cell-based data aggregation and packet relay is adopted. That is, a road segment is divided into several fixed cells, as shown in Figure 14.10. Herein, a road segment is defined as a section of a road between two intersections. All vehicles within a cell share similar traffic conditions. Each cell has a cell head: a car in a cell that is the closest to the cell center will be elected as a cell head (see Figure 14.11). A cell head is responsible for collecting the traffic information of a cell. To do so, all vehicles in a cell exchange their basic data (such as driving speed and driving direction) by periodic broadcasting hello messages. Then, a cell head can easily learn of the traffic condition in a cell, such as the number of cars in a cell and the average driving speed in a cell. To uniquely identify a cell, in the digital

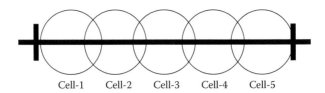

Cell-1      Cell-2      Cell-3      Cell-4      Cell-5

**FIGURE 14.10**
Driving a road segment into cells.

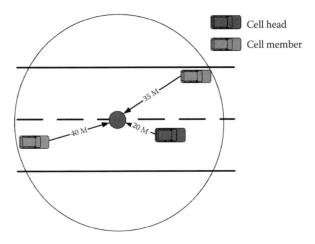

**FIGURE 14.11**
Cell head and cell members.

map, each road is assigned a road ID, a road segment is assigned a segment ID, and each cell is assigned a cell ID. Then, a vehicle can easily find out its current road ID, segment ID, and cell ID according to is current position obtained by the GPS receiver. With cell segmentation, only cell heads are involved in packet relay. Other cell members are not involved in packet relay, as shown in Figure 14.12. The packets used for collecting real-time traffic information usually contain some time fields to record the occurring time of different events. Compared to other wireless networks, it is easy to implement synchronization in VANET since all vehicles are automatically synchronized by the GPS system.

To find the quickest path for a guided vehicle, the guided vehicle will issue a route query packet. The query packet is a unicast packet. It is transmitted from the origin to the destination along the shortest path using multi-hop relay. After the query packet arrives at the destination cell, the destination cell head issues a route reply packet to the guided vehicle using flooding. By flooding we mean that the route reply packet will be duplicated at every road intersection, as shown in Figure 14.13. During the flooding process of the

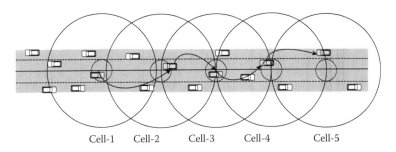

Cell-1    Cell-2    Cell-3    Cell-4    Cell-5

**FIGURE 14.12**
Cell-based data aggregation and packet relay.

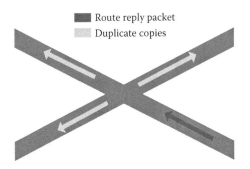

**FIGURE 14.13**
Flooding of route reply packet.

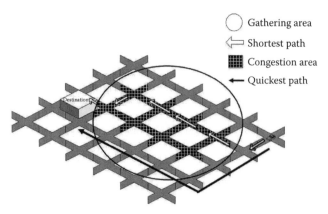

Gathering area
Shortest path
Congestion area
Quickest path

**FIGURE 14.14**
Overview of V2V-based vehicle navigation.

route reply packet, the route reply packet will record the traffic conditions of all traversed routes. After the guided vehicle receives multiple copies of the route reply packet from different directions, it can easily calculate the quickest path according to the real-time traffic information recorded in the route reply packets. Figure 14.14 shows the shortest packet taken by route query packet, and the flooding area traversed by the route reply packet, and then found the quickest route from origin to destination. Note that both route query packet and route reply packet are relayed by cell heads on a cell-to-cell basis.

The main problem with the route reply packet is that a simple unconstrained flooding scheme, unfortunately, will result in a broadcast storm in VANET. To solve this problem, three packet dropping mechanisms are developed [2] to limit the scope of flooding. The most important one is the long-route dropping scheme. During the delivery process of the route query packet, the query packet will record the traffic conditions of all traversed routes. After the destination cell head receives the route query packet, it can make use of this traffic information to estimate the travel time for a vehicle to traverse the shortest path. We call this time shortest-path travel time. Similarly, during the delivery process of the route query packet, the query packet will record the traffic conditions of all traversed routes. Due to flooding, the route reply packet will discover multiple potential routes for guided vehicles. When an intermediate vehicle in a potential route receives a route reply packet, it first estimates the travel time from the destination cell head to itself. If the estimated travel time has exceeded the shortest-path travel time, it knows that this potential route is useless since it results in longer travel time than the shortest path. Therefore, when such an event occurs, the intermediate vehicle drops the route reply packet and stops the flooding of the reply packet. It can be seen in Figure 14.14 that the flooding area

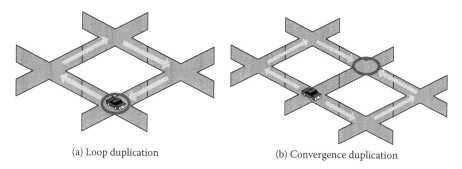

(a) Loop duplication            (b) Convergence duplication

**FIGURE 14.15**
Packet dropping schemes for two packet duplication problems.

is limited to a small scope. The other two packet dropping schemes are used to drop duplicate reply packets. The two schemes are the loop dropping scheme and the convergence dropping scheme. The loop dropping scheme is used to drop the duplicate route reply packet caused by loops in a road graph (see Figure 14.15a). The convergence dropping scheme is used to drop the duplicate route reply packet that is sent from different directions (see Figure 14.15b).

## 14.4 Comparison and Discussions

In this section, we compare V2R-based and V2V-based vehicle navigation systems from the following perspectives: network architecture, communication mode, and computing mode. Table 14.2 lists the summary of the comparison results.

From the point of view of network architecture, V2R-based systems are more complex and more expensive than V2V-based systems. A V2R-based system requires an infrastructure, which comprises a large number of roadside units (or sensors), BSs, and TIC. The communication between vehicles and roadside sensors usually employ VANET. The communication between BSs and TIC usually employ existing wired or wireless networks, such as GPRS/3G, WiMax, or frame relay. The total cost for deploying such an infrastructure is very high. Also, the total cost for maintaining the infrastructure is high. In contrast, for the V2V-based systems, there is no need to install and maintain an infrastructure.

From the point of view of the communication mode, V2R-based systems make use of V2R communication links of VANET to collect real-time traffic information. Since all roadside units (sensors) are fixed and all V2R links use one-hop communication, V2R links are quite stable. On the other hand, V2V-based vehicle navigation systems make use of V2V communication

**TABLE 14.2**

Comparison of V2R-Based and V2V-Based Vehicle Navigation Systems

|  | V2R-Based Navigation | V2V-Based Navigation |
|---|---|---|
| Network architecture | Infrastructure-based | Infrastructure-free |
| Required equipment | Roadside units (sensors), Base stations, Traffic information center, Onboard units in vehicles | Onboard units in vehicles |
| Backhaul connection | Required | No |
| Deployment cost | High | Low |
| Maintenance cost | High | Low |
| Transmission mode | One-hop transmission | Multi-hop relay |
| Transmission links | Vehicle-to-roadside links | Vehicle-to-vehicle links |
| Transmission reliability | High | Low |
| Network loading | Low | High |
| Computing mode | Centralized | Decentralized |
| Information management structure | Hierarchical | Flat |
| Real-time traffic information update | Periodic update | On-demand update |
| System robustness | Weak (single point of failure) | Robust |

links of VANET to collect real-time traffic information. That is, real-time traffic information is relayed from one vehicle to another using multi-hop relay. Since all vehicles are mobile and move fast, a long multi-hop relay link is vulnerable to link breaking. This means that the route query or reply packet issued by the V2V-based vehicle navigation systems may be lost during the packet relay process. Another important consideration is that the network load of V2V-based systems is higher than that of V2R-based systems because V2V-based systems employ multi-hop relay and flooding to discover potential paths, while V2R-based systems simply use one-hop communication to collect traffic information from passing vehicles. Therefore, from the perspective of communication, V2V-based systems are less reliable than V2R-based systems.

From the point of view of the computing mode, V2R-based systems employ a centralized and hierarchical architecture. All real-time traffic information is collected by roadside units (sensors), transmitted to BSs, and then transmitted to TIC to generate routing information for guided vehicles. However, since the TIC plays a vital role for the whole V2R-based system, it becomes a single point of failure. For any disastrous events (such as fire or network attack), the failure of the TIC will render the whole system useless. Form the point of view of system robustness, V2V-based systems are better than V2R-based systems.

## 14.5 Conclusions

Over the past few years, GPS-based vehicle navigation systems have received great interest and have been employed by more and more drivers. It can be expected that, in the near future, a vehicle navigation system will become an indispensable piece of equipment for drivers. Most current vehicle navigation systems use static routing algorithm without considering the real-time traffic information of roadways. Only a few navigation systems consider little real-time traffic information in making route decisions. By little real-time traffic information, we mean that only the traffic information of highways and some important trunk roads are taken into account. To implement an accurate and complete real-time vehicle navigation system, we can make use of a new network technology, VANET, to collect real-time traffic information of all roadways and to deliver the routing information to all guided vehicles. Currently, many VANET-based applications have been rapidly developed, such as emergency warning system for vehicles, intersection collision avoidance, and cooperative adaptive cruise control. It is reasonable to anticipate that VANET-based vehicle navigation systems will be turned into reality in the near future. In this chapter, we discussed the design principles of two types of VANET-based navigation systems: V2R-based and V2V-based. The two types of systems have been the subject of much research in the past few years. Due to the two types of approaches have their own advantages and disadvantages, it is important to develop a hybrid architecture that can combine their advantages, which still remains an open research issue that needs more investigation.

## References

1. Bose, A. and Ioannou, P. 1998. Evaluation of mixed semiautomated manual traffic. In Proc. of IEEE International Conference on Control Applications, vol. 2, 868–72, Trieste, Italy.
2. Ding, J. W., Meng, F. H., and Huang, Y. M. 2008. A real-time vehicle guidance system using P2P communication. In Proc. of IEEE First International Conference on Ubi-media Computing, 225–30, Lanzhou.
3. Ergen, M. and Varaiya, P. 2005. Throughput analysis and admission control for IEEE 802.11a. *Mobile Networks and Applications* 10 (5): 705–16.
4. Hartenstein, H. and Laberteaux, K. P. 2008. A tutorial survey on vehicular ad hoc networks. *IEEE Communications Magazine* 46 (6): 164–71.
5. Jerbi, M., Senouci, S. M., Rasheed, T., and Ghamri-Doudane, Y. 2007. An infrastructure-free traffic information system for vehicular networks. In Proc. of IEEE 66th Conference on Vehicular Technology, 2086–90, Baltimore, MD.

6. Schmitt, E. and Jula, H. 2006. Vehicle route guidance systems: Classification and comparison. In Proc. of IEEE Intelligent Transportation Systems Conference, 242–47, Toronto, Ontario.

7. Schrank, D. and Lomax, T. 2003. The 2003 annual urban mobility report. Texas Transportation Institute, The Texas A&M University System. Available at http://ntl.bts.gov/lib/24000/24000/24010/mobility_report_2003.pdf

8. Wischoff, L., Ebner, A., Rohling, H., Lott, M., and Halfmann, R. 2003. SOTIS: A self-organizing traffic information system. In Proc. of IEEE 57th Vehicular Technology Conference, vol. 4, 2442–46, Jeju, Korea.

9. Wolff, A. and Lee, C. C. 2006. A wireless mesh network for real-time vehicle guidance. In Proc. of the 9th International Conference on Applications of Advanced Technology in Transportation, 368–73, Chicago, IL.

10. Yamashita, T., Izumi, K., and Kurumatani, K. 2004. Car navigation with route information sharing for improvement of traffic efficiency. In Proc. of IEEE 7th International Conference on Intelligent Transportation Systems, 465–70.

11. Zhan, F. B. 1997. Three fastest shortest path algorithms on real road network: Data structure and procedures. *Journal of Geographic Information and Decision Analysis* 1: 69–82.

12. Zhan, F. B. and Noon, C. E. 1998. Shortest path algorithms: An evaluation using real road networks. *Transportation Science* 32: 65–73.

13. Zhang, J., Ziliaskopoulos, A. K., Wen, N., and Berry, R. A. 2005. Design and implementation of a vehicle-to-vehicle based traffic information system. In Proc. IEEE 8th International Conference on Intelligent Transportation Systems, 473–77.

14. Zhang, X., Hong, J., Fan, S., Wei, Z., Cao, J., and Ren, Y. 2007. A novel real-time traffic information system based on wireless mesh networks. In Proc. of IEEE Intelligent Transportation Systems Conference, 618–23, Seattle, WA.

15. Zhu, J. and Roy, S. 2003. MAC for dedicated short range communications in intelligent transport system. *IEEE Communications Magazine* 41 (12): 60–67.

# 15

## Patient Safety Management: RFID Technology to Improve Emergency Room Medical Care Quality

Chia-Chen Chao and Wen-Yuan Jen

## CONTENTS

Yearly, 44,000 to 98,000 patients die, $17 to $29 billion are lost due to treatment errors in the United States. Although the accuracy of the numbers has been challenged, medical disasters caused by treatment errors are serious. Many problems arise in emergency room operations such as severely ill patients waiting too long to receive treatment, unprofessional staff operations, improper drugs, injections, volume of anesthetics, surgeries, and so on. These drawbacks threaten emergency room medical care quality and endanger patient safety. The Joint Commission on Accreditation of Healthcare Organizations (JCAHO) urged that medical institutions pursue patient safety.

Improving patient safety with technology has thus become an urgent need. Hospitals need an innovative approach to improve the process of operations. The purpose of this paper is to apply radio frequency identification (RFID) to solve emergency room operational problems. The Wan-Fang Hospital is the only hospital implementing the emergency room RFID information system in Taiwan. This paper reports on the findings of a case-based research investigation. It introduces emergency room workflow, which monitors the triage patient's waiting time and the treatment process. The RFID system is applied in the emergency room workflow as a new service era to improve patient waiting time and accuracy for patient treatment and prevent human errors.

## 15.1 Introduction

There are 44,000 to 98,000 patients that die yearly, which leads to $17 to $29 billion in loss, and this is resulted from treatment errors in the United States [1]. Although the accuracy of numbers has been challenged, these medical disasters show the treatment errors are serious. The Joint Commission on Accreditation of Healthcare Organization (JCAHO) has proposed six major goals for medical institutions to pursue patient safety. These goals include accuracy of patient identification, effective communication between medical care providers and patients, improving medication safety of high-risk medication error, eliminating the occurrence of errors regarding patient's operation position, and cost-effectiveness of a clinical warning system. The Institute of Medicine (IOM) reports [2] its goal is to achieve a twenty-first century health care system that is evidence-based, patient-centered, and systems-oriented. Hence, using information technology to improve patient safety has become a primary trend [3], hospitals need innovative approaches to improve medical care and patient safety.

The emergency room is the place to fight for every second in order to save a patient's life. However, there are some crises contingencies in an emergency room, such as uncertainty of patient visits, shortage of medical staff, critically ill or trauma patient's long wait time before receiving treatment, and lacking managerial monitoring of the quality care mechanism. Basically, the hospital emergency room needs to audit and maximize its operation without having to dramatically increase employee numbers. It is required to apply information technology to achieve this goal [4]. Lately, hospitals are applying RFID to solve emergency room challenges [5] and drug safety [11], but there is the rare hospital applying RFID to manage emergency room workflow. Thus, this study focuses on a RFID-enabled information system, and the purposes of the system are preventing human errors, tracking every patient case, and pursuing patient safety.

This paper is divided into five sections. In Section 15.2, the mobile medical care information system literatures are viewed; meanwhile, the

RFID background is introduced as well. In Section 15.3, the deficiencies of emergency room workflow are depicted. In Section 15.4, the emergency room problems are presented based on a real case. In Section 15.5, we discuss the emergency room-radio freqency identification (ER-RFID) information system workflow, and emergency room prototyping functions of monitoring both patient's waiting time and medical treatment process.

## 15.2 Mobile Information System and RFID

### 15.2.1 Mobile Medical Care Information System

Health care facilities have implemented mobile medical care information system for managing organizational changes and challenges for years. For health care facilities, significant mobile advances have been made in devices, applications, and networking infrastructure, such as home care [6], a mobile clinical support system [7], user interface service [8], medical knowledge access [9], and mobile patient information [5,10,11] have been implemented and improved medical care service.

Some health care facilities install information systems in the emergency ambulance to enhance emergency prehospital care [12,13]. The Emergency Services Act defined the system, which provides the arrangement of personnel, facilities, and equipment for the effective and coordinated delivery of health care services under emergency conditions [14]. With the advantages of mobile technology, more mobile medical care information systems are implemented, such as a personal digital assistant (PDA) providing dedicated patient's information to physicians [15] and mobile physician order entry system [16]. We consider a mobile medical care information system an important role for managing clinical routine treatment. This study tends to believe that both an information system and RFID should have the potential to improve emergency room workflow and patient safety.

### 15.2.2 RFID (Radio Frequency Identification)

The RFID tag is an integrated circuit (IC) chip with an antenna attached by a printed conductor on a printed plastic sheet. The data can be stored inside the IC and communicate or transmit by the antenna. The size of the tag can be very small like a grain of rice or larger than a brick size. There are two kinds of RFID tags, one is passive (no battery) and the other is active. Passive tags are most popular for use. The RFID reader sends an energy field to wake up the tag and send energy to enable the processing of storing or sending data. The active tag has its own battery power. Data can be captured by processing readers. Encryption algorithms can ensure data security and protect transmission. Speed of data transmission and distance count on power output, antenna size, and interface. Tags have read-only, write-only, or both. Some memories can store ID permanently, others can update or use data storage later. Data is transmitted to read from RFID tags by RF signals. Therefore, in Table 15.1, various enterprises belonging to different industries have started to adopt RFID, such as Wal-Mart, Nokia, and U.S. Navy, for example. Many companies claim that the RFID improves organizational efficiencies and competitive advantages [17].

The advantage of the RFID tag is it can store a certain amount of data in the memory storage device, such as product identification number, price, cost, manufacture date, location, and product volume, and so on. This information can be read by a wireless scanner with fast speed. The RFID is able to process a large volume of multiple data at the same time. The RFID identifies its target within a short time, and its accurate matching competence will help organizational operations efficiently [22]. Some health care facilities have increasing need for RFID [23], and some hospitals have adopted RFID for solving emergency room challenges [5].

The health care industry is launching RFID applications to solve operational problems, such as the fight against drug counterfeiting [9], tracking blood products from donor to patient [25], and transmitting the correct medicine dosages to patients [11]. The RFID-enabled resource and workflow management solutions are designed to optimize asset utilization, reduce operating costs, and improve care quality in the health care industry [26]; there are more RFID-based system for various medical applications being developed in the United States [27]. Table 15.2 proposes health care applications suited

**TABLE 15.1**

Cases of RFID Being Adopted in Industries (RFID, 2004, 2005)

| Industry | Company | Application | Reference |
|----------|---------|-------------|-----------|
| Retailer | Wal-Mart | Tagging pallets and cases beginning in 2005 and to achieve internal efficiency material management, also to meet requirements from their customers for fast payment at cashier counter. | 18 |
| Cell phone | Nokia | It easily enables customers to initiate starting the mobile services, payment solutions, verification, authentication, ticketing, and exchanging business card. | 19 |
| Airline | Airport | RFID helps airlines to improve handling huge amounts of luggage items. | 20 |
| Pharmacy | The U.S. Food and Drug Administration | RFID tags attach at the genuine drug box. When the drug sells, it checks registration information to verify its authentication. | 21 |
| Navy | U.S. Navy | Track combat casualties and identify the wounded arriving for treatment at a field hospital in southern Iraq. | 21 |

**TABLE 15.2**

Cases of RFID Being Adopted in Healthcare Industry

| Hospital | Application | Reference |
|---|---|---|
| Johns Hopkins Medical Center | RFID monitors injection liquid bag for enough medicine to prevent air going into patient's body. | 24 |
| Georgetown University Hospital | RFID tracks blood products from donor to patient. | 9 |
| M D Anderson Cancer Center | Tagging drugs with radio chips, RFID ensures the cancer drug safety. | 11 |

to RFID. This study concludes the four traits for medical RFID application: (1) identification, tracking, and location in health care facilities; (2) identification of implantable medical devices; (3) access control in health care facilities; and (4) product packaging. From the literature review, it points out that some hospitals are utilizing RFID for tracking injection IV bags, blood bags, cancer medicine, and tracking wounded soldiers and treatments. But, there is a few hospitals that have applied RFID for emergency room workflow management. However, an emergency room is exactly the place that needs RFID-enabled resource to prevent medical errors and keep higher patient safety service quality. This study tends to explore the related issues of RFID-based information system managing both emergency room workflow and medical care service.

## 15.3 Emergency Room Problems: Case Description

Municipal Wan-Fang Hospital (WF hospital), which is a teaching medical center located in Taipei, Taiwan, and is affiliated with Taipei Medical University. This study portrays the emergency room problems as follows. When patients flood into the emergency room, waiting rooms for treatment, trauma, and critically ill patients need treatment in the shortest possible time. A large volume of unexpected patients arriving at the emergency room simultaneously would possibly cause medical staffs to commit negligence or malpractice by providing the wrong injection, medication, anesthesia, or even surgery. According to emergency room practices, this study proposed some deficiencies that need to be solved:

- Long waiting at emergency room causes patients delay in receiving prompt treatment and endangers their chances of recovery.
- Delay of medical staff in observation and treatment of critical triage level L1 or L2 patients due to the large volume of patients arriving at the emergency room for treatment.

- The fixed cost of the emergency room is too high to support its operation.
- The shortage of professional, well-trained staff and life-saving equipment.
- Providing wrong drugs, injections, volume of anesthetic, surgery, and so forth.

In accordance with the above deficiencies, we highlight two issues of emergency room as following: (1) monitor triage patient's waiting time. The emergency room has to monitor triage patients waiting time in order to prevent negligence of emergency room treatment follow up. (2) Monitor the treatment process. The emergency room has to monitor the treatment process in order to prevent human errors.

The RFID plays an important role to ensure the emergency room medical treatment process. With RFID functions, predefined waiting time can prevent patients from waiting too long to receive treatment, and it can insure the right person gets the correct treatment at the correct time to prevent human errors. The WF hospital implemented an emergency room RFID information system to monitor both patient's waiting time and treatment process. This study discusses the emergency room RFID information system as follows

- ER-RFID information system workflow
- Monitor triage patient's waiting time in emergency room
- Monitor the treatment process of emergency room

## 15.4 ER-RFID Information System

In order to monitor both triage patient's waiting time and medical treatment process for emergency room management, the Patient Safety Institute of the WF hospital implemented an emergency room RFID (ER-RFID) information system.

### 15.4.1 ER-RFID Information System Workflow

When the patient arrives at the emergency room of the WF hospital, the patient is triaged by identifying the injury level from level 1 to level 4, which are shown in Table 15.3.

Figure 15.1 shows four stages of the ER-RFID information system workflow as follows

**TABLE 15.3**

Triage Level of Patient's Injury

| Triage Level | Patient's Injury Identification |
| --- | --- |
| L1 | The injury threatening the patient's life that requires treatment immediately |
| L2 | The patient is in danger, but the injury will not immediately threaten the patient's life. |
| L3 | The injury qualifies for emergency room criteria, but is not life threatening. |
| L4 | The patient's injury is minor, which only needs outpatient or emergency room treatment. |

1. Triage: The patient has to complete his triage level identification at this stage.

2. Registration: When the patient completes the triage level identification, the nursing station inserts patient injury information and triage level into ER-RFID information system.

3. Assign: Appropriate specialization of physician to provide treatment, the patient receives treatment measure waiting, medical staff applies PDA to download patient treatment plan.

4. Monitor: Treatment mechanism or leave, RFID monitor patient's treatment process and waiting time before receiving treatment. Patient recovered, leaving emergence room, and returned RFID tag.

As Figure 15.2 shows, various hospital database applications work with the ER-RFID information system. Patient's treatment information, injured status, and test results are written into the database by reader. The patient's safety guidelines are embedded into the database to check treatment process following ER safety procedures. In accordance with the advantages of RFID, the medical care information systems work together to produce synergies.

### 15.4.2 Monitor Triage Patient's Waiting Time in Emergency Room

When the patient completes the triage level classification, the nursing station inputs patient injury information and triage level into the ER-RFID information system. Each triage level has its required safety time frame to monitor patient waiting time for receiving treatment. This study introduces the categories of the RFID wristband, and then discusses the alerting mechanism of monitoring wait time.

#### 15.4.2.1 The Categories of the RFID Wristband

After patient has been triaged, the patient is classified into an injury level. In Figure 15.3, different RFID wristband color indicator each patient's triage level and each triage level has a predefined waiting time schedule. In other

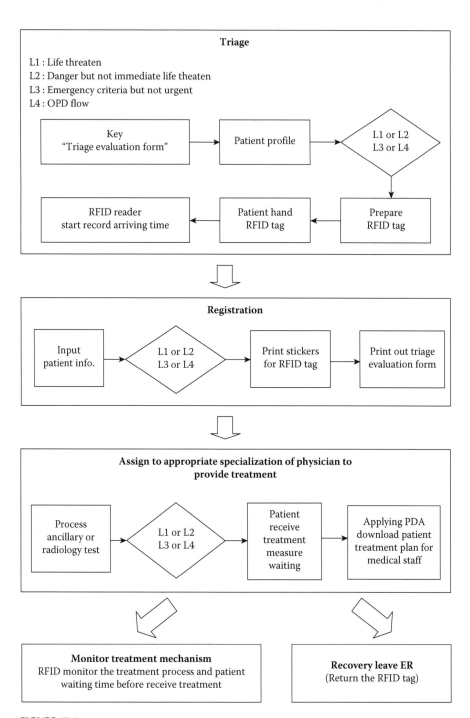

**FIGURE 15.1**
ER-RFID information system workflow chart.

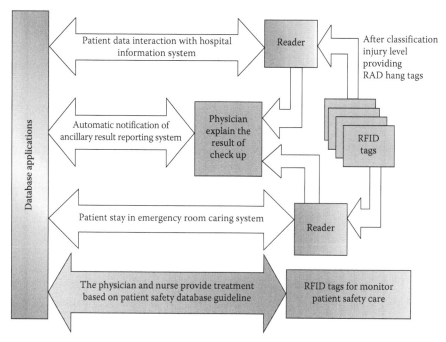

**FIGURE 15.2**
The infrastructure of ER-RFID information system.

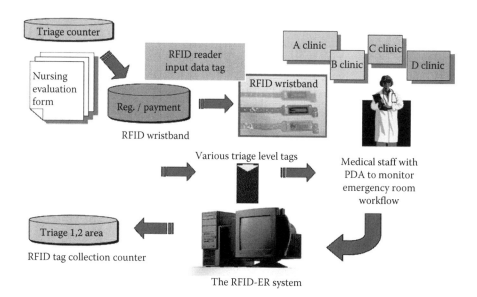

**FIGURE 15.3**
RFID-ER information system workflow.

words, a different RFID wristband means each patient's progression situation and treatment is different. The ER-RFID information system monitors patient treatment schedule and displays the patient's message on the screen based on the different RFID wristband colors. The objective of each RFID wristband color is defined as follows

- Red color: The red color of patient ID warns medical staffs that the patient should receive immediate treatment, test result reports, and patient progression evaluation.
- Yellow color: The yellow color is a warning message that the treatment should start, the report should be received, and the patient should be transferred to an inpatient bed for continuous treatment.
- Green color: The green color means medical staffs should observe the patient's progression and deliver treatment on the normal process track.

Figure 15.4 shows that medical staff need to take the PDA to scan the patient and the medical tray or ward RFID tag in order to record medical staff visitation time information into the PDA. Physicians can see the treatment plan and test results on the PDA. The information will transmit through the blue tooth into the medical information system database.

### 15.4.2.2 The Alerting Mechanism of Monitored Waiting Time

When a patient completes his or her triage level classification, the RFID first records the patient's waiting time. When a patient receives treatment, the RFID then records the start of treatment in the database. The ER-RFID information system analyzes the starting time minus the patient's completed triage level time to calculate the patient's waiting time in each triage level.

To address the long waiting time before receiving treatment, the RFID records the waiting time by tag after the patient's level is triaged. Later, the RFID compares the patient's waiting time before receiving treatment with each triage level standard safe waiting time. If the waiting time exceeded the standard waiting time, the abnormal record using a red color would be displayed in the alert fields to give notification and display on the ER-RFID information system to alert medical staff to provide treatment immediately. The three categories in the mechanism of the monitored waiting time are depicted as follows

- **Radiology and ancillary waiting time**
  When patients wait too long for radiology or ancillary test result reports, it may delay patient treatments and further endanger patients' lives. Therefore, applying RFID to monitor radiology and ancillary report waiting time helps patient safety.

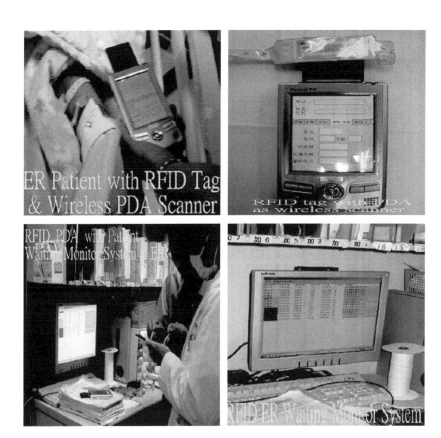

**FIGURE 15.4**
ER-RFID information system workflow pictures.

The method to measure waiting time for radiology and ancillary reports are to record the starting examination time and compare it with the report arrival time for the safety waiting time frame. If the waiting time was beyond the standard waiting time, ER-RFID information system displays an alert message on the computer screen, and its goal is to alert medical staff to get the test result reports.

- **Inpatients ward waiting time**
  In an emergency room, some severely ill patients are required to be transferred to inpatient status for continuous treatment. Waiting a long time for an inpatient ward at an emergency room threatens continuous treatment and patient recovery. When the physician issues a transfer record to transfer a patient from the emergency room to inpatient, the measurements of the patient waiting time for inpatient wards is monitored by the ER-RFID information system. The beginning of the transfer inpatient time is recorded into the database.

If the waiting time exceeds the standard predefined time schedule, the medical staffs solves the case after they get the alert message.

- **Triage level L1 and L2 waiting time**
  Severely ill patients of triage L1 or L2 require medical staff to check their medical progress within the required time frame. To prevent the triage level L1 or L2 patients from being neglected, the medical staff need to use a PDA to scan the ward tag, patient wristband tag and medical staff personal tag. Each visit time is recorded into the patient safety database. Patients at triage level L1 and L2 are monitored by the ER-RFID information system. If the medical staff did not observe patient progress or provide treatment for triage L1 or L2 patients, the ER-RFID information system displays the red abnormal patient's record warning message to alert the medical staff to observe the critically ill patient's progress and provide treatment.

### 15.4.3 Monitor the Treatment Process of the Emergency Room

In addition to monitoring a triage patient's waiting time, the other purpose of ER-RFID is to monitor the treatment process for emergency room management. The main solutions are designed to prevent human errors.

#### *15.4.3.1 Monitor Treatment Process*

The key to eliminate human errors in an emergency room is when there are an excessive amount of patients waiting for treatment. The ER-RFID plays an important role in monitoring the treatment processes based on the patient safety guideline. Hence, the ER-RFID functions are able to monitor the processes of drug, injection, anesthesia, and surgery for the right person, time, and place. The detail functions are listed below.

- Right drug: ER-RFID monitors the accuracy of matching the right drugs by comparing the patient's tag, medication tag, and treatment tray tag to assure distribution of the correct drug, dose, and time for each patient.
- Accurate injection and matching dose: ER-RFID monitors the accurate injection and matching dose for the right patient by the patient's tag, needle tag, and treatment tray tag to verify the correct treatment process for accurate treatment.
- Right volume of anesthetic: ER-RFID monitors the right volume of anesthetic by matching the right patient by the patient's tag, anesthetic needle tag, and treatment tray tag to assure the correct process and distribution.

- Right surgeon: ER-RFID monitors the right surgeon for the right patient by the patient's tag information verification.

The drug, injection, volume of anesthesia, and surgeon checking are monitored by the ER-RFID. The purpose of the ER-RFID assures that the right person is matched with the correct medical treatment distribution processes.

### 15.4.3.2 The Alerting Mechanism of Monitored Treatment Process

When a patient completes the triage classification, the RFID tag records the patient information in the patient's wristband tag. Three tags will be processed as follows

1. Tag for matching drug, injection, and anesthesia

The tag is attached to the patient medical tray to identify the tray that contains the injection, drug, dose of anesthetic, and so on, for the right patient. When the tray is delivered to the patient for treatment, the tag of the tray will be compared with the patient's wristband tag. This is a comparison and matching process. The purpose is to assure that drugs, injections, and anesthesia are chosen for the right person.

2. Tags for RFID reader and patient's identification

The other two tags information will go through the RFID reader to be input into the hospital patient safety information system to compare the patient identification number. As Table 15.4 shows, the ER-RFID information

**TABLE 15.4**

Items of ER-RFID Check List

| Patient Information | Matching Comparison | Normal/ Abnormal | Alert Notification |
|---|---|---|---|
| Patient tag no. | | | |
| Bed tag no. | | | |
| Tray/box tag no. | | | |
| Patient treatment plan | | | |
| Patient safety information system | | | |
| Prescription drug tag no. | | | |
| Injection tag no. | | | |
| Anesthesia injection tag no. | | | |
| Surgeon tag | | | |
| Nursing visit schedule (every 30 minutes) | | | |
| Physician visit schedule (every one hour) | | | |

system reads the tag to compare whether the treatment plan, such as drug, injection type, and dose of anesthesia information match the correct treatment plan. If it is mismatched, the system would enter an abnormal record into the database and warn medical staff by sound immediately. The alert warning system will provide an alarm signal and blink a warning light of a different color. The red color warning message is to alert medical staff to observe the condition of the critical patient and provide treatment instantly. The yellow color is to warn medical staff to visit patients based on scheduled time. The alerting mechanism controls medical staff following treatment safety procedures.

## 15.5 Conclusion

A successful information system should offer an efficient user interface to clinicians in order to get the most proper consultation results [28]. The RFID is applied for emergency room workflow, which monitors the patient's waiting time and reduces human errors. Before adopting ER-RFID, the medical staff identify patient's profile manually; in contrast, after adopting ER-RFID system, it automatically monitors patient's waiting time and medical treatment process. The contributions of ER-RFID information system is listed in Table 15.5.

### 15.5.1 Managerial Implications

The RFID system is one of the most important technologies of the twenty-first century. Enterprises apply RFID to improve its operation efficiency and

**TABLE 15.5**

Contributions of ER-RFID Information System

| Goal | Content | Contribution |
|------|---------|--------------|
| Monitor waiting time | Observe triage patient's waiting time | ER-RFID monitors each triage patient's waiting time and the alerting mechanism monitors the treatment process. |
| Prevent human errors | Prescription error | ER-RFID makes sure the prescription is correct and the patient is okay. |
| | Injection error | ER-RFID makes sure the injection is correct and the patient is okay. |
| | Anesthesia error | ER-RFID makes sure the dose is correct and the patient is okay. |
| | Surgical error | ER-RFID makes sure the surgeon is correct and the patient is okay. |

effectiveness [30–32]. Utilization of RFID systems can be for supply chain management, retailers, cell phone order payments, ticketing, airport luggage transportation, and the pharmaceutical fight against counterfeit drugs. For health care providers, RFID provides solutions. For instance, when medical staff look for medical devices, RFID directs where the device location is immediately [29]; for adopting RFID tracking injection IV bags, blood bags, cancer medicine, and tracking wounded soldiers and treatments and so on. From the businesses and hospitals utilizing RFID area analysis, the emergency room is one that needs the RFID system to improve its operational weaknesses. This study considers that the functions of monitoring patient's waiting time and preventing human errors could be a reference for health care providers [33].

This study suggests that medical staffs need a friendly and efficient user interface. The complicated user interface causes medical staff burdens and even disturbs their treatments. In addition, although there are many advantages to applying RFID, how to protect the patient's privacy and prevent anti-RFID devices will be the next issue to confront applying RFID in the emergency room.

# References

1. Naveh, E., Katz-Navon, T., and Stern, Z. 2005. Treatment errors in healthcare: A safety climate approach. *Management Science* 51 (6): 948–60.
2. Committee on Quality of Healthcare in America, Institute of Medicine. 2001. *Crossing the quality chasm: A new health system for the 21st century.* Washington, DC: Institute of Medicine National Academy Press.
3. Marin, H. F. 2004. Improving patient safety with technology. *International Journal Medical Informatics* 73 (7–8): 543–46.
4. Ferguson, T., Lin, B., and Chen, C. H. 2004. Leveraging the work force using information technology: A financial service case study. *International Enterprise Development* 1 (4): 316–31.
5. Yoon, S. C. 2005. Technological innovation as responding to business challenges: Case study and theorization. *International Journal of Technology Management* 29 (3/4): 295–307.
6. Ferris, M. T. 2005. Technology at work in home care marketing strategies. *Caring* 24 (1): 48–49.
7. Michalowskia, W., Rubinb, S., Slowinskic, R., and Wilkc, S. 2003. 'Mobile clinical support system for pediatric emergencies. *Decision Support Systems* 36:161–76.
8. Uslay, C., Malhotra, M. K., and Citrin, A. V. 2004. Unique marketing challenges at the frontiers of technology: An integrated perspective. *International Journal of Technology Management* 28 (1): 8–30.
9. Koleszar, A. J. 2004. On drugs and distribution. *Material Handling Management* 59 (1): 47–49.

10. Cooke, P. 2004. The role of research in regional innovation systems: New models meeting knowledge economy demands. *International Journal of Technology Management* 28 (3/4/5/6): 507–33.

11. Griffin, M., Stein, C., and Ray, W. 2004. Postmarketing surveillance for drug safety: Surely we can do better. *Clinical Pharmacology & Therapeutics* 75 (6): 491–94.

12. Anantharaman, V. and Lim, S. H. 2001. Hospital and emergency ambulance link: Using IT to enhance emergency pre-hospital care. *International Journal of Medical Informatics* 61 (2–3): 147–61.

13. Riediger, G. and Sperber, T. F. 1990. Efficiency and cost-effectiveness of advanced EMS in West Germany. *American Journal Emergency Medicine* 8:76–80.

14. Moore, L. 1999. Measuring quality and effectiveness of prehospital EMS. *Prehospital Emergency Care* 3 (4): 325–31.

15. Lin, B. and Vassar, J. A. 2004. Mobile healthcare computing devices for enterprise-wide patient data delivery. *International Journal of Mobile Communications* 2 (4): 343–53.

16. Gainer, A., Pancheri, K., and Zhang, J. 2003. Improving the human computer interface design for a physician order entry system. America Medical Informatics Association, Annual Symposium Proceeding, 847.

17. Bilgen, B. and Ozkarahan, I. 2004. Strategic tactical and operational production-distribution models: A review. *International Journal of Technology Management* 28 (2): 151–71.

18. Prater, E. and Frazier, G. V. 2005. Future impacts of RFID on e-supply chains in grocery retailing. *Supply Chain Management-An International Journal* 10 (2): 134–42.

19. Satoh, I. 2003. Spatial agents: Integrating user mobility and program mobility in ubiquitous computing environments. *Wireless Communications & Mobile Computing* 3 (4): 411–23.

20. Anon. 2005. ABI research sees RFID helping airlines track assets. *Microwave Journal* 48 (4): 47.

21. Yoshida, J. 2003, May 26. U.S. Navy uses RFID technology to track wounded in Iraq. *Electronic Engineering Times,* http://www.rfid-world.com/features/transportation/201805057.

22. McDonald, R. E. and Srinivasan, N. 2004. Technological innovations in hospitals: What kind of competitive advantage does adoption lead to? *International Journal of Technology Management* 28 (1): 103–17.

23. Miller, M. 1999. Tuning to future healthcare use of RFID. *Automatic I.D. News* 15 (2): 58–59.

24. Gell, G. 2002. Safe controllable technology? *International Journal of Medical Informatics* 66 (1–3): 69–73.

25. Gebhart. F. 2004. Hospitals start pilot testing RFID to curb Drug diversion: Drug topics. *Academic Emergency Medicine* 11:1155–61.

26. Kohn, C., and Henderson, C. W. 2004, May 10. RFID-enabled medical equipment management programs to reduce costs. *Managed Care Weekly Digest* 94–95.

27. Costlow, T. 2004. RFID extends to medical applications. *Design News* 59 (9): 43–44.

28. Chang, Y. J., Tsai, C. Y., Yeh, M. L., and Li, Y. C. 2004. America Medical Informatics Association Annual Symposium Proceeding, 808.

29. Roberts, S. 2003. Tracking medical devices via RFID. *Frontline Solutions* 12 (2): 54.
30. Lyon, D., Caplan, J., and Torpey, J. 2004. Review essay: Surveying surveillance Studies. *Canadian Journal of Communication* 29 (2): 1–6.
31. Henderson, J., McAdam, R., and Parkinson, S. 2005. An innovative approach to evaluating organizational change. *International Journal of Technology Management* 30 (1/2): 11–31.
32. Bennett, D. and Vaidya, K. 2005. Meeting technology needs of enterprises for national competitiveness. *International Journal of Technology Management* 32 (1/2): 112–53.
33. Shih, L. H., Huarng, F., and Lin, B. 1996. ISO in Taiwan: A survey. *Total Quality Management* 7 (6): 681–90.

# 16

## Multi-Agent Architecture and Location-Based Ubiquitous Learning Framework

Maiga Chang, Qing Tan, Fuhua Oscar Lin, and Tzu-Chien Liu

**CONTENTS**

This paper describes the MAA (Multi-Agent Architecture) and location-based learning framework for the support of learning in the ubiquitous-pervasive environment. A project plans to build a MAA-based learning environment with various agents. The agents have different abilities, for example, the agent can sense where the learner is located, the agent can plan the route for the learner, the agent can guide the learner traveling in the real world, and the agent can search and access the learning map for the learner. The agents communicate and cooperate with each other within the MAA framework.

## 16.1 Introduction

Many of the computer-based applications in e-learning focus on providing tools for teachers to either design courses (e.g., Macromedia Coursebuilder for Dreamweaver, Integrator2, and Toolbook 9) or do learning management (e.g., WebCT, Sakai, and Moodle), including some existing learner-centered educational applications, they usually have limitations in promoting adaptivity and context awareness and providing real-time support to learners from an educational point of view [17].

In the mobile learning environment, the learners could receive the learning materials provided by a system according to where they are when learning [4,5]. Mobile learning strategy can actually achieve the goal of learning at anytime and anywhere. Recently, ubiquitous learning has become an interesting and important issue in informal learning [19], focusing on learners being able to learn with their various devices whenever and wherever they want.

This paper reveals the project that plans to build a MAA-based ubiquitous learning environment and apply the positioning technologies to design location-aware learning services. For example, the dynamic grouping system will help learners find a collaborative group when they are doing learning quests, the learning path planner will arrange the route for learners when they arrive at a learning spot, and the location-based guidance system will navigate learners traveling from one learning spot to another.

Section 16.2 introduces the related research components, for example, situated learning theory, ubiquitous learning researches, MAA, positioning technologies, grouping technologies, spatial relation structures, and the landmark researches. Section 16.3 describes the main idea and design plan of the project's MAA-based platform. In Section 16.4, we are talking about the location-awareness application, the grouping system. Section 16.5 focuses on the learning path planner and the guidance message generator. Finally, Section 16.6 makes a conclusion and discusses the possible future.

## 16.2 Research Background

In a traditional learning environment, the education is held in the classroom, which find the teachers and learners face to face. The learners could only get the learning materials prepared in advance from the teacher. As a result, the learning activities are limited in what the teacher has arranged and then materials and courses are difficult to adjust immediately according to students' learning status and demands.

E-learning applies computer and the Internet technologies to assist teachers' teaching and learners' learning [2]. E-learning provides a new learning mechanism to run the education process with e-mail, web-camera, and web-based testing, the teacher and learners are not together [14]. However, some courses need students learning through observation and are not conducive to a web-based learning environment, for example, the butterfly watching or plant-learning experiences [4,5,20].

Mobile learning extends the learning from indoor to outdoor and gives learners opportunities to understand the materials via touching, watching, and feeling the objects in the real world [13,24]. Furthermore, mobile learning can stimulate and enhance learners to apply their knowledge to the real world [7]. However, there are still some unsolved issues, such as a flexible learning issue (i.e., the learners' learning activities will be limited in a specific and predefined learning environment). The learners are not just passive when receiving the learning materials from teachers but learn the concepts, knowledge, and skills via interacting with the real world [23]. Brown believes concepts and knowledge are situation-based [3], and learning is influenced by teaching activity, situations, and inter-actions called situated learning, which leads the ubiquitous-pervasive learning research [21].

There are four characteristics in a ubiquitous learning environment [11]: (1) context-aware, the learners' profiles and the learning environment are known; (2) personalization, the learning resources are provided according to a learners' profiles and learning status; (3) seamless, learning activities are not interrupted by place changes; and (4) calm, the learning materials are delivered to learners without interruption.

In the framework, with positioning technologies, online learners' positions are identified, and the grouping agent will provide the location based on grouping suggestion; the learning path planner will provide the learners personalized learning routes; the location-based guide agent will lead the learners from one learning spot to another; and the multi-agent architecture offers the learners a calm environment. The teaching material delivery is not included in the framework for now.

Java Agent Development Environment–Lightweight Extensible Agent Platform (JADE-LEAP), was chosen to develop FIPA-compliant MAS mobile devices because JADE has a run-time environment to support Connected Limited Device Configuration (CLDC) and Connected Device Configuration (CDC) specifications [12].

The Multi-Agent Architecture (MAA) allows great flexibility and scalability in the integration of components. It provides a simple yet extensible and powerful software layer to develop further pervasive learning environments, while simultaneously running multiple stationary and mobile agents on a Java enabled mobile device [1]. JADE-LEAP serves as the agent platform.

Ubiquitous learning can use a mobile device as the terminal to access digital contents via the cellular networks or other wireless communication. It offers a dynamic, anywhere and anytime accessible learning environment. In the cellular communication network, Location-Based Service (LBS) is one of its essential components. The location of each mobile device relative to the network is maintained by the LBS [15]. Mobile learning was born with an exciting characteristic, location awareness while inheriting most of the e-learning features. Integrating mobile device's location awareness provides unique and powerful potential in the online learning environment.

In this framework, the grouping agent gathers individual mobile learners based on their geographic locations and their learning behaviors. Therefore, mobile learners can take advantage of collaboration learning in a conventional education environment and enhance their learning in the ubiquitous learning environment.

In general, a LBS can be defined as triggered and user requested. Triggered LBS relies on a condition set up in advance that retrieves the position of a given device (e.g., passes across the boundaries, emergency services). With user-requested LBS, the user is retrieving the position once and uses it on subsequent requests for location-dependent information. This LBS type involves personal location (i.e., where you are) or services location (i.e., where the nearest is) [6]. The location-based optimal grouping service falls into the second category as the grouping service is offered on the basis of request.

The client agent has to be able to identify what location information is available, to acquire the location information, and to send the location information to a central server agent. It also has to collaborate with other agents in the multi-agent environment.

Regarding the guidance message generator in this framework, the spatial knowledge is an important research issue. Mohan and Kashyap have proposed an object-oriented model for representing the spatial knowledge in tree form [16]. All the objects in the real world can be represented in hierarchy form as Figure 16.1 shows. There are two benefits of using object-oriented representation to store the spatial knowledge: (1) it is easy to describe the relations among objects and (2) the structure can be transformed to various forms simply.

Del has discussed the spatial relationships in geometry and he has classified the relationships into directional relationships and topological relationships [8]. Directional relations include front, back, right, left, east, west, south, and north. Egenhofer and Fransoza have proposed a nine-intersection spatial model [9]. The nine-intersection model involves 29 topological relationships. The 29 topological relationships can be categorized into eight categories, including DISJOINT, MEET, INSIDE, EQUAL, CONTAINS, COVERS, COVERED BY, and OVERLAP.

Beside the spatial knowledge structure and relations, the landmark research is also necessary for developing the guidance message generator. There are three basic elements to guide a person moving in the real world: orientation,

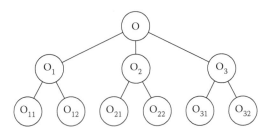

**FIGURE 16.1**
Spatial structure hierarchy.

actions, and landmarks [22]. Landmark is essential and it is more important than orientation and actions in guidance [10]. Sorrows and Hirtle [18] have proposed three landmark categories: visual, cognitive, and structure [18]. Each landmark category has to use a different navigation method.

## 16.3 Multi-Agent Architecture

Each learner is running the front-end of a learner agent on his or her own mobile phone or PDA supporting J2ME with the Mobile Information Device Profile (MIDP) 1.0 or greater APIs and Bluetooth JSR-82 or Wi-Fi APIs. A back-end mediator in the JADE-LEAP architecture, running on a host with a known, fixed, and visible address can automatically manage the back-ends of all front-ends active on the learner's device. Some types of agents located in a JADE agent platform are as follows:

- A main container, which stores default JADE agents.
- One or more containers, which execute system agents.
- Location-aware learner agents, one for each learner, which present each learner logged in within the area. The learner agent has capabilities, using the expressed preferences of learners or their previous behavior when navigating the system to provide information for the student model.
- Connection agents manage the connection between the mobile devices and agent platform.
- Service agents are available for each service such as location-aware service.
- Resource agents can manage resources, such as problem-based learning objects.

Figure 16.2 shows the Multi-Agent Architecture for ubiquitous learning.

## 16.4 Location-Based Optimal Grouping

The location-based optimal grouping service is to group, geographically, nearby students together to create ad hoc online learning groups. In this service, the positioning accuracy is not critical. Proposed conceptual system architecture for the grouping service is shown in Figure 16.3.

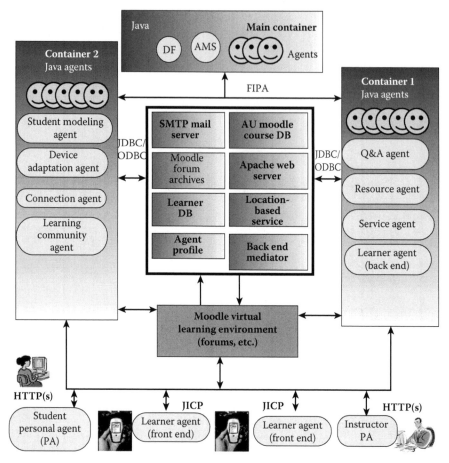

**FIGURE 16.2**
Multi-agent architecture for ubiquitous learning.

For the project, a client agent will be developed as a J2ME application to run on a Java enabled mobile device. The client agent needs to

- Acquire the location information on the mobile device from the cellular network.
- Identify if there is a GPS coordinate available on the mobile device.
- If the GPS coordinate is available, then obtain the GPS coordinates.
- Send the GPS coordinate with the location information from the cellular network to the location server agent.

The client agent has to be able to obtain the data from the mobile device as the initial location information for the grouping service. With all the location-based data, a simple matching method could be used to group the

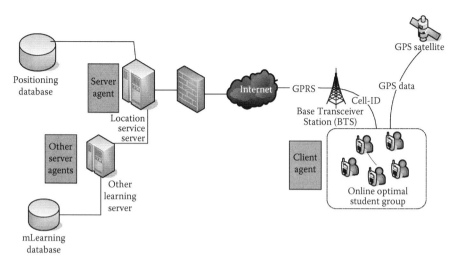

**FIGURE 16.3**
The conceptual system architecture.

mobile devices by determining the amount of shared similar location information. This approach could group online mobile learners within the radius of 500–1500 meters (almost one mile) in an urban area and 15 km (about 9 miles) in rural area.

Since GPS embedded mobile devices have become more and more popular, the client agent has to be able to work with this type of mobile devices. When the client agent is running on a mobile device, it will identify if the GPS coordinates are available (internally or externally). If yes, the client agent will periodically retrieve the GPS coordinates and send the position coordinates of the mobile device with its cellular network location data to a central location server agent to achieve high accuracy (within 3 meters or about 10 feet) location-based grouping.

However, the GPS technology has its own drawbacks. The GPS positioning is not available inside buildings or even between tall buildings, which restrict the view of the sky, and it also increases battery consumption and data-acquisition time. Therefore, it is very important to sufficiently use the GPS location data to update the cell geographic location information database and to enrich and correct the existing location database, eventually to enhance the over all location accuracy and the grouping performance.

The fundamental step of the client agent is to obtain the GPS coordinates from a mobile device. There are various software programs doing this job. The cellular phone manufacture also provides the data communication solutions, such as Nokia's PC Connectivity serial programs as high-level data communication development kits. Understanding the data communication protocol of a mobile device will be highly efficient for the client agent development

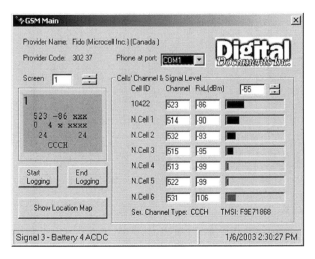

**FIGURE 16.4**
GSM cellular network data.

in respect to cellular location application system. Figure 16.4 shows cellular network data communication software, which works on Nokia GSM phones who is compatible with Nokia FBUS protocol, DAU-9P data cable. This application software was developed by one of authors earlier.

On the server side, an optimal grouping, algorithm-based location server agent will process the location data coming from a client agent with support of the existing location information database. The server agent will eventually group the nearby mobile learners with other grouping criteria. Finally, the grouping suggestion, inquiry, or automatic trigger will be sent to the group's client agents to deliver and implement the online optimal grouping service in the ubiquitous learning environment.

## 16.5 Learning Planner and Guidance System

This section focuses on the design of the learning path planner and the guidance message generator in the framework. First of all, the operation flow of the planner is described and is shown in Figure 16.5. Second, we design a two-phase way to apply the planner into the informal ubiquitous learning scenario, as Figure 16.6 shows. Figure 16.7 shows the flow of generating the guidance message in this framework.

Before the planner is starting to plan routes for learners, domain experts have to construct their own concept maps, which are so-called scaffolding concept maps (the step 1 in Figure 16.5). Moreover, experts have to produce

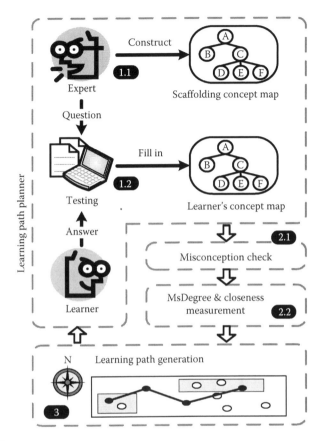

**FIGURE 16.5**
Flow of the learning path planner.

the related questions for each learning object for use in the paper-based or online testing.

As step 1.2 in Figure 16.5 shows, the planner can construct a learner's concept map according to his or her answers of the web testing. What the differences are between the expert and learner's concept map are easy to check by comparing these two concept maps and find out any misconceptions the learner has (the step 2.1 in Figure 16.5).

After the misconceptions are found, the planner uses a misconception degree and closeness as criteria to decide the feedback priority for the individual learner (step 2.2 in Figure 16.5). With the feedback priority, the planner is then able to generate the learning routes when considering both of the misconceptions and the learning objects in the real world (step 3 in Figure 16.5).

How to apply the learning path planner to the informal ubiquitous learning scenario? According to the operation flow of the planner, there are two main phases and five steps to learners as Figure 16.6 shows.

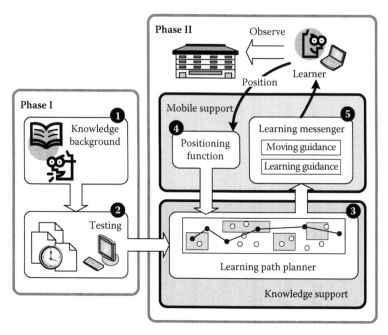

**FIGURE 16.6**
How to apply the learning path planner to the informal ubiquitous learning.

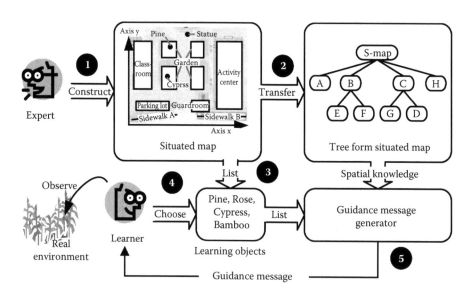

**FIGURE 16.7**
Generation and flow of guidance messages.

The two phases are the knowledge background and the testing phase (Phase I) and the knowledge and the mobile support phase (Phase II). Within the two phases, at first, learners have attained background knowledge as step 1 shows. They might have the knowledge via three ways: traditional classroom learning in the past; experiential learning in their daily life; and informal learning via Internet, newspapers, television, and books. Secondly, learners participate the testing as step 2 shows. The testing could be paper-based, web-based, or even mobile-based.

Thirdly, the learning path planner compares the concept maps and constructs the learning path for each individual as step 3 shows. Finally, we use positioning technologies and location-based services to get the learner's position as step 4 shows, and use the guidance message generator to give the learner either a moving guidance message such as "Please go to Area D in room 203, and find out the artifact No. 23!" or a learning guidance message such as "Please observe what special mark the vase has."

In general, the learning flow in the real world is: learners first have knowledge backgrounds and look at something they are interested in; then they will move to the next learning spot where they complete the observations; and repeat above steps as needed. Figure 16.7 shows how the location-based guidance message generator joins the ubiquitous learning framework.

First, domain experts construct the situated map according to the learning environment (e.g., a zoo; the step 1 in Figure 16.7). Step 2, the generator transforms the situated map into a tree-form situated map for generating guidance messages in the future. Step 3, the generator picks those learning objects from the situated map. Step 4, learners choose the objects they want to learn or they need to learn. Finally, the generator constructs the guidance messages to keep learners moving and studying in the real world (step 5 in Figure 16.7).

## 16.6 Conclusions

This paper reveals a MAA and location-based ubiquitous learning framework. The framework uses multi-agent architecture as the platform and designs different agents with different abilities. The agents can communicate and collaborate with each other via agent communication language.

The framework applies the positioning technologies to create several agents for ubiquitous learning, including dynamic grouping, learning path planner, and guidance message generator. The framework is still going on, which means there are more agents that might come out in the future and the current designs of the agents might also be improved in the future.

# References

1. Bellifemine, L., G. Caire, and D. Greenwood. 2007. *Developing multi-agent systems with JADE*. John Wiley & Sons, Ltd: Indianapolis, IN.
2. Brodersen, C., B. G. Christensen, Grønbæk, K., and C. Dindler. 2005, May 10–14. eBag: A ubiquitous web infrastructure for nomadic learning. In the Proceedings of the 14th International Conference on World Wide Web Conference, WWW2005, 298–306, Chiba, Japan.
3. Brown, J. S., A. Colins, and P. DuGuid. 1989. Situated cognition and the culture of learning. *Educational Research* 18 (1): 32–42.
4. Chang, A. and M. Chang. 2006, July 5–7. A treasure hunting learning model for students studying history and culture in the field with cellphone. In the Proceedings of the 6th IEEE International Conference on Advanced Learning Technologies, ICALT2006, 106–8, Kerkrade, The Netherlands.
5. Chen, Y.-S., T.-C., Kao, G.-J. Yu, and J. P. Sheu. 2004, March 23–25. A mobile butterfly-watching learning system for supporting independent learning. In the Proceedings of the IEEE International Workshop on Wireless and Mobile Technologies in Education, WMTE2004, 11–18, Chung-Li, Taiwan.
6. D'Roza, T. and G. Bilchev. 2003. An overview of location-based servers. *BT Technology Journal* 21 (1): 20–27.
7. Darmarin, S. K. 1993. School and situated knowledge: Travel or tourism? *Educational Technology* 33 (3): 27–32.
8. Del Bimbo, A. 1999. *Visual information retrieval*. San Francisco, CA: Morgan Kaufmann.
9. Egenhofer, M. J. and R. D. Franzosa. 1995. On the equivalence of topological relations. *International Journal of Geographical Information Systems* 9 (2): 133–52.
10. Golledge, R. G. 1999. Chapter 1: Human wayfinding and cognitive map. In *Wayfinding behavior: Cognitive Mapping and Other Spatial Processes*, 5–45. Baltimore, MA: The Johns Hopkins University Press.
11. Hwang, G.-J. 2006, June 5–7. Criteria and strategies of ubiquitous learning. In the Proceedings of the IEEE International Conference on Sensor Networks, Ubiquitous, and Trustworthy Computing, SUTC2006, vol. 2, 72–77, Taichung, Taiwan.
12. JADE: Java agent development framework. 2008. http://jade.tilab.com/
13. Kuo, R., M.-C. Wu, A. Chang, M. Chang, and J.-S. Heh. 2007, July 18–20. Delivering context-aware learning guidance in the mobile learning environment based on information theory. In the Proceedings of the 7th IEEE International Conference on Advanced Learning Technologies, ICALT2007, 362–66, Niigata, Japan.
14. Martin, B. L. 1994. Using distance education to teach instructional design to preservice teachers. *Educational Technology* 34 (3): 49–55.
15. Millar, W. 2003. Location information from the cellular network: An overview. *BT Technology Journal* 21 (1): 98–104.
16. Mohan, L. and R. L. Kashyap. 1988. An object-oriented knowledge representation for spatial information. *IEEE Transactions on Software Engineering* 14 (5): 675–81.
17. Rosatelli, M. C. and J. A. Self. 2004. A collaborative case study system for distance learning. *International Journal of Artificial Intelligence in Education* 14:1–29.

18. Sorrows, M. E. and S. C. Hirtle. 1999, August 37–50. The nature of landmarks for real and electronic spaces. In the International Conference on Spatial Information Theory, 37–50, Stade, Germany. *Lecture notes in computer science*, 1661:37–50. Berlin: Springer Verlag.

19. Syvanen, A., R. Beale, M. Sharples, M. Ahonen, and P. Lonsdale. 2005, November 28–30. Supporting pervasive learning environments: Adaptability and context awareness in mobile learning. In the Proceedings of the International Workshop on Wireless and Mobile Technologies in Education, WMTE2005, 251–53, Japan.

20. Thornton, P. and C. Houser. 2004, March 23–25. Using mobile phones in education. In the Proceedings of the IEEE International Workshop on Wireless and Mobile Technologies in Education, WMTE2004, 3–10, Chung-Li, Taiwan.

21. Thomas, S. 2005, March 8–12. Pervasive, persuasive elearning: Modeling the pervasive learning space. In the Proceedings of the 3rd IEEE International Conference on Pervasive Computing and Communications Workshops, PerCom2005, 332–35, Hawaii.

22. Tversky, B. and P. Lee. 1999, August 25–29. Pictorial and verbal tools for conveying routes. In the International Conference on Spatial Information Theory, 51–64, Stade, Germany. *Lecture Notes in Computer Science* 1661:37–50. Berlin: Springer Verlag.

23. Vygotsky, L. S. 1978. *Mind and society: The development of higher mental processes*. Cambridge, MA: Harvard University Press.

24. Yatani, K., M. Sugimoto, and F. Kusunoki. 2004, March 23–25. Musex: A system for supporting children's collaborative learning in a museum with PDAs. In the Proceedings of the IEEE International Workshop on Wireless and Mobile Technologies in Education, WMTE2004, 109–13), Chung-Li, Taiwan.

# 17

## Toward a Personalized E-Learning Environment

Elvis Wai Chung Leung

### CONTENTS

Presently, students can easily access the online course materials anytime and anywhere. However, students may have different linguistic, cultural, and academic backgrounds. So, diversity of student characteristics causes a significant issue. To address this issue, a personalized learning environment has been promoted recently. By providing a personalized learning environment, the major learning technologies and learning theories will be discussed in this chapter. In particular, Intelligent Tutoring System (ITS) and Adaptive Hypermedia (AH) are examined first. A personalization approach, e-learning standards, as well as cognitive theory and instructional design strategies will be discussed also.

### 17.1 Introduction

To cope with the increasing trend of learning demands and limited resources, most universities are taking advantage of Web-based technology for their distance education or e-learning. One of the reasons is due to the significant price drop of personal computers in recent years; the Internet and multimedia has penetrated into most households. Moreover, most students prefer to learn from an interactive environment through a self-paced style. Under the Web-based learning model, students can learn at anytime and anywhere because they are not required to go to school on schedule. Meanwhile universities also enjoy the economic benefit due to the large student base that can share the development cost of course materials and other operational expenses. Gradually, more and more universities are following this similar way to providing online education.

Presently, students can easily access online course material at anytime or anywhere. However, the diversity of student characteristics causes a significant issue. The students may have different linguistic, cultural, and academic backgrounds. It is difficult for the conventional e-course to meet student expectations. To address this issue, a personalized learning environment has been promoted recently. In this research area, some major research works [20,28,47] attempted to develop a learning system, taking into account psychological factors. At the same time, other research works [5,6,41,48] that focused on providing personalized learning facilities have also been developed. The key to a successful personalized learning environment for all students is to place students' needs into consideration. A reported survey [7] of student expectations has shown that students expect:

- Staff and instructors to be available at flexible times
- Good response times for enquiries and marking
- Easily accessible help with technology
- Easily navigated Web sites
- Course content based on relevant real-world situations
- Facilitation of collaboration in tasks

Clearly, students prefer instant feedback and more guidance rather than being totally free of learning directions. Providing a course syllabus or study guide is an effective learning strategy for online students [36]. Cordova and Lepper [18] evaluated the impact on personalization with respect to stimulating intrinsic motivation and learning in a computer-based educational environment. The finding was that students studying with the personalized contexts have better motivation, a higher degree of involvement and learned more than other students with traditional learning materials.

## 17.2 E-Learning Modeling

In response to the different needs and characteristics of a growing number of e-learning users, personalized e-learning systems have been exploited.

One strategy toward personalized tutoring systems is AH, which adapts both hypertext and multimedia to meet the individual student's needs. The primary goal of this approach is to provide adaptive presentation and adaptive navigation support [9]. Adaptive presentation refers to changing the content of Web pages to cater to individual needs. Adaptive navigation support refers to adding, hiding, or ordering the links presented on a Web page to better suit the user's needs.

Another approach uses agent-based technology to provide Web-based personalization in e-learning systems. Currently, providing student-centered learning contents and personalized guidance are the main research topics in intelligent tutoring systems. Ozdemir and Ferda [42] use interface agents to guide students through the course materials on the Web. Ankush et al. [3] proposes a content-based retrieval system to provide educational videos based on the individual student's needs.

The development of ITS, based on artificial intelligence and cognitive science researches, has accumulated considerable results in the past 20 years. At present, researchers attempt to deploy ITS onto the Web in view of the rapid development of the Internet and the World Wide Web.

### 17.2.1 Intelligent Tutoring System (ITS)

Most generic ITS architectures suggest building a good student model that reflects the system's beliefs on the mastery level of learners in particular concepts. Moreover, it is the driving force that enables the system to perform individualized tutoring to students [31]. To further extend the generic architecture, various types of ITS have been developed based on different domains, pedagogical strategies, and other affecting factors [37,46,56]. The major researches and developments of ITS are summarized in the following paragraphs.

Most researchers believe that the first ITS, SCHOLAR, was designed in the early 1970s, to teach South American geography [17]. Early ITS systems emphasized duplicating the structure of student-human-tutor interactions. In such systems, success has been limited because of the technological challenges of making a computer system sufficiently intelligent to field various questions in a pedagogically useful manner. The advent of the Web changed the pedagogical approaches to educational software, instead, it has less similarity to human tutoring models. These approaches are more focused on simpler instructional approaches and are easily computerized. Web-administered, multiple-choice tutors have shown promise because of their ease of development [29]. Other ITS systems create complex instructional strategies that require extensive expert system representations customized to the knowledge domain; representative models include geometry [1] and computer programming [38].

Other researchers have also attempted to create domain-independent standards for representing courseware. Information about the learning content, known as metadata, must be standardized to facilitate interoperability with other users. Samples are Dublin Core [55], Instructional Management Systems (IMS) project (www.imsproject.org), and Shared Content Object Reference Model (SCORM), created by Advanced Distributed Learning (www.adlnet.gov). The IMS project builds upon the Dublin Core to create the IMS metadata standard, while SCORM provides a single standard for educational content developers. To take advantage of the standards, ITS can focus more on student modeling. Meanwhile, the learning materials sequencing, which has been provided by ITS traditionally, is delegated to the standards.

The ITS system is also used to assess students' performance. Recent debates about what kinds of assessment data tutoring systems should be emphasized. The emphasis of one approach over the other reflects the designers' cognitive and pedagogical assumptions. Summative assessment and formative assessment are two major kinds of assessment methods being generally used. Summative assessments are used to formally assess whether students have achieved the learning targets; where formative assessments are used to help improve students' achievement of learning targets. There is a debate in the computerization of formative assessment in intelligent tutoring systems. It is argued that feedback must be given at an appropriate point in the learning process. According to Buchanan [15], to be useful in providing immediate feedback, computerized formative assessment must be gathered in an efficient and quantifiable manner.

In recent years, there have been various kinds of cognitive and pedagogical approaches to ITS design. These approaches not only address the issue of students versus tutor control, but also state the application of cognitive theories of learning and the role of assessment in intelligent tutoring systems. The Conceptual Helper, as part of the Andes project at the University of Pittsburgh, was designed to improve the instruction of physics [26]. By following the model-tracing paradigm, the tutor tried to model the thinking process of the student in the Conceptual Helper. However, the Conceptual Helper was not well adapted to use for other subjects. In another research, Aleven and Koedinger [1] found that students often lack the meta-cognitive awareness to seek help when required and thus, they suggest forcing students to use help even when they do not solicit it. Other researchers emphasized the importance of student control in the use of ITS [53,54].

### 17.2.2 Adaptive Hypermedia (AH)

In view of the rapid development of the Internet and World Wide Web, researchers attempted to deploy ITS on the Web. Most generic ITS architecture features are still retained in these systems [37]. Among these approaches, AH is a comparatively new approach that accounts for the hypermedia and user modeling fields together [10,11]. AH has always been applied in education. A number of non-Web-based stand-alone adaptive educational hypermedia systems were built between 1990 and 1996; with the early Web-based Adaptive and Intelligent Educational Systems (AIES) using AH technologies [14,21]. Since then the Web has become the primary platform for developing educational AH systems.

AH systems apply different forms of user models to adapt the content as well as the links of the hypermedia pages to the user. Brusilovsky further mentioned that AH is a field positioned on the crossroads of hypermedia and user modeling. Thus, "what can be adapted in terms of the content or structure of hypermedia" can be another considerable task. Brusilovsky also summarized the taxonomy of available AH methods today. As partially shown in Figure 17.1, two major technologies are identified: adaptive presentation and adaptive navigation support. The main purpose of all of these adaptive methods is to support users' performance on a particular task like "how to learn a topic."

The primary aim of the adaptive presentation technology is to adapt hypermedia page content based on the student's goals and knowledge as well as other information stored in the user model. Examples are ELM-ART [14], AST [50], and InterBook [13]. Instead, adaptive navigation support technology, by changing the appearance of visible links, supports the student in hyperspace orientation and navigation. Two well-known examples are ISIS-Tutor [12] and Hypadapter [27]. Direct guidance, adaptive link annotation, and adaptive link hiding are the three ways that are most popular in Web-based AIES and some of these features are also quite similar to ITS

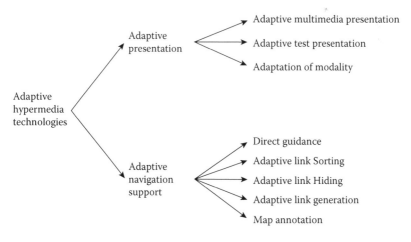

**FIGURE 17.1**
Brusilovsky's taxonomy of AH technologies.

adaptive sequencing. The difference between direct guidance and ITS adaptive sequencing disappears gradually in the Web context. As long as some types of educational material, such as presentations and questions, are represented as a set of nodes in hyperspace, the sequencing becomes indistinguishable from direct guidance. Popular examples are ELM-ART [14] and InterBook [13].

## 17.3 Personalization Approach

The goal of personalization is to provide users with what they want or need without asking explicitly [39]. In this section, different approaches are reviewed on how the personalization can be delivered to individual users.

### 17.3.1 Individual versus Collaborative

A personalization system may be built based on an individual user profile to predict and tailor future interactions. It is called the *individual approach*. This approach requires content descriptions and is often referred to as content-based filtering systems. NewsWeeder [32] is an example of using content-based filtering that automatically learns user profiles to recommend articles to the user. The major disadvantage of this approach is the filtering only relies on the users' previous interests for the recommendation.

There is an alternative approach that not only uses the individual user profile, but also takes care of other users who share similar preferences. It is called the *collaborative approach*. This approach is referred to as collaborative filtering. GroupLens [45] is an example that recommends articles to users based on a similar user profile. The major disadvantage of this approach is the reliance on the availability of ratings for any item prior to recommending it.

### 17.3.2 Reactive versus Proactive

The reactive approach to personalization refers to a conversational process that requires explicit interactions with the user in the form of queries or feedback. To provide the items of interest to the user, the feedback is also incorporated into the recommendation process for refining the search and suggestion. Most reactive systems for personalization have their origins in case-based reasoning research [16,34,35]. Reactive systems can also be classified based on the two major feedbacks: common feedback [34] and preference feedback [35]. In common feedback, the user must provide a rating for each recommendation based on the suitability to the user's needs. In preference feedback, the user is provided with a list of recommendations based on his or her best interest, a user is required to choose one recommendation out of the list to meet his or her requirement.

On the other hand, the proactive approach is to learn user preferences for providing recommendations without user enquiry. So, the user is not required to provide explicit feedback to the system to drive the current recommendation process. In particular, proactive systems provide users with recommendations that the user may choose to select or ignore. An example of the proactive system is the recommendation engine at Amazon.com [33].

### 17.3.3 Client Side versus Server Side

Personalization approaches can be classified based on the approaches that have been designed to run on the client side or on the server side. On the client side, data is only available to describe the individual user, thus only the individual approach can be applied on the client side. On the server side, the Web site has the ability to collect data from all users. Thus, both the individual and collaborative approaches can be applied.

In addition, server side approaches, in general, only have access to interactions of users with content being put on their own Web site. On the contrary, client side approaches can access data regarding the individuals' interactions with multiple Web sites. Given these characteristics, most client side approaches are aimed at personalized searches across multiple repositories [44].

To sum up the above analysis, personalization can be delivered in many different ways. Actually, a number of mixed approaches to personalization have also been proposed. These mixed approaches have been developed based on each of the recommended technology deficiencies that are difficult to overcome within the boundaries of a single recommendation approach.

---

## 17.4 E-Learning Standards

To provide personalized learning materials, e-learning standards play a significant role. E-learning standards not only enhance interoperability, but also provide a framework for designing personalized course materials. In this section, the major organizations for e-learning standards development and the standards' characteristics are discussed. Moreover, an overview of SCORM, a popular e-learning standard, is provided at the end of this section.

As e-learning has become popular in the past decade, recently researchers in this area are concerned about the issues of inflexibility of course development. The core problem is that the development of course materials is costly, time consuming, and not reusable. The researches showed that the most efficient way to address this problem is to break a course down into smaller, self-contained components that could be archived and reused whenever they are needed. Thus, there is a need to establish standard ways for the definition, representation, transmission, and reuse of course materials.

Currently, e-learning standards are mainly being developed by four main organizations, namely, the Aviation Industry CBT Committee (AICC), Instruction Management System Global Learning Consortium (IMS), Advanced Distributed Learning Initiative (ADL), and IEEE Learning Technology Standards Committee (IEEE LTSC).

AICC (www.aicc.org) is an international group of technology-based training professionals; AICC creates CBT-related guidelines for the aviation industry. IMS Global Consortium (www.imsproject.org) is a consortium of suppliers that work on specification development with a focus on the metadata used to address content packaging. ADL (www.adlnet.gov) is sponsored by the U.S. government; the main function of this organization includes the research and development of specifications that encourage the adoption and advancement of e-learning. The ADL Shareable Content Object Reference Model (SCORM) is the most widely accepted ADL publication. Some of the best elements of IEEE, AICC, and IMS specifications are combined by the SCORM specification into a consolidated document. IEEE's Learning Object Metadata (LOM) specification is the most widely acknowledged IEEE LTSC (www.ieeeltsc.org) specification, which defines element groups and elements that describe learning resources. The LOM elements and structures are used in the specifications of both IMS and ADL. Thus, SCORM is a major standard to be applied in Peels.

### 17.4.1 Sharable Content Object Reference Model (SCORM)

In order to provide a better definition of the different aspects of Learning Object (LOs), a number of standards organizations (IEEE, IMS, AICC) have joined forces with the initiative of the Advanced Distributed Learning consortium (ADL) and created a library of such standards as the SCORM. Moreover, SCORM is one of the most widely used standards and aims to provide [24]:

- Accessibility: It should be possible to identify, locate, and access LOs in one remote location and deliver LOs to other locations.
- Interoperability: LOs are developed by a set of tools on the platform and they should be accessible and reusable with different tools on different platforms.
- Durability: LOs should be able to cope with technology changes without redesigning, reconfiguration, or recoding.
- Reusability: LOs should be sufficiently flexible in multiple applications and contexts.

The SCORM specifications intend to make instructional components sharable and reusable by both systems and the human designers who assemble these components into instructional interactions. The SCORM specifications include the Content Aggregation Model, Run-Time Environment, and Sequencing and Navigation.

*The Content Aggregation Model* (CAM) describes (1) the types of content objects to be used in a content aggregation, (2) how to package the content objects for exchange from system to system, and (3) how to use metadata to describe the content objects for enabling search and discovery, and (4) how to define the sequencing rules for the content objects.

*The Run-Time Environment* describes the requirements of the Learning Management System (LMS) for interoperability of content across different LMSs. In particular, the requirements include (1) standardization of the content launch process, (2) standardization of communication methods between the content and LMSs, and (3) standardization of data model for passing information about the learner's interactions with the content.

The *Sequencing and Navigation* defines how SCORM materials may be sequenced via a set of learner-initiated or system-initiated navigation events or rules. Branch and flow may be defined by a predetermined set of activities, set of design time, or generated on demand.

## 17.5 Cognitive Theory and Instructional Design Strategies

The primary objective of a personalized e-learning system is to provide an environment where students are treated as a center of their own learning. Educators must know the principles of learning and how students learn before any learning materials are developed. Moreover, the development of effective online learning materials should be designed based on proven and sound learning theories. Thus, the major learning theories and learning material design are examined.

The major learning theories are discussed including behaviorism, cognitivism, and constructivism. Through discussion of the theories, it is a good opportunity to understand what kinds of factors should be considered in an effective learning materials design.

- **Cognitivism**
  It is a move from thinking only about overt behaviors to more about the inner workings of the mind. Cognitivists feel that behaviorism's emphasis on the link between a stimulus and a response was not sufficient to account for all human activities.

Cognitive psychologists explain that human learning involves the use of memory, motivation, thinking, and reflection. Craik and Lockhart [19] and Ausubel [4] also provided further explanation for learning as an internal process, and the learning outcome is affected by the following factors: (1) the amount learned depending on the processing capacity of the individual learner, (2) the amount of effort in the learning process, (3) the depth of the processing, and (4) the learner's existing knowledge structure.

To take into account the above factors, Anderson and Elloumi [2] suggested some guidelines for learning material design. First, the materials must be designed at difficulty levels that match the cognitive level of different learners, so that the individual learner can attend to and relate to the material for his or her learning. Second, it is proposed that prerequisite test questions be used to recall the prerequisite knowledge structure required for learning the new materials. With the flexibility of e-learning, students with diverse backgrounds and knowledge should be provided with the most appropriate path to review prerequisite learning before any new information is presented. Third, information should be presented in different modes (e.g., textual and visual) to accommodate individual differences in processing information and to facilitate putting the information into memory. According to the dual-coding theory [43], information received in different modes is better than in a single mode (textual only).

- **Constructivism**
  Constructivists believe learners are active rather than passive. They also feel that knowledge is not received from the outside or from someone else. Moreover, learning is a constructive process where learners build an internal illustration of knowledge and a personal interpretation of experience. Indeed, the position of the learner is the center of the learning and an instructor plays an advising and facilitating role. Students learn best when they actively construct their own understanding. They are encouraged to invent their own solutions and to try out ideas and hypotheses. As a result, learning is moving away from one-way instruction (teacher-centered) to construction and discovery of knowledge (student-centered) [22, 52].

  Jonassen [30], and Anderson and Elloumi [2] summarize the implications of constructivism for instructional design. The following principles illustrate how knowledge construction can be facilitated. First, learning should be designed as an active process. To facilitate the creation of personalized meaning, learners should be actively doing meaningful activities. For example, an active process may involve designing questions to help learners apply the learning contents in a practical situation. It also facilitates personal interpretation and relevance. Garrison and Shale [25] claim it has a significant impact on the learning experience if in the learning process students interact with the content, with other students, and with the instructors to test and confirm ideas and to apply what they have learned.

Second, learners organize the information they learn. The e-learning system can help students learn by presenting organized information. Learners should be given control of the learning process. There should be a form of guided discovery where learners are allowed to make decisions on learning goals, but with guidance from the instructor [40].

According to Ertmer and Newby [23], the above three theories can in fact be used as a taxonomy for learning.

- Behaviorist strategies [49,51] can be applied to teach the "what" (facts).
- Cognitive strategies can be applied to teach the "how" (processes and principles).
- Constructivist strategies can be applied to teach the "why" (higher level thinking that promotes personal meaning and contextual learning).

These educational research results and findings are taken into account in designing the instructional learning contents and tasks in different cases. In order to tailor the instructional contents to individual students effectively, a concept of delivery approach becomes critical. Bransford et al. [8] provided a comprehensive analysis on "how people learn" and proposed a conceptual framework for e-learning design. Indeed, the major elements in the conceptual framework include learner centered, knowledge centered, community centered, and assessment centered. To realize this conceptual framework in an e-learning system, Anderson and Elloumi [2] added a further enhancement on this framework based on agent technology. Bransford et al. proposed the "how people learn framework" and is summarized by Table 2.1 in his work [8], and the "affordances of the current web" and "affordances of the semantic web," as proposed by Anderson and Elloumi [2].

According to the above discussion on the learning theory, the design of learning materials plays a critical role that can directly affect the success of the e-learning system. Thus, the "how people learn framework" in Table 17.1 is also a good reference to design the personalized learning environment.

**TABLE 17.1**

Summary of the Proposed Conceptual Frameworks

| How People Learn Framework | Affordances of the Current Web | Affordances of the Semantic Web |
|---|---|---|
| Learner centered | Capacity to support individualized and community centered learning activities | Content that changes in response to individualized and group learner models |
| Knowledge centered | Direct access to vast libraries of content and learning activities organized from a variety of discipline perspectives | Agents for selecting, personalizing, and reusing content |
| Community centered | Asynchronous and synchronous; collaborative and individual interactions in varied formats | Agents for translating, reformatting, time shifting, monitoring, and summarizing community interactions |
| Assessment centered | Multiple time and place shifted opportunities for formative and summative assessment by self, peers, and teachers | Agents for assessing, critiquing, and providing "just in time feedback" |

## 17.6 Conclusion

In the modern world, e-learning has become a fact of life. No one disputes whether e-learning should be applied or not, the remaining questions are how and when it should be applied. In response to the different needs and characteristics of a growing number of e-learning users, the major technologies and strategies for a personalized learning environment have been examined. In the future, more and more learners will not only go to schools for courses, learning experiences will come to learners in response to their strengths and prior learning, interests, and aspirations.

## References

1. Aleven, V. and Keodinger, K. 2000. Limitations of student control: Do students know when they need help? In Intelligent Tutoring Systems, Proceedings of the 5th International Conference, ITS, Montreal, Canada. *Lecture Notes in Computer Science*, 1839, eds. G. Gauthier, C. Frason, and K. VanLehn, Berlin: Springer, 292–303.
2. Anderson, T. and Elloumi, F. 2004. *Toward a Theory of Online Learning, Theory and Practice of Online Learning*. Edmonton: Athabasca University, 33–60.
3. Ankush, M., Kanishka, R., and Sung, W. K. 2003. Content-based retrieval systems for personalization of educational videos. *Artificial Intelligence in Education*, Sydney: IOS Press, 289–96.

4. Ausubel, D. P. 1974. *Educational Psychology: A Cognitive View*. New York: Holt, Rinehart and Winston.
5. Balabanovic, M. 1997. An adaptive web page recommendation service. In Proceedings of the First International Conference on Autonomous Agents, Marina del Rey, CA.
6. Balabanovic, M. and Shohan, Y. 1997. Fab: Content-based, collaborative recommendation. *Communications of the ACM* 40:66–72.
7. Bolliger, D. U. and Martindale, T. 2004. Key factors for determining student satisfaction in online course. *International Journal on E-Learning*, January–March, 3:61–67.
8. Bransford, J., Brown, A., and Cocking, R. 1990. *How people learn: Brain, mind experience and school*. Available from: http://www.nap.edu/html/howpeople1
9. Brusilovsky, P. 1999. Adaptive and intelligent technologies for Web-based education. *Künstliche Intelligenz* 4:19–25.
10. Brusilovsky, P. 2000. Course sequencing for static courses? Applying ITS techniques in large-scale web-based education. In Proceeding of International Conference on Intelligent, 73 Tutoring Systems, 625–34.
11. Brusilovsky, P. 2001. Adaptive hypermedia. *User Modeling and User-Adapted Interaction* 11:87–110.
12. Brusilovsky, P. and Pesin, L. 1994. An intelligent learning environment for CDS/ISIS users. In eds. J. J. Levonen, and M. T. Tukianinen. Proceeding of the Interdisciplinary Workshop on Complex Learning in Computer Environments (CLCE94), Joensuu, Finland, EIC, 29–33.
13. Brusilovsky, P., Eklund, J., and Schwarz, E. 1998. Web-based education for all: A tool for developing adaptive courseware. In Computer Networks and ISDN Systems, Proceedings of 7th International World Wide Web Conference, 291–300.
14. Brusilovsky, P., Schwarz, E., and Weber, G. 1996. ELM-ART: An intelligent tutoring system on World Wide Web. In *Intelligent Tutoring Systems, Lecture Notes in Computer Science*, 1086, eds. C. Frasson, G. Gauthier, and A. Lesgold. Berlin: Springer Verlag, 261–69.
15. Buchanan, T. 2000. The efficacy of a World-Wide Web mediated formative assessment. *Journal of Computer Assisted Learning* 16:193–200.
16. Burke, R. 2000. Knowledge-based recommender systems. *Encyclopedia of Library and Information Systems*, 69:180–200.
17. Carbonell, J. R. 1970. AI in CAI: An artificial intelligence approach to computer-assisted instruction. *IEEE Transactions on Man-Machine Systems* 11 (4): 190–202.
18. Cordova, D. L. and Lepper, M. R. 1996. Intrinsic motivation and the process of learning: beneficial effects of contextualization, personalization, and choice. *Journal of Educational Psychology* 88 (4): 715–30.
19. Craik, F. I. M. and Lockhart, R. S. 1972. Levels of processing: A framework for memory research. *Journal of Verbal Learning and Verbal Behavior* 11:671–84.
20. Cristea, A. 2004. Adaptive and adaptable educational hypermedia: Where are we now and where are we going? In *Proceedings of Web-based Education*, Innsbruck, Austria, 16–18.
21. De Bra, P. M. E. 1996. Teaching hypertext and hypermedia through the Web. *Journal of Universal Computer Science* 2 (12): 797–804.

22. Duffy, T. M. and Cunningham, D. J. 1996. Constructivism: Implications for the design and delivery of instruction. In *Handbook of Research for Educational Communications and Technology,* ed. D. H. Jonassen. New York: Simon & Schuster Macmillan, 170–98.

23. Ertmer, P. A. and Newby, T. J. 1993. Behaviorism, cognitivism, constructivism: Comparing critical features from an instructional design perspective. In *Performance Improvement Quarterly* 6 (4): 50–70.

24. Fletcher, J. D. 2006. *The ADL Vision and Getting from here to there, Web-Based Learning: Theory, Research, and Practice.* Mahwah, NJ: Lawrence Erlbaum, 31–53.

25. Garrison, D. R. and Shale, D. 1990. A new framework and perspective. In *Education at a Distance: From Issues to Practice,* eds. D. R. Garrison, and D. Shale, D. Malabar, FL: Robert E. Krieger, 123–33.

26. Gertner, A. and VanLehn, K. 2000. Andes: A coached problem solving environment for physics. In Intelligent Tutoring Systems, Proceedings of the 5th International Conference, ITS, Montreal, Canada. *Lecture Notes in Computer Science,* 1839, eds. G. Gauthier, C. Frason, and K. VanLehn. Berlin: Springer, 133–42.

27. Hohl, H., Bocker, H. D., and Gunzenhauser, R. 1996. Hypadapter: An adaptive hypertext system for exploratory learning and programming. In *User Modeling and User-Adapted Interaction,* 6 (2–3): 131–56.

28. Hong, H. and Kinshuk, H. 2004. Adaptation to student learning styles in web based educational systems. In eds. L. Cantoni and C. McLoughlin, Proceedings of ED-MEDIA 2004: World Conference on Educational Multimedia, Hypermedia & Telecommunications, June 21–26, Lugano, Switzerland, 491–96.

29. Hoole, D., Yogendran, N., Thavachandran, S., Priyatharshan, P., and Hoole, S. R. H. 2002. A bank of chemistry questions on an on-Line server. In *Journal of Science Educational and Technology* 11 (1): 9–13.

30. Jonassen, D. 1994. Thinking technology. In *Educational Technology* 34 (4): 34–37.

31. Lajoie, S. and Derry, S. 1993. *A Middle Camp for (Un)Intelligent Instructional Computing: An Introduction, Computers as Cognitive Tools.* Hillsdale, NJ: Erlbaum, 1–11.

32. Lang, K. 1995. Newsweeder: Learning to filter netnews. In Proceedings of the 12th International Conference on Machine Learning, Tahoe City, CA, 331–39.

33. Linden, G., Smith, B., and York, J. 2003. Amazon.com recommendations: Item-to-item collaborative filtering. In *IEEE Internet Computing,* 7:76–80.

34. Lorenzi, F. and Ricci, F. 2005. Case-based recommender systems: A unifying view. In *Intelligent Techniques in Web Personalisation, LNAI.* Berlin: Springer-Verlag.

35. McGinty, L. and Smyth, B. 2005. Improving the performance of recommender systems that use critiquing. In *Intelligent Techniques in Web Personalisation. LNAI.* Berlin: Springer-Verlag

36. Mehrotra, C., Hollister, C. D., and McGahey, L. 2001. *Distance Learning: Principles for Effective Design, Delivery, and Evaluation.* Thousand Oaks, CA: Sage.

37. Melis, E., Andres, E., Budenbender, E., and Frischauf, A. 2001. ActiveMath: A generic and adaptive web-based learning environment. *International Journal of Artificial Intelligence in Education* 12:385–407.

38. Mitrovic, A. and Suraweera, P. 2000. Evaluating an animated pedagogical agent. In Intelligent Tutoring Systems, Proceedings of the 5th International Conference, ITS, Montreal, Canada. *Lecture Notes in Computer Science,* 1839, eds. G. Gauthier, C. Frason, and K. VanLehn. Berlin: Springer, 73–82.

39. Mulvenna, M., Anand, S. S., and Buchner, A. G. 2000. Personalization on the net using web mining. *Communication of ACM* 43 (8): 123–25.

40. Murphy, K. L. and Cifuentes, L. 2001. Using web tools, collaborating, and learning online. In *Distance Education* 22 (2): 285–305.

41. Ooney, R. J. and Roy, L. 2000. Content-based book recommending using learning for text categorization. In Proceedings of the 5th ACM Conference on Digital Libraries, San Antonio, TX, 195–204.

42. Ozdemir, B. and Ferda, N. A. 2000. An intelligent tutoring system for student guidance in web-based courses. In 4th International Conference on Knowledge-Based Intelligent Engineering Systems and Allied Technologies, 2, 835–39.

43. Paivio, A. 1986. *Mental Representations: A Dual Coding Approach*. Oxford: Oxford University Press.

44. Parent, S., Mobasher, B., and Lytinen, S. 2001. An adaptive agent for web exploration based on concept hierarchies. In Proceedings of the 9th International Conference on Human Computer Interaction, New Orleans, LA.

45. Resnick, P., Iacovou, N., Sushak, M., Bergstrom, P., and Riedl, J. 1994. Grouplens: An open architecture for collaborative filtering of netnews. In Proceedings of the 1994 Computer Supported Collaborative Work Conference, Chapel Hill, NC, 175–86.

46. Rickel, J. and Johnson, W. 1999. Animated agents for procedural training in virtual reality: Perception, cognition, and motor control. *Applied Artificial Intelligence* 13:343–82.

47. Rumetshofer, H. and Wöß, W. 2003. XML-based adaptation framework for psychological-driven e-learning systems. *Educational Technology & Society* 6 (4): 18–29. Available from http://http://www.ifets.info/journals/6_4/4.pdf

48. Rusilovsky, P. 2001. Adaptive hypermedia. *User Modeling and User-Adapted Interaction* 11, (1–2): 87–100.

49. Skinner, B. F. 1974. *About Behaviorism*. New York: Knopf.

50. Specht, M., Weber, G., Heitmeyer, S., and Schoch, V. 1997. AST: Adaptive WWW-courseware for statistics. In eds. P. Brusilovsky, J. Fink, and J. Kay, Proceeding of Workshop "Adaptive Systems and User Modeling on the World Wide Web" at 6th International Conference on User Modeling, UM97, Chia Laguna, Sardinia, Italy, Carnegie Mellon, 91–95.

51. Standridge, M. 2002. Behaviorism. In *Emerging Perspectives on Learning, Teaching, and Technology,* ed. M. Orey, University of Georgia. Available from http://www.coe.uga.edu/epltt/Behaviorism.htm

52. Tapscott, D. 1998. *Growing Up Digital: The Rise of the Net Generation*. New York: McGraw-Hill.

53. Thomas, J. and Rohwer, W. 1993. Proficient autonomous learning: Problems and prospects. In *Cognitive Science: Foundations of Instruction*, ed. M. Rabinowitz. Hillsdale, NJ: Lawrence Erlbaum Associates, 1–32.

54. Vogel, D. and Klassen, J. 2001. Technology-supported learning: status, issues, and trends. *Journal of Computer Assisted Learning* 17:104–44.

55. Weibel, S., Kunze, J., Lagoze, C., and Wolf, M. 1998. Dublin core metadata for resource discovery. IETF #2413. The Internet Society.

56. Zayas, B. 2001. Learning from 3D VR representations: Learners-centered design, realism and interactivity. In Proceedings of Workshop on External Representations in AIED, San Antonio, TX. Available from http://www.psychology.nottingham.ac.uk/research/credit/AIED-ER/zayas.pdf

# Contributors

**Chin-Chen Chang, BS, MS, PhD,** received his BS in applied mathematics in 1977 and his MS in computer and decision sciences in 1979, both from the National Tsing Hua University, Hsinchu, Taiwan. He received his PhD in computer engineering in 1982 from the National Chiao Tung University, Hsinchu, Taiwan. During the academic years of 1980–1983, he was on the faculty of the Department of Computer Engineering at the National Chiao Tung University. From 1983 to 1989, he was on the faculty of the Institute of Applied Mathematics, National Chung Hsing University, Taichung, Taiwan. From August 1989 to July 1992, he was a professor and also served as the head of the Institute of Computer Science and Information Engineering at the National Chung Cheng University, Chiayi, Taiwan. From August 1992 to July 1995, he was the dean of the College of Engineering at the National Chung Cheng University. From August 1995 to October 1997, he was the provost at the National Chung Cheng University. From September 1996 to October 1997, Dr. Chang was the acting president at the National Chung Cheng University. From July 1998 to June 2000, he was the director of the Advisory Office of the Ministry of Education of the Republic of China. From 2002 to 2005, he was a chair professor of National Chung Cheng University. Since February 2005, he has been a chair professor of Feng Chia University. In addition, he has served as a consultant to several research institutes and government departments. His current research interests include database design, computer cryptography, image compression, and data structures.

**Han-Bin Chang, PhD,** is currently an assistant professor of the Department of Computer Science and Information Engineering at Yuanpei University, Taiwan. He received his PhD from the Department of Computer Science and Information Engineering at Tamkang University, Taiwan, in 2008. His current research interests are in the scope of multimedia systems, video processing, image processing, human computer interaction, ubiquitous computing, and distance education technologies.

**Hsuan-Pu Chang** received his PhD from the Department of Computer Science and Information Engineering at Tamkang University, Taiwan, in 2008. He is currently an assistant professor in the Department of Digital Game Design at Kao Fong College, Taiwan. His research interests include game-based learning, multimedia applications, and ubiquitous computing technologies.

**Maiga Chang** is an assistant professor in the School of Computing Information and Systems, Athabasca University, Athabasca, Alberta, Canada. He is the local chair of the Second Institute of Electrical & Electronics Engineers Digital Game and Intelligent Toy Enhanced Learning (DIGITEL 2008) and general co-chair of the Fourth International Conference on e-Learning and Games (Edutainment 2009). He serves on the Athabasca University Press and on nine peer-reviewed journals, including Springer Publishing's *Transaction on Edutainment* as an editorial board member and executive peer reviewer. He has participated in 87 international conferences and workshops as a program committee member and has authored or coauthored more than 99 book chapters, journal articles, and international conference papers. He has been a member of the International Who's Who of Professionals since 2000. In September 2004, he received the 2004 Young Researcher Award in Advanced Learning Technologies from the Institute of Electrical & Electronics Engineers Technical Committee on Learning Technology. He has been a valued Institute of Electrical & Electronics Engineers member for 12 years and a member of the Association of Computing Machinery, American Association for Artificial Intelligence, the INNS, and the Phi Tau Phi Scholastic Honor Society. His research mainly focuses on mobile and ubiquitous learning, museum e-learning, game-based learning, educational robots, learning behavior analysis, and data mining in e-learning.

**Chia-Chen Chao, MS, PhD,** is an assistant professor in the Information Science Department at the National Taipei University of Education, Taiwan. He received one MS from New York University, one from New York City University, and one from the National Chen Chi University and received his PhD from the National Chen Chi University. Dr. Chao's research is published in the *International Journal of Medical Informatics, The International Journal of Technological Innovation, Entrepreneurship and Technology Management, International Journal Management and Enterprise Development, International Journal Electronic Healthcare,* and a number of national and international conference proceedings. His current research interests include a marketing force strategic information system, RFID-based mobile service, and health informatics.

**Louis R. Chao, BS, MS, PhD,** received his BS and MS from the National Taiwan University in 1965 and 1968, respectively, and the PhD from Duke University, Durham, North Carolina, in 1971, all in electrical engineering. From July 1971 to February 1972, he was a research associate in the Statistics Department, the University of North Carolina at Chapel Hill, working mainly on sonar detection systems. He was then appointed as chairman of the Computer Science Department by Tamkang University and returned to his home country, Taiwan. In 1974, he was promoted to dean of the College of Engineering, and then dean of Academic Affairs and vice president of Financial Affairs, respectively, in 1978 and 1984 and became the president of Tamkang University in 1989. He was actively exploiting the bilateral programs between institutions of higher education and has organized a number of international conferences held in Taiwan. In 1993, he was appointed as the member of the Control Yuan by the President of the Republic of China and was reappointed in 1999. Since 2005, he returned back to Tamkang University and was then appointed as chair professor of Computer Science and Information Engineering. Dr. Chao was reappointed to Control Yuan in 2008 and retired from Tamkang University. He is now an honorable professor with Computer Science and Information Engineering. Dr. Chao is a member of IEEE, and the former president of the Systems Analysis Society and of the e-Learning Society of Taiwan. His long-time main interests include digital communication systems, operations research, and e-learning.

**Jen-Wen Ding, BS, MS, PhD,** is currently an assistant professor in the Department of Information Management, National Kaohsiung University of Applied Sciences, Kaohsiung, Taiwan. He received his BS, MS, and PhD in engineering science from the National Cheng Kung University, Tainan, Taiwan, in 1996, 1998, and 2001, respectively. His research interests include multimedia communications, peer-to-peer computing, and mobile computing.

**Yufei Du** is currently an MPhil student at City University of Hong Kong. His research interests include universal mobile telecommunications systems, wireless sensor network technology, and mobile video quality predictions and estimations.

**Victor Gau, BS, MS,** received his BS in electrical engineering from the National Taiwan University, Taipei, Taiwan, in 1993, and the MS in biomedical engineering from National Yang-Ming University, Taipei, Taiwan, in 1999. He is currently a PhD candidate at the Department of Electrical Engineering at the University of Washington at Seattle. His research interests include peer-to-peer streaming systems, multimedia networking, wireless networks, and computer vision.

**Sami J. Habib, BS, MS, PhD,** received his BS in computer engineering from Iowa State University in 1993. After graduation, he spent a year working as a lab engineer in the Department of Electrical and Computer Engineering at Kuwait University. Then, he pursued graduate study at the University of Southern California, where he earned an MS and PhD in computer engineering in 1995 and 2001, respectively. Currently, he is an associate professor in the Computer Engineering Department at Kuwait University. Dr. Habib has served on several technical program committees, and a reviewer for a number of international conferences and journals. He has published many journal and conference papers. He is a member of the Institute of Electrical & Electronics Engineers, Institute of Electrical & Electronics Engineers Computer Society, and Association of Computing Machinery. His current research focuses on developing computer-aided design methodologies and performance analysis techniques for designing—redesigning distributed systems, especially the physical network topology, data management system, and sensor network.

**Masahito Hirakawa, MS, PhD,** received his degrees from Hiroshima University, Japan. He is a professor of computer engineering at Shimane University, Japan. His research interests include visual interfaces, multimodal interfaces, augmented reality, and multimedia computing.

**Hui-Huang Hsu, PhD,** is an associate professor in the Department of Computer Science and Information Engineering at Tamkang University, Taipei, Taiwan. He received his degree from the Department of Electrical and Computer Engineering at the University of Florida in 1994. Dr. Hsu is a senior member of the Institute of Electrical & Electronics Engineers. His current research interests are in the areas of machine learning, data mining, ambient intelligence, bioinformatics, and multimedia processing.

**Jiung-Yao Huang, PhD, MS,** is a professor of the Department of Computer Science and Information Engineering at National Taipei University, San Shia, Taipei Hsien, Taiwan. Previously, he taught in the Department of Communication Engineering at National Chung Cheng University, Taiwan, between 2003 and 2006. Before that, he served in the Department of Computer Science and Information Engineering at Tamkang University, Taiwan. He actively served as the director of the Advanced Digital Research Section in the Information Processing Center, and the executive editor of the *Tamkang Journal of Science and Engineering* at Tamkang University, Taiwan from August 2001 to July 2003. He received his degree in electrical and computer engineering from the University of Massachusetts at Amherst in 1993. Prior to receiving his MS in computer science from Tsing-Hua University, Taiwan, in 1988, he graduated in 1983 from the Department of Applied Mathematics at Chung-Hsing University, Taiwan. He has been a member of the Institute of Electrical & Electronics Engineers and Association of Computing Machinery since 1990 and joined the Society for Computer Simulation in 1996. His research interests include pervasive computing, computer graphics, networked virtual reality, and multimedia systems.

**Ray Yueh-Min Huang, MS, PhD,** is a distinguished professor and chairman of the Department of Engineering Science, National Cheng-Kung University, Taiwan. He received his MS and PhD in electrical engineering from the University of Arizona in 1988 and 1991, respectively. He has coauthored two books and has published about 200 refereed professional research papers. Dr. Huang has received many research awards, such as the Best Paper Award of 2007 IEA/AIE Conference, Best Paper Award of the Computer Society of the Republic of China in 2003, the Awards of Acer Long-Term Prize in 1996, 1998, and 1999, and the Excellent Research Awards of

National Microcomputer and Communication Contests in 2006. Dr. Huang has been invited to give talks and has served frequently in the program committees at national and international conferences. Dr. Huang is on the editorial board of the *Journal of Wireless Communications and Mobile Computing, Journal of Security and Communication Networks,* and *International Journal of Communication Systems.* Dr. Huang is a member of the Institute of Electrical & Electronics Engineers as well as the Institute of Electrical & Electronics Engineers Communication, Computer, and Circuits and Systems Societies. His research interests include multimedia communications, wireless networks, artificial intelligence, and e-learning.

**Runhe Huang, BSc, PhD,** is a professor on the faculty of Computer and Information Sciences at Hosei University, Japan. She received her BSc in electronics technology from the National University of Defense Technology, Changsha, China in 1982, and her PhD in computer science and mathematics from the University of the West of England, United Kingdom in 1993. She worked at the National University of Defense Technology from 1982 to 1988. In 1988, she received the Sino-Britain Friendship Scholarship for her PhD study in the United Kingdom. She worked in the Computer Science and Engineering Lab at the University of Aizu from 1993 to 2000. She joined Hosei University in 2000. Dr. Huang has been working in the field of computer science and engineering for more than 25 years. Her research fields include computer-supported cooperative work, artificial intelligence, computational intelligence, multiagent systems, ubiquitous computing, and mobile computing.

**Jenq-Neng Hwang, PhD,** received his degree from the University of Southern California; he then joined the Electrical Engineering Department of the University of Washington at Seattle, where he has been a professor since 1999. He also served as the associate chair for research and development in the Electrical Engineering Department from 2003 to 2005. He has published more than 250 journal articles, conference papers, and book chapters in the areas of image—video signal processing, computational neural networks, multimedia system integration, and networking. Dr. Hwang received the 1995 Institute of Electrical & Electronics Engineers Signal Processing Society's Annual Best Paper Award and is a fellow of Institute of Electrical & Electronics Engineers.

**Azizan Ismail, BSc, MS,** received his degree from the University Utara Malaysia and his MS in multimedia from the University Putra Malaysia. He has worked as a lecturer at the University Tun Hussien Onn Malaysia and is now on study leave pursuing his PhD in the School of Computing and Mathematical Sciences at the Liverpool John Moores University. His research interests are image processing, indexing and retrieval, and managing digital life memories.

**Wen-Yuan Jen, MS, PhD,** is an associate professor and chairman of the Institute of Information and Society at National United University, Taiwan. She received her MS from Texas A&M University and her PhD from the National Central University, Taiwan. Dr. Jen's research is published in *International Journal of Medical Informatics, The International Journal of Technological Innovation, Entrepreneurship and Technology Management, Lecture Notes in Computer Science, International Journal Management and Enterprise Development, International Journal Electronic Healthcare,* and a number of national and international conference proceedings. Her current research interests include mobile service, electronic commerce, and information society.

**Weijia Jia, BSc, MsC, PhD,** is currently a full professor in the Department of Computer Science and the director of Future Networking Center, ShenZhen Research Institute of City University of Hong Kong. He received his BSc and MSc from Center South University, China in 1982 and 1984 and MAppSc and PhD from Polytechnic Faculty of Mons, Belgium in 1992 and 1993, respectively, all in computer science. He joined the German National Research Center for Information Science in Bonn, St. Augustine from 1993 to 1995 as a research fellow. In 1995, he joined the Department of Computer Science, City University of Hong Kong as an assistant professor.

**Kamen Kanev, MS, PhD,** received his MS in mathematics and his PhD in computer science both from Sofia University, Bulgaria in 1984 and 1989, respectively. He is a professor with the Research Institute of Electronics and Graduate School of Informatics, Shizuoka University, Hamamatsu, Japan, where he teaches and supervises students majoring in computer and information science and electronics. On this and related topics he has authored and coauthored more than 70 scientific journal and conference papers and patents. Dr. Kanev is a member of the Institute of Electrical & Electronics Engineers, the Association of Computing Machinery, and the Asia-Pacific Society for Computers in Education. His main research interests are in interactive computer graphics, user interfaces and surface-based interactions, and in vision information processing.

**Takayuki Koyama, BS,** received his degree from Shimane University in 2009. He is working for a software vendor, ICR, Japan. His research interests include human–computer interaction and image processing.

**Sinjae Lee, BS, MS, DSc,** received his degree in mathematical education from Kookmin University, Seoul, Korea in 1990. He also received his second BS, MS, and DSc in computer science from Towson University in 1999, 2001, and 2008. He is a research professor in the Department of Computer Science and Engineering at Korea University, Seoul. His research interests include a mobile wireless network, peer-to-peer reputation systems, network simulation, and network security.

**Wonjun Lee, BS, MS, PhD,** received his degrees in computer engineering from Seoul National University, Korea in 1989 and 1991, respectively. He also received his MS in computer science from the University of Maryland at College Park in 1996 and his PhD in computer science and engineering from the University of Minnesota at Minneapolis in 1999. In 2002, he joined the faculty of Korea University, Seoul where he is currently an associate professor in the Department of Computer Science and Engineering. He is also adjunct professor in the Department of Electrical Engineering, State University of New York at Stony Brook. He has held a faculty position at the University of Missouri at Kansas City. He has authored or coauthored over 100 papers in refereed international journals and conferences. He served as TPC member for the Institute of Electrical & Electronics Engineers INFOCOM 2008–2009, Association of Computing Machinery MOBIHOC 2008–2009, Institute of Electrical & Electronics Engineers ICCCN 2000–2008, and over 90 international conferences. He currently serves as an editor for *Journal of Information Processing Systems* in the area of communication systems and security. He is a senior member of the Institute of Electrical & Electronics Engineers. His research interests include mobile wireless communication protocols and architectures, wireless sensor networking, wireless mesh network protocols, and internet architecture technology.

**Elvis Wai Chung Leung, PhD,** received his degree in computer science from City University of Hong Kong. He is a course leader for the Bachelor's Degree Program jointly offered by De Montfort University, United Kingdom and City University of Hong Kong. Dr. Leung has published several book chapters and papers for technical journals and conferences. He has been actively involved in the research community as a reviewer for technical journals, as a workshop co-chair of WBL2008, and as a publicity chair of HSI2005. His current research interests include information technology in education, artificial intelligence, multimedia systems, and internet applications.

**Qing Li** is a professor at the Department of Computer Science, City University of Hong Kong. Concurrently, he is a guest professor at the University of Science and Technology of China and Zhong Shan (Sun Yat-Sen) University, and a guest professor (Software Technology) of the Zhejiang University (Hangzhou, China)—the leading university of the Zhejiang province where he was born. Professor Li has edited many books and published over 280 technical papers and book chapters in these areas. In addition, he has been serving as an associate editor and reviewer for technical journals, and as an organizer/co-organizer of numerous international conferences. He is an associate editor of the *ACM Transactions on Internet Technology, World Wide Web* (Springer Publishing), and the *IEEE Transactions on Knowledge and Data Engineering*. Dr. Li is the chairperson of the Hong Kong Web Society, and serves as a councillor of the Database Society of Chinese Computer Federation, a councillor of the Computer Animation and Digital Entertainment Chapter of Chinese Computer Imaging and Graphics Society, and is a Steering committee member of DASFAA, ICWL, and the international WISE Society. His current research interests include *object modeling, multimedia databases, web services, and e-learning.*

**Chia-Chen Lin, BS, PhD** (also known as Min-Hui Lin), received her degree in information management in 1992 from the Tamkang University, Taipei, Taiwan. She received both her MS in information management in 1994 and PhD in information management in 1998 from the National Chiao Tung University, Hsinchu, Taiwan. Dr. Lin served as visiting associate professor at the Business School of the University of Illinois at Urbana Champaign during the period of August 2006 to July 2007. The visiting scholarship was appointed and sponsored by the Ministry of Education, Taiwan. During her visit, she worked with Professor Klara Nahrstedt at the Department of Computer Science on copy detection and the collaboration continues. Dr. Lin is currently a professor of the Department of Computer Science and Information Management, Providence University, Sha-Lu, Taiwan. Since August 2008, she is the associate dean of Academic Affairs at Providence University. She is also a member of the Institute of Electrical & Electronics Engineers and a member of the Association of Computing Machinery. Her research interests include image and signal processing, image hiding, mobile agents, and electronic commerce.

**Fuhua Oscar Lin, PhD,** is a full professor and graduate program director of the School of Computing and Information Systems. Dr. Lin obtained his degree from Hong Kong University of Science and Technology in 1998. Prior to working at the Athabasca University, Dr. Lin was a research officer with the Institute for Information Technology of National Research Council of Canada. He did postdoctoral research at the University of Calgary, Canada. Dr. Lin has more than 70 publications, including edited books, journal articles, book chapters, conference papers, and reviews. Dr. Lin served as invited reviewers for many international journals including *Institute of Electrical & Electronics Engineers Transactions on Multimedia, Transactions on Knowledge and Data Engineering,* and *Institute of Electrical & Electronics Engineers Transactions on Learning Technologies.* His current main research topics include reasoning capability of intelligent agents, immersive and intelligent educational environments, and complex systems modeling.

**Tzu-Chien Liu** is now an associate professor in the Graduate Institute of Learning & Instruction and the director of the Center for Teacher Education, National Central University. Dr. Liu is an associate editor of the *Journal of National Taiwan University* and special issue co-editor of the *International Journal of Engineering Education.* He has participated in more than 15 international conferences (or workshops) as a co-chair or program committee member. He has authored or coauthored more than 125 book chapters, journal articles, or international conference papers. His research mainly focuses on ubiquitous learning, instruction and learning science, and educational application of innovative technology.

**David Llewellyn-Jones, PhD,** is a research fellow at Liverpool John Moores University working on projects in computer security. He received his degree in model theory from the University of Birmingham in 2002 before joining the school as a research assistant to work on the EPSRC-funded PUCSec project. This project looked at developing solutions for software service composition within pervasive computing environments, including trust and privacy aspects. He is involved in various international conferences, journals, collaborative industry projects and initiatives, including the Networked European Software and Services Initiative, and the United Kingdom CyberSecurity KTN Secure Software Development Special Interest Group. He is currently working in areas of computer security, including secure component composition, software security modeling—evaluation, digital rights management, systems—networks interoperation, security protocols, and network security.

**Jianhua Ma** is a professor at Hosei University since 2000. Previously, he had 16 years of teaching–research experience at the National University of Defense Technology, Xidian University, and the University of Aizu, Japan, respectively. His research from 1983 to 2003, covered coding techniques for wireless communications, secure data-audio-video transmissions, speech recognition and synthesis, multimedia QoS, hyper-interface, graphics ASIC, e-learning and virtual university, CSCW, multiagents, mobile Web service, and peer-to-peer network. Since 2003 he has been devoted to what he called smart hyperspace and smart worlds pervaded with three essential kinds of smart/intelligent u-Things: u-object, u-environment, and u-system, toward ubiquitous intelligence or pervasive intelligence based on u-science for u-services with a Ubisafe guarantee.

**Fa-Hung Meng, BSB, MBA,** is a software engineer at Wistron Information Technology & Services (WistronITS) Corporation. He received his BSB degree in information management from WuFeng Institute of Technology, Taiwan in 2002, and his MBA in information management from the National Kaohsiung University of Applied Sciences, Taiwan in 2008. His research interest lies in internet technology, peer-to-peer computing, and mobile computing.

**Madjid Merabti** is a professor of networked systems and director of the School of Computing & Mathematical Sciences, Liverpool John Moores University, United Kingdom. A graduate of Lancaster University, he has over 20 years experience in conducting research and teaching in Distributed Multimedia Systems (networks, operating systems, and computer security) and over 100 publications in these areas. He leads the Distributed Multimedia Systems and Security Research Group, which has a number of government and industry-supported research projects in the areas of network security, instrusion detection system (IDS), multimedia networking, secure component composition, differential services networking, mobile and ad hoc networks, networked appliances, and sensor networks. He is collaborating with a number of international colleagues in these areas, he is editor of a number of related journals, and he chairs several annual international conferences.

**Nikolay Mirenkov, PhD, DSc,** received his degrees in computer science and engineering from the Novosibirsk Technical University and the Union of Soviet Socialist Republics Academy of Sciences in 1972 and 1993, respectively. He had been the head of the Supercomputer Software Department at the Computing Center, Novosibirsk, Russia from 1984 to 1993, the head of the Graduate Department of Information Systems from 1997 to 2001, and the head of the Computer Software Department from 2001 to 2004 at the University of Aizu, Japan. He is now vice president and dean of the Graduate School of Computer Science and Engineering at the University of Aizu. His research interests include self-explanatory components; human–computer interface for people with special needs; visualization, sonification, and filmification of methods and data; as well as parallel programming and high performance computing.

**Takeshi Nakaie, MS,** received his degree from Shimane University, Japan in 2008. He is now working for Dai Nippon Printing Co., Ltd. His interests include human–computer interaction and audio processing.

**Batu Sat, MS, BS,** is a PhD candidate in the Department of Electrical and Computer Engineering at the University of Illinois at Urbana–Champaign. He received his MS in electrical engineering from Illinois in 2003 and his BS in electronics and telecommunications engineering from Istanbul Technical University in 2001. He is a student member of the Institute of Electrical & Electronics Engineers and Association of Computing Machinery. His current research interests are in the development of evaluation and design methods for real-time multimedia communication systems.

**Yuanchun Shi, PhD, MS, BS,** is a professor of the Department of Computer Science, the director of the Institute of Human Computer Interaction & Media Integration of Tsinghua University, and the director of the Pervasive Computing Division of Tsinghua National Lab of Information Science and Technology, China. She received her degrees in computer science from Tsinghua University. She was a senior visiting scholar at Massachusetts Institute of Technology Artificial Intelligence Lab during 2001–2002. She is an Institute of Electrical & Electronics Engineers senior member. Professor Shi is the chairperson of the Technique Committee of Pervasive Computing, China Computer Federation. Her research interests include pervasive computing, human computer interaction, network multimedia, and e-learning technology.

**Yue Shi** is a PhD student of the Department of Computer Science, Tsinghua University. His research interests focus on tabletop interaction, multimodal interface, and adaptive user interface.

**Timothy K. Shih** is a professor and the Dean of College of Computer Science, Asia University, Taiwan. He is a fellow of the Institution of Engineering and Technology. In addition, he is a senior member of ACM and a senior member of IEEE. Dr. Shih also joined the Educational Activities Board of the Computer Society. Dr. Shih has edited many books and published over 430 papers and book chapters, as well as participated in many international academic activities, including the organization of more than 60 international conferences. He was the founder and co-editor-in-chief of the *International Journal of Distance Education Technologies,* published by Idea Group Publishing, USA. Dr. Shih is an associate editor of the *ACM Transactions on Internet Technology* and an associate editor of the *IEEE Transactions on Learning Technologies.* He was also an associate editor of the *IEEE Transactions on Multimedia.* Dr. Shih has received many research awards, including research awards from the National Science Council of Taiwan, IIAS research award from Germany, HSSS award from Greece, Brandon Hall award from USA, and several best paper awards from international conferences. Dr. Shih has been invited to give more than 30 keynote speeches and plenary talks in international conferences, as well as tutorials in IEEE ICME 2001 and 2006, and ACM Multimedia 2002 and 2007. His current research interests include multimedia computing and distance learning.

**Sud Sudirman, BEng, PhD,** received his BEng from the University of Sheffield, United Kingdom and his PhD in computer science from the University of Nottingham, United Kingdom His doctorate thesis was titled "Colour Image Coding, Indexing and Retrieval Using Binary Space Partition Tree." He is currently working as a lecturer at Liverpool John Moores University. He is a member of the Institute of Electrical & Electronics Engineers and its affiliated Signal Processing and Computer Societies. His research interests are image indexing and retrieval, pattern and object recognition, and digital image watermarking.

**Wei-Liang Tai, BS, MS, PhD,** received his BS in computer science and information engineering from Tam Kang University, Taipei, Taiwan in 2002. He received his MS and PhD in computer science and information engineering in 2004 and 2008, respectively, from National Chung Cheng University, Chiayi, Taiwan. His current research interests include data hiding, watermarking, and information security.

**Qing Tan, PhD,** newly joined the School of Computing Information and Systems, Athabasca University as an academic coordinator with his 15 years research and industry experiences. Dr. Tan earned his degree in cybernetics engineering for robotics from the Norwegian Institute of Technology in 1993. Dr. Tan took several senior and technical management positions during his industrial working period. He led and was involved in various software application developments and cutting-edge technology system research and development and integrations. Dr. Tan has been a program committee member, reviewer, and guest editor of several international conferences and journals since he has come back to his academic career. His research interests are mainly location-based technologies, mobile computing and technologies, mobile learning, network technologies and security, enterprise modeling and information management system, and robotics and sensory systems.

**Chung-Hsien Tsai**, MS, received his degree in telecommunication management from Polytechnic Institute of New York University in 2001. He is currently a PhD student in the Department of Computer Science and Information Engineering at National Central University, Taiwan. His current research interests include context awareness, mobile computing, mobile augmented reality, and GPS technology.

**Fung Po Tso, BEngCE, MPhil,** received his degrees from City University of Hong Kong in 2005 and 2007, respectively. His research interests include 3G technology, wireless mesh network, and embedded system design.

**Ming-Chih Tung, MS, PhD,** is an assistant professor in the Department of Computer Science and Information Engineering at Ching Yun University, Taiwan. He received his degrees in computer science from Tamkang University, Taiwan in 1996 and 2004. His research interests include modeling and simulation, military simulation, mobile augmented reality, and networked virtual environment.

**Benjamin W. Wah, PhD,** is currently the Franklin W. Woeltge endowed professor of Electrical and Computer Engineering and professor of the Coordinated Science Laboratory of the University of Illinois at Urbana-Champaign. He serves as director of the Advanced Digital Sciences Center in Singapore, a research center between the University of Illinois at Urbana-Champaign, and the Agency for Science, Technology and Research (A*STAR). He received his degree in computer science from the University of California at Berkeley in 1979. He served on the faculty of Purdue University from 1979 to 1985, as a program director at the National Science Foundation during 1988–1989, as Fujitsu visiting chair professor of Intelligence Engineering, University of Tokyo in 1992, and McKay visiting professor of Electrical Engineering and Computer Science, University of California at Berkeley in 1994.

In 1989, he was awarded a university scholar of the University of Illinois; in 1998, he received the Institute of Electrical & Electronics Engineers Computer Society Technical Achievement Award; in 2000, the Institute of Electrical & Electronics Engineers Millennium Medal; in 2003, the Raymond T. Yeh

Lifetime Achievement Award from the Society for Design and Process Science; in 2006, the Institute of Electrical & Electronics Engineers Computer Society W. Wallace-McDowell Award and the Pan Wen-Yuan Outstanding Research Award; and in 2007, the Institute of Electrical & Electronics Engineers Computer Society Richard E. Merwin Award and the Institute of Electrical & Electronics Engineers–CS Technical Committee on Distributed Processing Outstanding Achievement Award. Dr. Wah cofounded the *Institute of Electrical & Electronics Engineers Transactions on Knowledge and Data Engineering* in 1988 and served as its editor-in-chief between 1993 and 1996, and is the honorary editor-in-chief of *Knowledge and Information Systems.* He currently serves on the editorial boards of *Information Sciences, International Journal on Artificial Intelligence Tools, Journal of VLSI Signal Processing,* and *World Wide Web.* He has chaired a number of international conferences, including the 2000 IFIP World Congress and the 2006 Institute of Electrical & Electronics Engineers / WIC / Association of Computing Machinery International Conferences on Data Mining and Intelligent Agent Technology. He has served on the Institute of Electrical & Electronics Engineers Computer Society in various capacities, including vice president for publications in 1998 and 1999 and president in 2001. He is a fellow of the AAAS, Association of Computing Machinery, and Institute of Electrical & Electronics Engineers. His current research interests are in the areas of nonlinear search and optimization, multimedia signal processing, and computer networks.

**Yi-Hsien Wang, PhD,** received his degrees in computer science and information engineering from National Chiao-Tung University, Hsinchu, Taiwan in 2000 and 2002, respectively. He was also a visiting scholar at the Department of Electrical Engineering, University of Washington at Seattle from 2007 to 2008. He received his PhD from the Department of Computer Science in National Chaio-Tung University, Hsinchu, Taiwan, in 2009. His research interests include internet gaming technology and software engineering.

**Peng-Jung Wu, PhD,** received his BS in computer information science from Soochow University, Taipei, Taiwan in 2001, and his MS in computer science and engineering from Sun Yat-Sen University, Kaohsiung, Taiwan, in 2003. He received his PhD from the Department of Computer Science and Engineering in Sun Yat-Sen University, Kaohsiung, Taiwan, in 2009. He was also a visiting scholar at the Department of Electrical Engineering, University of Washington at Seattle from 2008 to 2009. His research interests include wireless network and peer-to-peer streaming system.

**Lei Ye, BEng, PhD,** received his degrees from Xidian University, China, in 1982 and 1989, respectively. He joined the University of Wollongong, Australia in 2004 after working in both industry and academia with Motorola Australian Research Centre, Nanyang Polytechnic, Singapore and the University of Electronic Science and Technology of China. His work in JPEG and MPEG standardization was awarded Significant Participation of Global Standards Award by Motorola. He is also the recipient of several other awards including First Award of the Progress of Science and Technology from Chinese Ministry, Best Paper Award from Chinese Institute of Electronics, and Distinguished Doctoral Dissertation Award from Xidian University. His recent research outcomes in image retrieval have led to the creation of a spin-off company of the University of Wollongong. Dr. Ye has published several dozen conference papers and journal articles in his academic career. He is a senior member of the Institute of Electrical & Electronics Engineers and active in the professional community serving as conference organizer, invited speaker, and paper reviewer. His current research interests include image processing, retrieval and annotation, multimedia communication and computing, multimedia content management, rights management, and security.

**Chun Yu** is a PhD student of the Department of Computer Science, Tsinghua University. His research interests focus on tabletop interaction, multimodal interface, and adaptive user interface.

# Index